THE
ACCIDENTAL GOD FALLACY

A Primer on the Deception
of Scientific Atheism

GEORGE COLEMAN CONRAD

ISBN: 1-4392-0127-7
ISBN-13: 9781439201275

Visit www.booksurge.com to order additional copies.

For my wife Deneise

and

my daughter Melissa

TABLE OF CONTENTS

THE GREATEST FALLACY EVER TOLD

PART I
THE STORY OF CREATION AND EVOLUTION WITHOUT GOD
- THE BIG PICTURE -

PART II
ASSEMBLING A LIFE-FRIENDLY WORLD WITHOUT GOD
- A CLOSER LOOK AT 'ACCIDENTAL' NECESSITIES -

PART III
ASSEMBLING LIFE WITHOUT GOD
- A CLOSER LOOK AT 'ACCIDENTAL' NECESSITIES -

PART IV
DIVERSIFYING LIFE WITHOUT GOD
- A CLOSER LOOK AT 'ACCIDENTAL' NECESSITIES AND
THE AMAZING RESULTS -

PART V
THE SIGNIFICANCE OF INFORMATION

PART VI
SCIENTIFIC BELIEFS

PART VII
THE ATHEISTIC RELIGION OF
SCIENCE HARMS SOCIETY

THE GREATEST FALLACY EVER TOLD

Throughout recorded history, the elite scientists of each era have adamantly believed in fallacies that they proclaimed to be 'scientific truth'.

For nearly 2,000 years Aristotle's universe, with the Sun revolving around a stationary Earth, was the 'scientific truth' describing our solar system. Today, the evidential discoveries of physics clearly reveal that the Earth orbits the Sun. The 'scientific truth' of an Earth-centered universe has been proven to be a fallacy.

Charles Darwin and his scientific colleagues of the 19th century were absolutely convinced of the 'scientific truth' that the 'pangenes' of a mother and father 'preformed' a very small person who simply grew larger and larger within the mother's womb. Today, expectant parents can actually see the evidence for themselves that such, indeed, is a fallacy. On the first visit to the obstetrician's office they can view ultrasound images of the tiny forming embryo that looks nothing at all like Mom and Dad.

Until the 20th century, elite scientists proclaimed the 'scientific truth' of the spontaneous generation of life from non-living matter, whereby frogs grew out of mud, rats grew out of garbage, and flies were born from rotting meat. Today, scientific discoveries of the life sciences clearly reveal that such a 'scientific truth' was really a fallacy.

These and many other 'scientific truths' of bygone days seem almost comical to us today. Scientists have discovered so much about the universe and life itself that such fallacies seem to be just remnants of the past. For, after all, today's elite scientists are the individuals who have discovered the evidence that disproves such fallacies.

Today, elite scientists proclaim as 'scientific truth' only those things that have been actually proven to be true by the evidential discoveries of science, don't they?

Actually, no. The elite scientists of 21st century America today proclaim as 'scientific truth' a fallacy that is far more insidious than any of the scientific fallacies past. This modern fallacy undermines the very foundation of western civilization.

Our most elite scientists do not believe in God. And they maintain that their belief in scientific atheism has been 'proven' to be true based on the factual evidence discovered by science. That is an outrageous misrepresentation. The discoveries of science have most certainly not revealed the non-existence of God.

God is simply the term used for ultimate reality. Everyone believes in an ultimate reality of some kind. For atheists, the ultimate reality is the 'Accidental God'. For theists, the ultimate reality is a 'Purposeful God'. No one on Earth can know for certain which of these two ultimate realities is true. Each of us must determine our foundational belief concerning which of the two is the most probable belief based on incomplete information. That requires faith. Atheism and theism are both faith-based religions.

Faith for atheists is based on an abiding belief in the power of chance and accident. Theistic faith is based on a belief in the power of purpose and design. Both beliefs are faith-based and neither should be taught as 'scientific truth' in science class. Yet, our elite scientists, who are our preeminent scientific authority figures, insist that the tenets of their faith in scientific atheism must be taught as 'scientific truth'. Such religious instruction in the faith of scientific atheism should be banned from the science classes of our public schools.

Atheism and theism are both faith-based religions. This primer was written to enable each of us to decide for ourselves, which of these two faith-based religions is better supported by the factual discoveries of science. Thoughtful examination will thereby focus the light of reason and common sense on the greatest fallacy ever told.

CHAPTER 1
The Fallacy and Deception of Scientific Atheism

Do not be deceived by the claim made by America's most elite scientists that they have discovered the **ultimate 'scientific truth'.** These brilliant scientists now proclaim that they have compelling evidence **that the universe and life itself began and evolved without purpose and without intention.** All things just happened by chance, without design and without meaning. All that exists is the result of a cosmic accident. Their claim is both erroneous and deceptive. **It is the greatest fallacy ever told,** and it is a fallacy that is harmful for American society.

When these brilliant scientists proclaim that the discoveries of science evidence that the universe and life itself began and evolved with no need of a creative, intelligent agent, they are not stating a scientific fact. They are expressing their metaphysical belief, called **'scientific atheism',** and are paying homage to an **Accidental God. The discoveries of science show no such thing.**

My intention in writing the book, that you have now begun to read, is to subject the fallacy and deception of scientific atheism to the tests of scientific fact, reason and common sense.

On July 23, 1998 an article appeared in the distinguished science publication *Nature* that was entitled "Leading Scientists Still Reject God". That article reported the findings of a survey of the membership of the National Academy of Sciences (NAS), the most distinguished group of scientists in America. Over half of the membership responded to the survey. The findings were quite telling. On average only 7% of those eminent scientists believe in a purposeful God. Mathematicians had the highest rate of belief (14.3%); physicists and astronomers much less (7.5%); and biological scientists had the lowest rate of belief in God (5.5%). Put another way, over 90% of the most eminent scientists in America do not believe in God.

The members of the National Academy of Sciences are the really smart guys of science. They are the elite scientists in America. They set the stage for telling other scientists and ordinary folks the 'truth' about scientific discoveries. They are a whole lot smarter than I am, and their brilliance is quite intimidating.

The world of science and mathematics has always intimidated most people, including me. That world has always seemed to possess an aura of great mystery. Hidden answers to the mysteries of the universe seem to dwell within. The hidden answers seem to remain forever unknown except to the brilliant minds that can penetrate that secret world.

There must be a compelling reason for these brilliant scientists to not believe in God. I thought that they must have discovered scientific evidence that unravels the mysteries of the universe and life itself that clearly reveals that God does not exist, for why else would they so overwhelmingly support the belief that God does not exist? It would certainly be good for folks like me to understand the factual basis for their belief. For the existence of God is a real important issue. If God exists there is a higher purpose and meaning for the universe and for life itself. If God does not exist there is no purpose and meaning for the universe, and the only purpose and meaning in life is that which we make up. That's a real important difference.

As a society, we empower these elite scientists as authority figures. We place our trust in them and expect them to tell us the truth. We become accustomed to accepting their knowledge as 'scientific truth' based solely on the brilliance of their authority. Yet, the question of ultimate purpose and meaning in life is far too important to everyone - scientist and layman alike - to simply concede that these gifted scientists actually know the true nature of ultimate reality. While we may not be able to 'do the math', we should be able to comprehend the factual basis for their belief.

So, I began to read and research a lot of the stuff of science on my own. I am very glad that I did. What I discovered for myself was most compelling. I found that, while I may not be able to penetrate the secret mathematical world of science, I am able to comprehend the wonders of the natural and living world that science has discovered. And, most importantly, I discovered that the actual evidence adduced by science directly contradicts the belief held by these most-brilliant scientists that God does not exist.

The belief of brilliant scientists that God does not exist is not based on the discoveries of science. That belief is solidly founded on the 'idea' that nothing beyond nature can possibly exist. The belief is called **scientific atheism** and it has become a fervent religious belief. It is different and more insidious than all other religious beliefs for it is cloaked in the 'authority of science'.

The National Academy of Sciences first provided a bright light to illuminate their belief through the publication *Science and Creationism: A View from the National Academy of Sciences*. The 2nd edition was published in 1999. Then, in 2008 the Institute of Medicine of the National Academy of Sciences published a book entitled *Science, Evolution and Creationism*. These publications provide the 'scientific evidence' in support of the atheistic belief maintained by the overwhelming majority of the NAS membership.

These publications provide a clarion call from the elite scientists of America for teaching 'scientific atheism' in the science classrooms of our public schools. These books are a not-very-subtle effort to require that the 'scientific truth' be taught that the universe and life itself is nothing other than the result of chance and accident. They make it very clear that any mention of planning or design should be banned from the classroom, for it carries religious overtones that are contrary to science.

Since the National Academy of Sciences is regarded as the highest scientific authority in America, their teachings are widely held to be most factual. Those teachings are regarded by most mainstream scientists and laymen alike to be 'scientific truths'. In these books the NAS far more than infers that atheism is a factual 'truth', they provide 'scientific evidence' in support of the atheistic belief maintained by the overwhelming majority of the NAS membership.

Simply put, I found the pronouncements of the National Academy of Sciences to be contrary to my innate 'common sense'. But, if science has in fact made discoveries that evidence that the universe and life itself are the result of unintentional cosmic accidents, so be it. I can change my worldview based on the evidence presented. That is what reasonable, intelligent people do. So, I undertook to look behind the cloak of scientific authority to determine for myself if the discoveries of science actually support the claims of the National Academy of Sciences.

I am writing this book because I have found that the NAS publications are simply an outrageous attempt to justify the teaching of atheism in public schools. They undermine the search for truth and meaning that is the birthright of every human being. When anyone says that they have the definitive answers to the ultimate truths of the universe and life, they become most suspect as preaching their beliefs. And, indeed, the NAS publications preach the scientific atheistic beliefs of the vast majority of NAS scientists.

By publishing these books the NAS denigrates the beliefs and the work of other gifted scientists who do not conform to the party line. The NAS is clearly placing anyone who may believe in 'intelligent design' based on the evidence discovered by science into the camp of the 'unscientific'. To the NAS the work of scientists who believe in 'intelligent design' has no value and any hint of planning or design should be forbidden in scientific journals. Banning the consideration of planning and intelligence from the disciplines of science in this manner is simply a travesty. It is an affront to the search for truth and it is the imprimatur of scientific bigotry.

I have no religious credentials. NONE. I am not a born-again, evangelical Christian. I am not a member of any church or any organized religion. I do, however, believe in a purposeful God. And, quite relevantly, many of the underpinnings of that deep belief are based on the discoveries made by these brilliant scientists.

I have no scientific credentials. NONE. No scientific credentials are necessary to challenge the atheistic beliefs of these most-eminent scientists. They have every right to believe in atheism, even if the facts discovered by science undermine that belief. But, they have no right to contend by virtue of their scientific authority that their atheistic belief has been confirmed by scientific evidence if no such evidence in fact exists.

I wholly accept the scientific discoveries made by science as true. But, I wholly reject the contorted conclusion that science has discovered that the mechanism of chance and accident is the operative principle of nature that has produced the wonders discovered by science. My rejection of that contorted conclusion is based on logic and reason applied to the compelling evidence discovered by science.

Compelling evidence discovered by science supports the conclusion that the universe was created from a singularity, an infinitesimally tiny point wherein the laws of nature did not obtain, about 14 billion years ago. Thereafter the matter and energy of the universe evolved in strict accordance with the laws of physics and chemistry that emerged from that creation event. Yet, no scientific evidence has been discovered to explain the actual 'mechanism' through which the matter and energy and forces and laws of the universe were created. Scientific atheists believe that the creation and evolution of the universe occurred by chance, not design. Yet, there is absolutely no scientific evidence in support of the conclusion that the universe began and evolved as the result of pure chance, with no planning involved. There is no evidence that the universe was simply the result of an unplanned cosmic accident.

Compelling evidence discovered by science supports the conclusion that all biological organisms living on Earth today evolved from common ancestors by means of 'natural selection'. Yet, no scientific evidence has been discovered to explain the actual 'mechanism' through which natural selection operates. Scientific atheists believe that the creation and evolution of life occurred by chance, not design. Yet, there is absolutely no scientific evidence in support of the conclusion that life began and evolved through the process of undirected natural selection of 'random mutations' working on groupings of non-living molecules, with no planning involved.

Atheists Believe in an 'Accidental God'
In common parlance one who believes in the existence of God is known as a theist, while one who does not believe in the existence of God is called an atheist. Such an atheistic belief is really a misnomer. God is a metaphysical concept, defined quite simply as the supreme or ultimate reality.

Even the staunchest atheist believes in an ultimate reality of some kind. For atheists that ultimate reality is unplanned. Ultimate reality is chance. Chance is the operative principle whereby things happen without intention or purpose. An accidental event is one that occurs by chance, without intent, without purpose. The product of pure chance. The atheist's God is an 'Accidental God'.

Theists Believe in a 'Purposeful God'
Purpose is something set up as an object or end to be attained. A purposeful event is one that occurs as an object or end to be intentionally attained. It is the product of intention and planning. The theist's God is a 'Purposeful God'.

The term 'Purposeful God' as used in this book connotes absolutely no theological meaning. It simply refers to a creative intelligent agent. The existence of a creative intelligent agent necessarily implies that there is real purpose and meaning in universal development and life. It in no way implies that anyone is capable of scientifically determining what that purpose and meaning may be. This book is not about what the purpose and meaning of life is. This book is about whether there is any purpose and meaning in life at all, other than that which we invent. If there is a 'Purposeful God', there is. If there is not a 'Purposeful God', there is not. That is a really important point.

Is There Purpose and Meaning in Life?

For each of us the first step toward the discovery of purpose and meaning in life is to determine whether or not there is in fact any real purpose and meaning at all. That determination is the very foundation of our overall worldview.

It turns out that a scientist's answer to the question: "Is there real purpose and meaning in the universe and in life?" is not so clear-cut today. Most eminent scientists say 'no'. There is no 'Purposeful God'. There is no purpose and meaning in life other than what we make up. However, a minority group of scientists still believe in a 'Purposeful God'.

We live in what has been termed a post-modern world. One hallmark of this post-modern world is the widely-held belief that science evolves eternal truths. Most people in western society accept that today. Yet, few of us are scientists. For that matter, few of us have even the foggiest of ideas of what science really is.

The post-modern world began following the last great epiphanies of science of the 20th century, namely:

- Einstein's theories of relativity;
- quantum mechanics of particle physics; and
- unveiling the secrets of DNA and mapping the human genome.

We have thereby learned and demonstrated, quite literally, how to fly to the moon, blow up the world, and clone ourselves. But the knowledge of how to do those things is possessed by a minuscule number of people.

The world we live in today is one of immense complexity and specialization. Not only a world of complexity and specialization, but one wherein the busyness of multi-tasking is the norm and 'speed is life'. We don't have time to think, and we certainly don't have the time, or the energy, to think about such things as purpose and meaning. Even if we did have the time and the energy to think about such things, where would we even begin to formulate a factual basis for such an inquiry? Information in today's world is simply overwhelming. We have to rely on expert authority to tell us the 'truth' about things. Don't we?

The answer is both yes and no.

The Yes Answer

We do have to rely on expert scientists to tell us the truth about the wonders of the physical world. Such truth must be based on evidence that has been discovered by brilliant scientists.

We don't often think about it, but there are two, and only two, ways that we gain knowledge about stuff:
- what we come to know through our personal experience; and
- what we are told based on the authority of other people.

And we rely very, very heavily on authority figures.

Most things that we know are derived from the authority of others. That is actually a good thing, because much of what we think we know from our personal experience turns out to be not true.

Our personal experience is based on our five senses and can tell us things that are in fact not true. Aristotle reasoned from his personal experience that the Earth was the center of the universe and that the Sun revolved around the Earth. Personal experience would make me an Aristotelian. What I personally observe is that the Earth stands still and that the Sun travels each day through the sky above from east to west. Only the authority of science tells us this true reality: that the Earth is rotating on its axis at a speed of over 900 mph; that the Earth is orbiting around the Sun once each year; and that the Sun and the Earth and the other planets in our solar system are traveling around the Milky Way galaxy at a speed far in excess of a million miles a day in an orbit that takes over 200 million years to accomplish. Brilliant scientists have made factual discoveries of these things and we rely on the authority of science to tell us these factual truths.

My personal experience tells me that the chair that I am sitting on is made of stable, solid wood. The authority of science tells us that the stable solid wood is in reality composed of various molecular compounds that are bonded together through electrical connections occurring in the outermost shells of the atoms composing those compounds. And, the components of the atoms themselves are anything but stable. They are spinning, and the electrons are revolving around the tiny, tiny nucleus of each atom, and the atom itself is composed almost entirely of open space. Brilliant scientists have made factual discoveries of these things and we rely on the authority of science to tell us the true reality of the chair I am sitting on. And, that knowledge is quite different than what I gain through personal experience.

The point is quite simple. We trust in science. And, we rely on scientists to tell us the truth about the factual nature of things based on the evidentiary discoveries that they have actually made. We as a modern society rely on authority to tell us the truth and we go to school for a long time to learn these truths. The scientific truths that we learn in science class should be based on facts, not conjectures.

The No Answer

We do not have to rely on expert authority to tell us the truth about purpose and meaning in life. Each of us must discover our own core belief concerning purpose and meaning in life. And, that discovery should be based on the best 'truth' that we can adduce. Of necessity, that discovery will be greatly influenced by others. The facts discovered by authority figures will be extremely useful in enabling us to discover the 'truth'. But, our individual human intelligence and dignity is ennobled when we examine the bases for our belief through our own human reason, independent of the human bias of authority figures.

Life is much simpler for us if we rely on authorities to supply us with easy answers. Answers that we feel we can trust. For answers to the ultimate questions of purpose and meaning we, as a society, have always relied on the answers supplied by religion and science. Most fortunately for human civilization, science and religion have largely been in lock-step regarding these questions throughout history. **Religious authorities and scientific authorities traditionally have shared a common worldview that served them both.** The foundation of this worldview has been cloaked in different garbs in many different societies, but has, at core, always been the same. In essence, this worldview provides that universal world order was created and maintained by the laws established by a creator who intelligently designed the universe and all therein. Both clergymen and scientists largely agreed on this worldview that proclaimed the universe and the lives of mankind to have real purpose and meaning.

During the past few decades **a new worldview** has grown to become the mainstream view of the modern scientific community. It is **a worldview based on 'naturalism' which maintains that scientific laws of nature alone are adequate to account for all phenomena, and that the laws themselves need no explanation. All phenomena are, at core, the result of the laws of nature working on blind chance. There is no need for a 'Purposeful God'. Since there is no need for a 'Purposeful God', there is no purpose and meaning in life other than what we make up.**

A real conundrum now presents itself to us. The 'truth' of mainstream religion is now opposed to the 'truth' of mainstream science. There is no easy answer to that conundrum (an intricate, perplexing, baffling, and mysterious problem). One of the 'truths' has to be wrong. We may not know a lot but, for sure, we can figure that one out. Something is fundamentally wrong.

Religious leaders don't talk about the polar differences in the worldviews between mainstream science and mainstream religion. They don't want to acknowledge that the rift with science is so large.

Of course, elite and mainstream scientists never state the matter so bluntly. They insist that they have done no such thing as to devoid life of purpose and meaning. All they have done, they say, is to show us the 'truth' of their discovery that a 'Purposeful God' is not necessary to explain either the creation or evolution of the natural universe and life itself. They say that everything, at core, has simply evolved through chance alone. A 'Purposeful God' was and is not necessary. If we ordinary folks want to believe in a purposeful, intelligent God, that's just fine with them, but we are pretty stupid if we do. The really smart guys know the real 'truth' that no such God exists. That 'truth' must be acknowledged by any scientist who does not want to be ostracized from the respected scientific community, and that 'truth' must be presented in the science classes of public schools.

Most of us have a great reverence for science and scientists. Little wonder in that. We believe in science for we see daily the benefits and the marvels that scientists have brought into our lives. Household computers, cell phones and plasma TVs have become part of our daily utensils. Surgeons are using lasers to remove tumors from our bodies and correct lifelong eyesight deficiencies to leave us free of eyeglasses forever. Organ transplants are everyday occurrences, wonder drugs ease our pain and suffering, DNA matches are used to convict criminals, and the human genome has been mapped. The scientists are the smart guys who are doing all this incredible stuff. No one thinks they are making this stuff up.

We see the truthful results of science each and every day of our lives. We trust science. Scientists base their worldview on facts, don't they? If these really intelligent guys have factually determined that everything in the universe is the result of mere chance, that there is no 'Purposeful God', and if I trust their authority in revealing that to me as 'truth', then the fact is that there is no real purpose and meaning in life.

If there is no real purpose and meaning in life, it would really be stupid to revere a 'Purposeful God', for science has determined that there is no such God. It would really be stupid to believe in any objective moral standards to live by. It would be really stupid to be hard-working or kind or generous or forgiving or loving or caring or anything else except a self-centered pleasure-seeker. Not much of a basis for a civilization.

Some of the underlying unrest in America today is due to the queasy uncertainty that most intelligent people feel because we trust science to tell us the truth but, at the same time, we want to reject the worldview of the elite and mainstream scientific community that there is no purpose and meaning in life. We yearn to be purposeful. We yearn to be special in the eyes of God. We yearn to be rewarded for living a moral and righteous life. And, we want our yearning not to be stupid. Our most-human answer to the queasiness is to ignore it. We exhaust ourselves with the busyness of living and ban the conundrum from our minds. The queasiness remains.

The Ultimate Question

Life either has real purpose and meaning or it does not. One or the other. Queasy uncertainty about whether there is real purpose and meaning in life provides for a fretful existence.

In order to get rid of that queasy uncertainty, it is necessary to examine a great deal about the fundamentals of both living and non-living things in our universe. That examination is intended to provide the perspective necessary to allow the reader to make an informed judgment concerning the fundamental underlying basis for the most truthful worldview. Did all this stuff occur by chance or not-by-chance? 'Accidental God' or 'Purposeful God'?

Chance or not-by-chance? 'Accidental God' or 'Purposeful God'? That is the bottom-line question concerning worldviews. It is the basis of whether there is real purpose and meaning in the universe and in life.

Why This Book Was Written
This book was written to make one point and to examine a second point:

* Point to be made: **No religion, neither theism nor atheism, should be taught in science class.**
* Point to be examined: **Do the discoveries of science better support:**
  (a) accident, or
  (b) intention and planning
  as the best explanation for the creation and evolution of the universe and the creation and evolution of life on Earth?

Point # 1
All Religions Should be Banned from Science Class
This country's most brilliant scientists are the elite members of the National Academy of Sciences. They are the guys who can really 'do the math'. They thereby access a world hidden from the rest of us, a world of mathematical precision and harmony and beauty that serves as the very foundation for the reality of the world we live in. The members of the NAS are the role models and teachers of mainstream scientists and are the acknowledged authority figures for all things scientific.

These scientists employ the 'scientific method' to rigorously examine the natural world in which we live. That method is necessarily limited to natural phenomena.

These scientists deal exclusively with natural phenomena. They exclusively use the scientific method to examine those natural phenomena. They thereby came to believe that nothing beyond the natural world can possibly exist.

It is quite understandable that they became wedded to the belief that nothing beyond the natural world can exist. That belief is called scientific atheism. Yet that belief is not a factual discovery of science. It is not a 'scientific truth'. It is a metaphysical belief that is founded on a firm conviction that ultimate reality is unplanned. These scientists believe that ultimate reality is chance. Chance is the operative principle whereby things happen without intention or purpose. That ultimate reality is thereby accidental. Behold the Accidental God.

No one knows if the universe began as an accident or as a planned event. But these brilliant scientists are committed to the belief that the universe began and evolved as a cosmic accident.

No one knows if life on Earth began and evolved as an accident or as a planned event. But these brilliant scientists are committed to the belief that the creation and evolution of life was the result of chance, accidental and unplanned events.

In America everyone is entitled to believe in any God that they choose, whether Purposeful God or Accidental God. But, no God, neither Purposeful nor Accidental, should be glorified as the 'truth' in the science classes of our public schools. Belief in an Accidental God is the religion of scientific atheism. It is a metaphysical religious belief and should not be taught in science class.

Point # 2
Purposeful God or Accidental God?

Do the discoveries of science better support atheism or theism as an overall worldview? If you have read this far, you know that there is no suspense about my conclusion. But, your reasoned conclusion may be different from mine. Each of us should make a reasoned decision based on the best evidence available. We should each base our core beliefs on our individual reasoning, and not simply adopt the beliefs held by authority figures or anyone else.

This primer is an honest effort to identify and examine the scientific fundamentals of the world we live in, as revealed to us through the painstaking efforts of brilliant and dedicated scientists. It is acknowledged at the outset that this will be an elementary work. That is why it is called a primer. The enormity of the subjects to be examined dictates that. However, an attempt will be made to weave the fundamentals of science into a meaningful fabric of basic knowledge about the world we live in. If the attempt is successful it may result in a fuller perspective regarding the world we live in and ourselves. As you proceed through the chapters ahead you should gain your own perspective of whether the discoveries of science better support atheism or theism as an overall worldview.

The focus of this book centers on the elite and mainstream scientific worldview of atheism. The first four parts of the primer consist of basic background information. If you have a background in basic cosmology, biology, and the physical sciences, feel free to turn directly to begin with Part V. If you do not have a background in science, you may find the materials contained in the first four parts to be helpful for gaining some rudimentary scientific knowledge necessary to gain perspective.

The chapters in Part I will recount the epic scientific story of creation and evolution of the universe and life without God, without will and purpose. What was necessary for those events to happen by accident, without will or purpose. Part II will take a closer look at the details of the creation and evolution of the universe without God. Parts III and IV will take a closer look at the details of the creation and evolution of life without God.

Part V will review the recent discoveries of science concerning the information found within all living things and examine the scientific significance of the information in life.

Part VI will review the history of scientific beliefs and the discoveries of science that resulted in changing scientific beliefs. Finally, Part VII will examine why the elitist religion of scientific atheism is harmful for society.

The Purpose of This Primer

The purpose of this primer is to directly challenge the bold assertions of the scientific elite of this country that:

- the universe began and has evolved without planning, without intention and without purpose; and
- life itself began and has evolved without planning, without intention and without purpose.

These assertions are presented as scientific facts by our most gifted scientists. And these most-gifted scientists insist that these assertions must be taught as 'scientific truth' in the science classrooms of our public schools.

These assertions are not supported by factual discoveries that have been made through painstaking scientific research. To the contrary, the evidential discoveries of science blatantly contradict such assertions. Yet, our most brilliant scientists seem to be oblivious to these most-obvious contradictions. They steadfastly maintain that no planning, no intention, and no purpose is ever involved in the creation and evolution of the natural world and the creation and evolution of the living world.

A pariah is an outcast. The elite scientists of this country have deemed any contrarian scientists who discern any actual design or planning in the natural and living worlds to be pariahs. These scientists become outcasts from the mainstream scientific community and their works are denied publication in peer-reviewed scientific journals. It seems that the party line of scientific atheism must be maintained at all costs, even at the expense of the reputations of distinguished fellow scientists who refuse to follow the party line. And, that is just plain wrong.

The Thesis of This Primer

The thesis of this primer is quite simple. It is a thesis of common sense, as follows:

> If intelligent people are armed with basic factual knowledge about the many wonders of the world we live in, the genius of common sense can determine the most probable 'truth'. If the fundamental basics can be extracted from the overwhelming mass of information detail and can be presented in a straight-forward manner, then intelligent people can discover and decide their own core beliefs concerning purpose and meaning in life based on the best facts available. Human intelligence and dignity are ennobled when we examine the bases for our beliefs through our own human reason, independent of the bias of authority figures.

We have more information available to us today at the beginning of the 21st century than any society has ever before dreamed of. But, the mass of information available is really useless unless it leads us to knowledge. And, knowledge is never fully useful unless it leads us to wisdom.

This primer is intended to challenge the reader to engage in independent thought. Most appropriately, it is dedicated to two of the most thoughtful people on the planet, with whom I am privileged to share my life: my wife Deneise and my daughter Melissa.

If others have the patience to read this book, it is my hope that the reader finds it to be of some small assistance in the quest for purpose and meaning in life that we all pursue.

It is hoped that upon completing this book both the author and the reader become a little wiser.

PART I

THE STORY OF CREATION AND EVOLUTION
WITHOUT GOD - THE BIG PICTURE -

Over 90% of the most eminent scientists in America do not believe in a Purposeful God. They believe that the scientific laws of nature themselves are adequate to account for all phenomena, and that the laws themselves need no explanation. To them the supreme or ultimate reality is chance alone - accidental. Belief in such an Accidental God underlies their story of creation and evolution with no need for a real God.

CHAPTER 2
The Start of It All

The biggest thing discovered to date by science is the universe. It is absolutely immense. Cosmologists estimate that the universe began about 14 billion years ago and continues to expand at about the speed of light. That's a speed of 186,000 miles per second, or roughly 11 million miles per minute, 660 million miles per hour, 16 billion miles per day. Light is composed of photons, particles of light energy rocketing through the universe at this great speed in electromagnetic waves.

We can try to imagine the size of the universe as an area comprised of 16 billion miles covered each day for a period of 14 billion years. We can try, but we really can't grasp it. It is certainly a big thing. And it has gotten several million miles bigger since you started to read this paragraph. The universe is not just big. It is beyond human scale or comprehension.

The Big Bang

Science tells us that at some precise moment some 14 billion or so years ago, our universe quite literally began with a bang. Extreme space-time curvature had resulted in all matter and energy and forces of the universe being compressed into a single point of infinite density. The immense compression (we are not exactly sure how that happened) centered in a hot point of infinite density, a single point called a **singularity**, where the laws of physics did not exist. **The 'Big Bang'** start of the universe occurred when, from that singularity, a rapid expansion of hot dense gas was emitted (we are not exactly sure how that happened either). Everything that exists anywhere in our universe today is a remnant of that initial explosion.

The universe began at a single point - a singularity. The singularity preceded known physical laws. The singularity preceded matter, energy, space, and time.

Dr. Arno Penzias, recipient of the Nobel Prize for Physics, explains ultimate scientific beginnings in this manner:

> "In order to achieve consistency with our observations we must, according to Einstein's General relativity, assume not only creation of

matter and energy out of nothing, but creation of space and time as well. Moreover, this creation must be very delicately balanced. The amount of energy given to the emerging matter must be enough to move it fast enough to escape the bonds of gravity, but not so fast that the particles lose all contact with each other. Enough of the initially-created matter must pull together under gravity to form galaxies, stars, and planetary systems which allow for life. Thus the second 'improbable' property of the early universe, almost as improbable as creation out of nothing, is an exquisitely delicate balance between matter and energy. Third - and this one puzzles scientists at least as much as the first two - somehow all these pieces, each without having any proper contact with the others, without having any way of communication, *all* must have appeared with the same balance between matter and energy at the same instant."

What preceded the huge explosion, the 'Big Bang'? The scientific answer is simple: We don't know. Some scientists do, however, wonder.

Renowned scientist, Heinz Pagels, provides this:

"The nothingness 'before' the creation of the universe is the most complete void that we can imagine - no space, time or matter existed. It is a world without place, without duration or eternity, without number - it is what mathematicians call 'empty set.' Yet this unthinkable void converts itself into the plenum of existence - a necessary consequence of physical laws. Where are these laws written into that void? What 'tells' the void that it is pregnant with a possible universe? It would seem that even the void is subject to law, a logic that exists prior to space and time."

And further wonder is provided by the notable theoretical physicist, Steven Hawking, author of *A Brief History of Time* and *The Theory of Everything*:

"What is it that breathes fire into the equations and makes a universe for them to describe? The usual approach of science of constructing a mathematical model cannot answer the question of why there should be a universe for the model to describe. Why does the universe go to all the bother of existing?"

In short, no one scientifically knows why. But it does. And, here we are.

The First Ten Billion Years

At a very brief instant after the 'Big Bang', the physical laws of the universe sprang into being. Scientists have denoted the exact instant as a mere fraction of a second following the 'Big Bang'. The fraction is indeed minuscule, calculated as 0.0000000000000000000000000000000000000000001 (10 to the minus 43rd power) of a second. That's pretty quick. The moment is known as the Planck Epoch in honor of quantum physicist Max Planck. Before the Planck Epic the known physical laws of the universe that cover the interactions of all of the fundamental particles of matter and energy and forces, and time and space itself, did not exit.

How those laws happened to spring into being we don't scientifically know. But science has discovered more and more of and about those amazing physical laws. Science does not invent those laws. Science just discovers them. They are 'out there'. And, the same seems to be true for one of the most important tools of science, mathematics. Mathematics seems to be 'out there' as well, and waiting for our most brilliant scientific minds to make further discoveries about mathematics and about our universe.

Let's review the fundamental forces and particles of matter that were instantly created out of the 'Big Bang'. The 'Big Bang' created:

- the force of gravity;
- the electromagnetic force (that is the force that creates electrical and magnetic fields);
- the strong nuclear force (that is the force that holds the nucleus of an atom together); and
- the weak nuclear force (that is the force that causes radioactive decay of neutrons).

The 'Big Bang' created the initial and irreducible particles of ordinary matter. Those particles are called quarks and electrons. The quarks combine together to make the protons and neutrons which comprise the nucleus of each atom. The nucleus of each atom has a positive electrical charge. The electrons are the irreducible particles that orbit around the nucleus of each atom. The electrons have a negative electrical charge that exactly off-sets the positive electrical charge of the nucleus.

Those are the fundamental forces of the universe and the irreducible elementary particles that make up all ordinary matter in the universe. Those are the basics.

As the physical particles of matter and energy and the fundamental forces appeared, all that followed the 'Big Bang' proceeded in accordance with the physical laws of the universe that are commonly called the laws of nature. The laws of nature govern the interactions of the fundamental particles and the fundamental forces that somehow emerged from the singularity that preceded the 'Big Bang'.

The cooling of the universe allowed particles to come together, but the new physical laws of the universe dictated that the unbelievable heat from the creation event would keep particles of matter whizzing about without taking much shape for some time to come. Science provides the following rough outline of early universal development.

On Your Mark, Get Set - Go

At the Planck Epoch all matter, energy, forces, space and time were combined in an unimaginably hot and unimaginably dense point (less than 10 to the minus 33rd power of a centimeter). The laws of nature did not exist. Then, in less than a billion trillionth of a second - Bang.

At 10 to the minus 35th power of a second after the 'Big Bang' the universe burst forth with incredible and rapid expansion. Each of the fundamental forces had its turn at center stage, if only very briefly, as the universe unfolded. In the first instant the strong nuclear force prevailed, followed quickly by electromagnetism. Then the weak nuclear force had a few seconds of predominance before the force of gravity became the macroscopic winner for all time thereafter.

At 10 to the minus 5th power of a second, the irreducible quarks began to stick together and to thereby form protons and neutrons. The other irreducible particles of ordinary matter, electrons, whizzed about wildly. The intense nuclear furnace of creation also emitted vast numbers of unstable high-energy sub-atomic particles that rapidly decayed and did not survive to become part of ordinary matter. However, their initial role of bombarding the quarks and electrons affected the early formation of the universe.

At extremely hot temperatures, particles move about so fast that the electromagnetic and nuclear forces are not strong enough to attract the particles so that they can stick to each other. When temperatures cool enough, the fundamental forces can then attract those atomic particles together.

About two minutes after the 'Big Bang' things had cooled to the point where protons and neutrons could stick together and thereby form nuclei of the fundamental elements of the universe, hydrogen and helium. (Note: The nucleus of a hydrogen atom contains but a single proton, the nucleus of a 'heavy' hydrogen atom, called deuterium, contains one proton and one neutron, while the nucleus of a helium atom contains two protons and two neutrons). At this two-minute mark the ratio of element nuclei was about 75% hydrogen and 25% helium. That was roughly the state of things from about two minutes out until about half-a-million years after the 'Big Bang'.

All the ordinary matter that existed in the universe at that time were free electrons, atom nuclei, and electromagnetic radiation photons of light that had been generated by the 'Big Bang' and which spewed throughout the cosmos.

The electrons could not yet get into orbit around the nuclei and thereby form atoms, because those unstable sub-atomic particles and other whizzing electrons or photons from the electromagnetic radiation would knock them out of orbit.

This was the time before atoms were formed. And in that unformed state, the electromagnetic radiation got absorbed by a proton or electron that vibrated in response to the wave of the electrical field. This made the universe opaque. Matter and photons of electromagnetic radiation were as one.

Then, about half-a-million years after the start of it all, the universe had sufficiently cooled to the point that atoms could form. That was a really big deal. Once atoms had formed, matter could begin to condense into galaxies and stars.

The electrons now joined in orbit around the nuclei, thereby **forming atoms with a neutral electric charge.** The reactions with the electromagnetic waves of radiation were thereby dramatically reduced. Photon radiation and atoms of matter parted company. **The universe became transparent, and a gigantic burst of electromagnetic radiation was thereby released.** That resulted in another explosion with a 'Great Flash' of light spreading throughout the universe.

The photons that were released continued to lose energy as the universe further cooled, and, as a consequence, their frequencies became lower. The release of that electromagnetic radiation spread uniformly throughout the cosmos and now comprises the cosmic microwave background radiation that exists throughout the universe today. It results in outer space having a temperature of about 3 degrees above absolute zero.

The Birth of a Galaxy

For untold ages after the formation of atoms of hydrogen and helium, the universe just grew colder and darker, and colder and darker. If the universe had spread out with total uniformity following the 'Big Bang', no galaxies or stars would have ever developed. The only matter in the universe would be hydrogen and helium.

But, there is some scientific evidence that shock waves from the 'Big Bang' and the subsequent 'Great Flash' affected the newly forming matter of the universe by creating **irregularities**. Because of those irregularities, matter now began to clump together in different areas of the heavens. Galaxies were the first heavenly bodies to form.

The force of gravity brought gas particles together in greater and greater numbers. Vast gas clouds began slowly spinning and getting hotter. As gravitational collapse continued the early galaxies increased their rate of rotation. Many varieties of galaxies thereafter emerged in the universe under the force of gravity working on irregularities: ring galaxies, pinwheel galaxies, elliptical galaxies and spiral arm galaxies took shape.

The Birth of A Star

About a billion years after the start of it all, gasses in galactic areas began to swirl together into denser and denser structure through gravitational attraction. The first stars were thereby formed. As gravity further condensed the particles of gas, they collapsed further under the force of gravity into spinning eddies. Because there was no other force to act on it, the ever-growing mass continued to spin and was destined to continue to spin throughout its life.

The predominant atoms of hydrogen gas began to aggregate in large enough clusters that their mutual gravitational attraction caused them to begin to further collapse in upon themselves to form a core. In this way, one star, then another, and another…began to form.

The hydrogen gas gravitational contraction in a star's core causes the hydrogen atoms to collide faster and faster with each other. That motion causes - heat. The hydrogen gas continues to heat up more and more. When the heat becomes intense enough the atoms stop bouncing off each other. They now merge, or fuse, together with each other. This thermonuclear reaction creates a super-intense heat. It is called **nuclear fusion**. This super-intense heat increases the pressure of the gases to the point that the gas pressure exactly balances the pressure of the gravitational contraction. The gases then stop contracting and will remain in a steady state (or stasis), steadily burning hydrogen for a very long period of time. A star is born. And the heat and light that is released by these steady and stable atomic reactions produces sunshine.

The Life and Death of a Star and A Galaxy

Once a star has reached its stable state of stasis it will burn its hydrogen fuel to make helium for a long time, in fact, until it runs out of it. Science has discovered that our Sun has been doing this for about five billion years and should continue to do so for yet another five billion. Four protons of hydrogen atoms fuse to produce the nucleus of a helium atom. The process seems simple, but actually consists of five discrete steps. And, the result has been amazing for planet Earth and its inhabitants.

If a star is large enough, when all of the hydrogen has been converted to helium, the zone of atomic fusion expands outward. As that happens, the helium nuclei are pressed closer and closer together and a second stage fusion reaction begins to convert helium into lithium, then beryllium, boron, then carbon and nitrogen and oxygen. The thermonuclear reaction continues through the process of adding further helium nuclei to create successively heavier elements, like sodium, magnesium, silicon, sulfur and potassium until finally manganese burns to make iron. Iron is a stable element. You cannot extract further energy from iron by the star's normal thermonuclear process. The star has become a heavy mass of heavy elements. And, it has stopped burning and no longer gives off sunshine. What happens then?

What happens to a star after it has completed its thermonuclear reactions, has run out of fuel, and become 'element heavy', depends largely on its size and density when the burning process has been completed. Paradoxically, a large star will have a shorter life than some smaller stars because it needs to burn more and more fuel more rapidly in order to counter-balance the greater force of gravity applied to its greater mass.

When a star runs out of fuel, it will again start to contract. The gas pressure produced by the burning of fuels no longer counter-balances the pressure of gravity. Gravitational pressure thereby becomes more and more intense. The increased pressure compresses the particles of matter in the star's core more and more.

Brilliant quantum physicists have discovered that two particles of matter cannot have both the same velocity and occupy the same spatial position. Therefore, under the pressure of such immense compression they develop very different velocities. The difference in their velocities makes the star expand. This repulsive force of expansion now momentarily counter balances the gravitation attraction force. The result is a balance very similar to when the star was producing steady sunshine.

What happens after that balance is reached really depends on the mass and density that the star has upon reaching that point of potential equilibrium.

If the size and density are just right, the equilibrium will continue indefinitely and the star becomes a 'white dwarf star'. A '**white dwarf**' will have a radius of several thousand miles and a density of hundreds of tons per cubic inch.

Other stars, with a mass of about three to five times the size of our Sun, continue to contract after they reach the point of potential equilibrium. In less than one second, the star's final phase of gravitational collapse is reached. Iron atoms are further crushed together as the core temperature raises to over one billion degrees. A shock wave proceeds outward from the core. As the wave encounters the material in the star's outer layers, the superheated material is fused into new elements and radioactive isotopes heavier than iron. The extremely heavy stuff of the universe like gold and the radioactive elements are thereby created. The shock wave propels the matter from this 'exploding' star outward into space in what is called a '**supernova**'. This process occurred over and over in the first 9 or 10 billion years of the universe that elapsed before our Sun was born. Following the explosion, **the core becomes a 'neutron star'**. It will finish life with a radius of only about ten miles, and will have a density of hundreds of millions of tons per cubic inch. But, its death has provided a great service by spreading much of the mineral wealth of the universe.

Finally, the mass and density of some unfortunate stars will doom them to an even more inelegant fate. Their density and mass ratio condemns them to continue to compress and compress and get denser and denser until they reach a point (quite literally) wherein they are so dense that nothing, not even light, can escape their gravitational force. They become a 'black hole' in space.

On a more friendly note, the lonely and gloomy 'black hole' fate of a star may, at times, be shared by its neighbors. The gravitational collapse of all the stars and matter in an entire section of a galaxy is called a 'quasi-stellar object' or 'quasar'.

The same universal force of gravity that created the stars grouped the stars into galaxies, galaxies into galaxy clusters, and clusters into superclusters. And, all of these large structures are separated by vast voids of inter-galactic space. And, the voids become larger and larger between the galaxies as the universe continues to expand and expand.

An Expanding Universe

In 1929 Edwin Hubble discovered through his telescope at Mount Wilson near Los Angeles that the stars in distant galaxies are **red-shifted**. That discovery was profound because it provided proof of a non-static universe. Hubble had discovered an **expanding universe.** The following digression may prove useful to understanding Hubble's discovery.

Isaac Newton had done a lot of work with optics. He observed that when white light passes through a prism it breaks-up into the colors of the rainbow. The colors are arranged from low-frequency on one end (red) to high-frequency on the other end (blue, shading off into violet and ultra-violet). The blue light frequency is about double the frequency of red light.

In the early 19th century Christian Doppler, an Austrian physicist, discovered and documented what has come to be called the Doppler effect. He noted that the pitch and frequency of sound waves of an approaching train increases as the train nears and diminishes as it speeds past you. The change in pitch and frequency of sound waves can be used to accurately measure the speed of the train (or your car, as you may recall from your last speeding ticket).

Doppler extended the effect from sound waves to light waves. When a luminous object approaches you at high speed, the frequency of its light increases and it appears bluish. Its light is said to be blue-shifted. When a luminous object moves away from you at high speed the frequency of its light decreases and it appears reddish. Its light is said to be red-shifted. What Hubble observed atop Mount Wilson was that all the light from the stars above was red-shifted. He learned how to judge the distances to various galaxies and how to use the Doppler effect to measure their speed. He discovered Hubble's Law that the speed of the galaxies is in proportion to their distance. In other words, the further a galaxy is away from Earth, the faster it is moving away from us. This discovery was precisely what would be expected if all of the universe had started out at a single point at a single instant, with a 'Big Bang'.

As Edwin Hubble discovered, ever since the 'Big Bang' the universe has been expanding.

The idea of an expanding universe was anathema to Albert Einstein. The mathematics of his theory of relativity predicted that the universe was not static, that it had to be either expanding or contracting. But, he so disliked the idea of a non-static universe that he proposed a '**cosmological constant**' which would serve as a kind-of 'anti-gravity' force necessary to maintain a static universe. He had no evidence in support of such an anti-gravity force, so after Hubble's discovery he declared that the cosmological constant was the greatest scientific mistake he had ever made. Einstein was a scientist who followed the evidentiary trail wherever it led.

So, we know that the universe is expanding based on scientific evidence. Do we know if it will expand forever? That depends on how much 'mass' is in the universe.

Hidden Stuff

Since the universe is expanding the actual mass of the entire universe, the 'whole', is unknown. An estimate of the whole must suffice. Since inter-galactic space is scientifically viewed as a 'void', the total mass of all the observed galaxies in the universe is used to estimate the 'whole'. Back to Doppler.

The **Doppler shift** is used not only to determine the rate of acceleration of galaxies; it is **also used to determine the rate of rotation**. And, once the rate of rotation is known, the 'mass' of a galaxy can be mathematically calculated with great accuracy.

As brilliant mathematicians made these detailed calculations a haunting fact was discovered. The calculations didn't make sense. The problem was that the speed of rotation did not correspond to the computed mass of the stars and planets in the galaxy. The amount of mass was so relatively small that every observed galaxy would fly apart at the speed of rotation observed. The discrepancy between the mass necessary to maintain stable galaxies observed and the calculated mass is not small. Not by a long shot. The discrepancy is roughly 90%. Ninety percent.

Scientific evidence now revealed that only 10% of the known universe consists of stuff that we can see. The other 90% has never been seen. This 90% of the universe does not emit detectable electromagnetic radiation, or else we would see it. But, we know it is there because of its gravitational effects on the stars and galaxies that we do see. It is elusively called both **Dark Matter** and **Dark Energy**.

Dark Matter

What about the planets? Can't the missing mass be accounted for by them? After all, they don't emit electromagnetic radiation that we can observe. Might that not account for the missing mass? Science tells us - no. The planets in our solar system amount to far less than 1% of the Sun's mass. So planets orbiting other stars in the heavens above and throughout the universe wouldn't make a significant dent in accounting for the missing mass. What other kind of matter could it be?

Most scientists believe that most dark matter is not ordinary matter, but consists of sub-atomic particles that are 'inferred' from scientific calculations. **These particles may or may not exist**. They are affectionately called **'wimps'** (weakly interactive massive particles). Science has not proven that they exist. Wimps would pass through ordinary matter, including the Earth and our bodies, without detection. **Wimps would only effect ordinary matter gravitationally**.

Dark Energy

Hubble discovered that the **universe is expanding**. As scientific instruments have improved, scientists have been able to make better and better observations of galaxies in the distant universe and have concluded, based on the brightness of the most distant supernovas, that the rate of expansion is accelerating.

Some scientists believe that the only explanation for an expanding and accelerating universe is that the **energy content of a vacuum**

(in inter-galactic space) is non-zero with a negative pressure. This unknown vacuum energy is called **dark energy**. It is believed that dark energy may be an intrinsic, fundamental energy that uniformly fills otherwise empty inter-galactic space. Dark energy may be viewed as a 'cosmological constant' with anti-gravitational properties. Albert Einstein may not have been mistaken after all.

It is postulated that the inflationary outward pressure of dark energy more than offsets the deflationary pressure of gravity. So, the universe expands at an accelerating rate. The June, 2003 issue of *Science* reported that such dark energy, that no one has ever seen and which no one understands, may dominate the energy density of the universe.

The dark energy theory not only accounts for an accelerating expanding universe; it may also account for the lion's share of the missing mass of the universe. A rough approximation of the best scientific estimate for the composition of the universe as of this writing is:

- 10% ordinary matter;
- 20% dark matter; and
- 70% dark energy.

The Fate of the Universe

Cosmologists generally agree on what is known as the **Cosmological Principle**, to wit: matter in the expanding universe is homogeneous and isotropic. In essence, on a large scale, the universe is the same everywhere and the same in all directions.

And the fate of the universe will depend on what scientists have calculated as the required **critical density**. The value of the critical density is calculated to be quite small, on the order of six hydrogen atoms per cubic meter of space. But most important.

As the universe expands, its fate ultimately will be determined by its density. If the amount of matter in the universe is less than or equal to the critical density, the force of gravity will not be strong enough to stop the expansion and it will continue to expand as an 'open universe' forever. If there is more matter than the calculated critical density, then the force of gravity will eventually result in the universe collapsing back upon itself, a predicted feature of a 'closed universe'. Things would then run backwards, ultimately resulting in a 'Big Crunch'.

Theoretical physicist Steven Hawking has concluded that the expansion rate of the universe is incredibly fine-tuned and the fine-tuning had to occur within the first second of creation. In his book, *The Theory of Everything*, he provides this perspective.

> "If the rate of expansion one second after the big bang had been smaller by even one part in a hundred thousand million, million, the universe would have recollapsed before it ever reached its present size. On the other hand, if the expansion rate at one second had been larger by the same amount, the universe would have expanded so much that it would be effectively empty now."

Hawking notes that scientific evidence now suggests that the universe will expand forever, but his calculations show the universe to be on the very cusp of collapse and expansion. As he puts it:

> "The most remarkable thing about the universe is that it is so close to the borderline between open and closed. The probabilities against it being on such a borderline are enormous. Yet it is still so close that we haven't been able to decide which side it is on."

The prospect of a 'closed universe' that ultimately collapses back upon itself in a 'Big Crunch' only to be born again in another 'Big Bang' is known as an 'oscillating universe'. If the universe does oscillate and run backwards, some speculate that causality may become inverted, with effects preceding causes.

On a more comforting note, science predicts that whatever the ultimate fate of the universe happens to be, that fate won't happen for a long, long time. Billions and billions of years from now.

Science has discovered that after the universe evolved over a period of some nine or ten billion years, the Sun and the planets of our solar system began to take shape. So let's proceed now to the story of how our solar system was formed and how our home planet, Earth, was born and developed.

CHAPTER 3
How Planet Earth Began and Evolved

About 14 billion years ago an immense explosion emanated from an infinitesimal point of compression, a singularity. The 'Big Bang' released a blinding flash of light and created the first particles of matter in the universe along with the fundamental forces and the laws of nature. As the universe cooled the first atoms were formed creating the simplest elements of hydrogen and helium. The atoms were accelerated outward in all directions at the speed of light.

Our Galaxy and Our Sun

As the universe continued to expand, in time it began to cool enough for matter to begin to aggregate due to the universal force of gravitation. (Note: Isaac Newton discovered the universal law of gravity which provides that each body in the universe is attracted by every other body by a force that is stronger the more massive the bodies are, and the closer that they are to each other.) About one billion years following the 'Big Bang', clusters of hydrogen and helium molecules were mutually attracted to each other by gravitation and began to form first galaxies and then stars. As stars were formed they were attracted together to remain in galaxies. Galaxies, in turn by the same universal force, grouped together into galaxy clusters, and clusters, in turn, grouped into super-clusters. All the while, the universe continued to expand ever-outward in all directions from the infinitesimal point of the singularity of universal origin.

One of the galaxies that so formed is our galaxy, the Milky Way galaxy. Our Sun is a medium-sized star located in an outer spiral arm of the Milky Way.

The Milky Way galaxy is part of a Local Cluster of about twenty galaxies located in a spatial area of several million light years across. The **Milky Way galaxy** alone contains about **four hundred billion stars**. It spans an area of about 130,000 light years wide. It is pancake-shaped with a bulge in the middle about 10,000 light years thick. Our Sun is located in a quiet neighborhood about 30,000 light years from galactic central point, far away from the dangers of deadly radiation that has been detected in the center of the galaxy, and far away from potential new supernova explosions. Our Sun is located in one of the 'spiral arms' of the Milky Way galaxy in an outer area that is about 1,000 light years thick. Our Sun and all the other stars in the Milky Way slowly rotate around the center of the galaxy at a rate of about once every two hundred and fifty million years or so.

While the Sun and the planets and all of the stars are moving around the Milky Way, the Earth and other planets in our solar system are circling the Sun in elliptical orbits at various speeds. As the Earth spins on its axis at a speed of about 1,000 miles an hour it travels in orbit around the Sun at a speed of about 1½ million miles each day in a circuit that takes just about 365 days to complete. And what most of us simply observe is the Sun rising each day in the east and setting each day in the west.

Stars are classed by color and luminosity. About **95% of the stars in the Milky Way are smaller than our Sun** and are called class M red dwarf stars. These red dwarfs burn fuel more slowly than our Sun and, therefore, have a life-span longer than our Sun. It is most unlikely that we would be writing or reading this if our Sun was a red dwarf. Since the red dwarfs generate less heat, in order for a satellite planet to have liquid water, necessary for life, it would have to have an orbit so close to the red dwarf sun that the gravitational force would 'lock' the planet's revolution so that one side always faces the red dwarf sun (like one side of our Moon always faces the Earth). The result would be an uninhabitable planet, boiling on one side and freezing on the other.

The usual distance between stars in a galaxy is a few light years. The closest star to our Sun is Proxima Centauri, located about 4.2 light years away.

Our Sun is in the grouping of stars called the **main sequence**. These stars fuse hydrogen to make helium. Main sequence stars that are more than ten times larger than our Sun burn hydrogen at such a rapid rate that they may run out of fuel after several million years. By contrast, our Sun is 'right-sized' so that it should enjoy a life span of about ten billion years. Our Sun is a slow and steady burner, a **yellow dwarf** of the main sequence, evolved to last a long time.

Our Sun and our solar system were born about four to five billion years ago, when the universe was about ten billion years old. If calculations are correct and all goes well, it should have enough fuel to last for about another five billion years. Science tells us that the heavier elements, essential for life, like oxygen and carbon, could come into being only after several generations of large stars had elapsed. Fortunately, our Sun is a third generation star, born from the remnants of a supernova explosion.

The supernova explosion created a dense cloud of interstellar gas and dust of the heavier elements created through that explosive process. Then, with gravitational attraction, concentration of this interstellar matter formed a flattened, rotating cloud, called the **solar nebula**. While 98% of the mass of the solar nebula consisted of hydrogen and helium gas, the remaining 2% contained the chemical elements essential for life. Over a period of about ten million years, the matter in that cloud condensed more and more as the Sun and the planets were formed. Over eons of time, the force of gravity brought some of the particles closer and closer together to form the Sun, while other particles came together to form the planets that orbit their common star.

The Sun is huge. Its diameter is over 100 times that of the Earth's and its mass is over 300,000 times greater. It provides all of the light and heat energy for us and for all of our sister planets in the solar system. That energy originates in the Sun's core.

The core of the Sun comprises about 10% of the orb's mass and is the location of its nuclear furnace. The core is unimaginably dense and unimaginably hot, more than 20 million degrees F. The immense density and heat drive the nuclear fusion process that supplies all of that light and heat energy.

In a five-step process, atoms of hydrogen fuse together to create the element helium and release the energy held within matter in the process. About **4 million tons of mass are converted into heat and light energy by this thermonuclear fusion every second.** The energy release is done in strict accord with Einstein's $E = mc^2$. (Note: Albert Einstein discovered the equivalence of matter and energy and that energy and mass can be converted, one into the other.) As 600 million tons of hydrogen mass is converted into helium, less than 1% of the mass disappears. The electromagnetic radiation that is released through that conversion then heads toward the surface.

Extending outward from the core's thermonuclear furnace is the **radiative zone** that includes about 85% of the Sun's radius and which is composed of transparent matter. Electromagnetic radiation does not travel through transparent matter at the speed of light. Far from it. It takes the photons of radiant energy as long as a million years to reach the Sun's surface, called the **photosphere**. But when they complete that journey they are radiated out into space as sunshine, at the speed of light.

The Sun's atmosphere, called the **corona**, lies beyond the photosphere. While the temperature at the Sun's surface is only about 10,000° F, the corona atmosphere heats up again to well over one million degrees. No one knows why.

While most of the Sun is composed of hydrogen and helium (by mass 78% and 20% respectively) the remaining 2% is spread between oxygen and carbon with traces of metals, like iron, silicon, and magnesium.

Our Sun has about twice the amount of carbon than other stars in our corner of the Milky Way. And, the stars and planets in our corner of the galaxy are comparatively rich in other mineral elements which have proved to be so useful in building complex living things. Our thanks goes out to the supernova parent of our solar system.

So, our Sun was born from a cloud of stardust containing a relatively abundant share of the heavier elements necessary for complex life to survive. It is located in a hospitable neighborhood in a quiet corner of the Milky Way. And, it is of the right size to burn a steady-state, yet abundant, amount of fuel for a long, long time - about 10 billion years. It is good fortune for us indeed that our Sun is: just the right size to provide a long and stable energy source for us; was born at just the right time to provide us with the minerals that we need; and is located at just the right distance from Earth to provide us with a benign energy source.

Our Sister Planets in Our Solar System

Galileo's observations in the 17th century confirmed the theory advanced by Copernicus in the 16th century that our Sun is the center of our solar system. Our solar system has nine satellite planets that orbit our Sun in a continuous cycle, each in accord with the universal law of gravitation.

Our solar system was formed when a swirling cloud of gas and mineral particles became denser and denser through gravitational attraction and then collapsed to form the nebula from which our Sun and solar system were formed. The collapse was not uniform. Some of the materials were attracted by gravity to aggregate into the mass of the Sun. Other materials became confined through gravity to a spinning disk around the Sun. Further work of aggregation and gravity created the solar system of nine planets out of that spinning disk. Our planet Earth is the third-closest planet to the Sun. And, it just so happens that such a location is just about perfect for us.

It is most likely that our Sun was formed from condensation of a supernova's nebulous dust cloud. The nine planets, in turn, were most likely formed that same way, as a consequence of the initial spin of the dust cloud, followed by gravity's compaction. This conforms with the fact that the Earth and all the planets, and the asteroid belt, revolve around the Sun in the same direction that the Sun is spinning. The Sun and all the planets keep forever spinning because there is no outside force that acts upon them to stop the spin.

The four 'rocky' planets, in order of proximity to the Sun, are Mercury, Venus, Earth, and Mars. These planets, and the asteroid belt, just beyond Mars, are composed of a large variety of mineral elements. The asteroid belt, located between Mars and the giant gas planet, Jupiter, is composed of thousands of 'rocks' that orbit the Sun in the same plane as the planets do. It is believed that the 'rocks' of the asteroid belt are the remnants of a planet that never really came together.

Next beyond the asteroid belt are the giant gas planets of Jupiter and Saturn and beyond them Uranus and Neptune. These planets have a very high percentage of the initial elements of the universe, hydrogen and helium, and, as a consequence, do not have the rich rocky density of the inner planets.

The ninth planet, and the furthest from the Sun, is little Pluto, estimated to be about two thousand times smaller than the mass of the Earth. Pluto actually may be just one of the many 'planetoids' of the Kuiper Belt, which is located just beyond the orbit of Neptune. The Kuiper Belt contains both comets (up to 20 kilometers in size) and small 'planetoids' (up to about 2,000 kilometers in size). They all cycle around the Sun in a period of 200 years or less. Comets also come from the much-more-distant Oort Cloud, and they cycle around the Sun in a period of 1,000 years or more. Comets contain frozen water, ammonia, carbon monoxide, and methane.

The very early history of our solar system was characterized by violent impacts and collisions of heavenly bodies. The impacts of early comets and asteroids that did strike the Earth in fact served to enrich our planet with abundant mineral elements and life-giving water. But, too many impacts could have doomed our planet as has been the case with others in our solar system. Scientists believe that the location of the giant gas planets has served to protect Earth from many violent collisions by either swallowing or diverting many comets and asteroids away from us. It is estimated that this 'protection' offered by Jupiter and Saturn may have drastically reduced the probability of devastating comet or asteroid impact with Earth from once every 10 thousand years to once every 100 million years, with highly significant evolutionary results.

As it turns out, the size and location of our sister planets, the frozen constituents of the comets, the mineral content of the asteroids, along with the proper number of more-than-close encounters in the past, have a great deal to do with our good fortune here on Earth. We will see even more facets of this phenomenon as we review that most-beneficent satellite, our Moon.

Our Moon

As the Earth was in a fairly early stage of planet formation, about 4.5 billion years ago, scientists theorize that our new planet got hit hard by a heavenly body about the size of the planet Mars. And, again as it turns out, that **'big whack'** was a most propitious event for us. The impact vaporized much of the heavenly body that hit the Earth. The vaporized matter was thrown back into Earth's orbit and condensed to become our Moon. The impact increased the mass of our planet and added an abundance of heavy metals to the Earth's core. The 'big whack' also increased the Earth's spin so that we now have a 24-hour day, and likely created the tilt of the Earth that provides us with our four seasons, and cleansed our atmosphere of many poisonous gasses. If you have to take a 'big whack', its fortunate when it proves to be good for your health.

Our Moon is located 239,000 miles from the Earth and is absolutely essential for life on this planet. The Moon provides a highly stabilizing influence for us. The gravitational attraction provides us with only a slight wobble along our vertical axis, which is called **'precession'**. The current tilt of the Earth away from vertical is 23.4 degrees (during a 41,000-year period we wobble between 21.8 to 24.4 degrees). If we did not have the Moon, we would wobble greatly more, between a tilt of 0 and 85 degrees, which would dramatically change our climate patterns and make the likelihood of terrestrial life on Earth very small.

The Moon's gravitational pull creates a tidal drag in our oceans which helps slow down our rate of spin. The Sun's gravitational pull also regulates our rate of spin. The combined effect of the Sun and the Moon's gravitational attraction produces bulges in the oceans pointing away from and also toward the Moon and the Sun as the Earth spins. The result is the daily reoccurrence of both solar and lunar tides. These combined tidal forces of the Sun and Moon together slow our planet to a spin rate resulting in a 24-hour day. If Earth's spin were slower, temperatures would vary by hundreds of degrees. If faster, hurricane-force winds would be virtually constant. And, rainfall patterns would be limited to very narrow bands around the planet.

The Moon is roughly one quarter of the Earth's diameter and it is located roughly one quarter of a million miles from our planet. The size and location of the Moon have had even more profound effects on our ability to learn about the wonders of the world we live in. The Sun is 400 times further away from Earth than the Moon, and it is also 400 times larger than the Moon. This amazing fact allows for us on Earth to experience nearly perfect total solar eclipses.

Solar eclipses have enabled scientists to make great discoveries about the nature of the Sun's atmosphere, the nature of the stars, the different wavelengths of light, and the speed at which the Earth rotates, as well as providing evidence to confirm Einstein's theory of General Relativity. In short, discoveries at the very core of modern astrophysics and cosmology have resulted because of this amazing phenomenon. If the positioning of the Sun and Moon had been only slightly less-perfect than they are, these revelations may never have occurred.

Without our Moon and our sister planets, it is highly unlikely that organic life would have developed on Earth at all. And, the chances of us being here to write or read this paragraph are slim to none.

Our Beautiful Planet Earth

Planet Earth was formed from the solar nebulae, which in turn was born of a supernova, that also gave birth to our Sun. The mass of our planet grew through the process of **accretion**. Accretion occurs when solids collide and stick to one another and thereby grow larger and larger. As accretion progressed over millions of years and as gravity attracted particles of the elements closer and closer together into a spinning mass, the planet Earth was formed. The solid inanimate structure of the Earth is called the **geosphere.**

Geosphere Formation

At first Earth was a totally molten planet, with a surface of magma. Then, as the planet began to cool over time, a thin skin, or **crust**, began to form over the surface as the magma solidified on top into heavy rock known as **basalt**. Today the solid outer crust is about 25 miles thick and consists of oxygen, silicon and other metals, like iron.

Below the crust is the mantle that extends about 1800 miles below the crust and consists of dense rock and metal oxides. It accounts for about four-fifths of the volume of the planet.

Deepest within the Earth is the **core.** The core is of two parts, is **made of iron and nickel,** and maintains in a fiery furnace of unimaginable heat. An outer core is molten liquid and the inner core is solid iron due to the unimaginable density and pressure. The core of the Earth has a large magnetic field which creates the planet's **magnetosphere**, and a magnetic north and south pole.

The Earth's **inner core** is solid iron, due to the unimaginable pressure and heat. The **outer core** is made of liquid iron and nickel, flowing within a furnace of about 4,000 degrees. The planet rotates fast enough to create eddies in this outer metallic core, and the circulation of the liquid creates convection currents that conduct electricity. The outer core thus sets-up a dynamo generator and a magnetic field. This causes the Earth to be surrounded by a gigantic magnetic field created by the convective movement in the liquid core. That is the magnetosphere.

The Earth's interior is continuously heated by the radioactive decay of the elements uranium, thorium and potassium. And the heat energy within is unbelievable. This interior energy, that is the product of radioactive particle decay, maintains the planet's magnetic fields and also serves as the energy source to thrust-up new land areas to the Earth's surface.

Above the core is a layer of inner-Earth, the radioactively-heated **mantle**. The mantle lies directly beneath the Earth's crust. The mantle is a fluid, but not like water. It is a **fluid like glass**. It seems to be solid, but it behaves as **a viscous fluid**. As uranium and thorium and potassium decay deep within the Earth, their radioactive heat is shed as they breakdown into isotopic forms. By the laws of thermodynamics, the heat rises toward the surface. The rising heat creates huge convection cells of liquid-rock heat in the mantle. As the viscous upper mantle rises it moves parallel to the surface and then moves back down deeper through a series of dynamic planetary processes. 'Plates' are thereby formed as a composite layer of crust and mantle.

The mantle is a semi-molten layer that serves to create the mountains and trenches of the surface. The magma rises, spreads-out and releases heat at the surface by both volcanism and plate tectonics. We intuitively figure-out that **volcanism** is produced by the pressures deep-within that erupt through the surface as volcanos. But, what is **plate tectonics** and where does it come from?

As rock is heaved-up from the radioactive mantle, great upward pressure creates ridges and trenches in the sea floor above. A series of 'plates' are thereby created. Each plate is similar to a huge conveyor belt. And that conveyor-belt-effect causes continental drift. The plates beneath the oceans are about 30 miles thick while those beneath the continents are about twice that depth.

Where the tectonic plates go, the continents go. The continents float on the heavier viscous, glass-like plates below. Where plates are shoved together mountains are formed. But, the plates move the continents very, very slowly, at most a few centimeters a year. In this way the radioactive interior energy of the planet continuously thrusts-up new land to the surface above and moves the continents about the globe.

According to the theory of **plate tectonics**, the top portion of the crust of the planet is comprised of a number of 'plates' that float and travel over the plastic-like mantle. Volcanos, earthquakes and mountain-building occur from the interaction of these tectonic plates where they come together.

Plate tectonics ultimately causes mountain chains and terrestrial life. Without radioactive decay within the planet's interior there would be no plate tectonics, and the surface of the Earth would be covered only with water, in an ocean world.

Basalt rock provides the linings for the bottom of the world's oceans. Basalt is of a lower-density than the rocks of its mantle parent below that pushed it up toward the surface, but is more dense than the silica rock of which the continents are constructed.

As the Earth matured, water pooled over the lowest surface points on the planet that had been formed by the heavy basalt rock. The world's oceans rest atop the high-density basaltic rock. The world's continents are composed of low-density **silica** rock (like granite). The silica rock is loosely-embedded atop the basaltic rock below. The silica continents thereby 'float' on the basaltic rock bed below.

So, the initial state of the Earth was simply a fiery mass, core to surface. In 4 ½+ billion years the planet has evolved from a totally molten mass to a sphere with a fairly rigid skin on the surface. The effect of gravity gave the Earth its spherical shape and layered density, and also separated the magma, bringing the lighter elements to the surface. As cooling occurred over time, the continents were formed from the cooled silica matter.

While all of the 92 naturally-occurring elements exist on Earth they are far from being equally proportioned. By weight, **only eight elements provide for over 99% of the mass**: oxygen (46.4%); silicon (28.2%); aluminum (8.2%); iron (5.6%); calcium (4.1%); magnesium (2.3%); sodium (2.4%); and potassium (2.1%). But the other 84 elements that account for only 0.7% of the total mass are far from unimportant. Without light hydrogen we, of course, would have no water. And, many of the essential processes of life are catalyzed by enzymes that are configured around the heavy metals like copper and manganese and zinc.

Only four elements provide for 96% of the mass of our bodies: oxygen (65%); carbon (18%); hydrogen (10%); nitrogen (3%). But, without sodium and potassium, for example, we would have no neural activity in our brains. Many times small things produce big and unexpected results.

Another producer of big and unexpected results is the **magnetosphere**, which affects both the biosphere and the atmosphere of the Earth. You will recall that the inner-Earth's magnetism causes the planet to be surrounded by a gigantic magnetosphere. The magnetosphere swirls out into space to a distance of up to ten times the radius of the planet. The structure and behavior of the magnetosphere is affected by the outflow of hot **plasma** from the Sun, called the **solar wind**. The solar wind is powered by the immense million-degree-temperature within the Sun's corona.

The intense heat within the corona strips free electrons from atoms (just as it does within the interior of the Sun itself) and thereby fills the Sun's atmosphere with an electrically-neutral medium of positive protons (ions) and free electrons. As that medium radiates into space it becomes a plasma of ionized gas called the 'solar wind'. The 'solar wind' responds strongly to electromagnetic fields. When the swirling 'solar wind' hurls through space and interacts with the Earth's magnetosphere, it ultimately determines the overall configuration of the magnetosphere. If we are lucky, we then can see the resultant awesome beauty of the aurora borealis (northern lights) dancing in the northern skies. It is an amazing atmospheric light show.

Atmosphere Formation

Gasses from the Earth's interior rose to the surface to form an atmosphere, and the oceans were thereby ultimately formed. The oceans are actually a product of that out-gassing. As the steamy water vapor cooled over time it condensed to form the oceans that today comprise about 70% of the surface area of the Earth. Over further time the oceans became 'salty' through interactions of water with the other elements composing the Earth's crust.

The initial period of Earth's formation is called the **Hadean Eon** (a hellish-like period) and lasted from about 4.5 billion years ago to 3.8 billion years ago. Gasses of hydrogen, methane, ammonia, nitrogen, carbon dioxide and water formed the Earth's first atmosphere. The planet remained covered for a long period in the darkness of burning clouds of gasses that were infused with steamy water vapor from volcanic activity. As cooling occurred, rain fell from this steamy atmosphere and the first fledgling oceans were formed.

During the period from 3.8 to 2.5 billion years ago (the **Archean Eon**), **the entire planet was covered with ocean water**. But, beneath the waves the continental land masses were forming. The Earth's surface crust formed about 3.5 billion years ago. Volcanic emissions of steamy water vapor, ammonia and carbon dioxide provided a second atmosphere that was devoid of oxygen but had about 100 times as much gas as today's atmosphere. The **high concentrations of carbon dioxide produced a greenhouse effect**. As the rains ensued and the Earth became wholly covered with water, about one-half of the carbon dioxide in the atmosphere was absorbed into the world ocean. The first bacteria used methane or hydrogen for metabolism, a much less effective metabolic process than respiration. Those bacteria began to eat each other to survive, and higher life may not have evolved if bacteria had not developed a better metabolic system. Thankfully, as bacteria evolved they began to convert carbon dioxide into oxygen through **photosynthesis**. Thus primitive blue-green algae evolved and began to clump together into large units called **stromatolites**. They now did not need to feed on each other. They could now feed on sunshine.

The atmosphere dramatically changed as the result of microbial photosynthesis as free oxygen was emitted into the ocean and the sky. During the **Proterozoic Eon** (2.5 billion years ago to 570 million years ago) more and more photosynthesis took place and more and more carbon dioxide was cleansed from the air. Our current atmosphere then took shape throughout the **Phanerozoic** Eon (570 million years ago to present) as green plants and land animals arrived on the world stage.

Today's atmosphere is composed of 78% nitrogen, 21% oxygen, and 1% other gases (including carbon dioxide and ozone). Beyond the four layers of our atmosphere lies the exosphere, where the atmosphere thins out as it melds into outer space. Not surprisingly, the exosphere is composed of hydrogen and helium for, after all, that's the way things started out with the 'Big Bang', with just hydrogen and helium.

Sunlight illuminates and warms the planet and provides the source of energy for all plant and animal life. But, without our atmosphere sunlight would kill us. Based on the reflective characteristics of the planet and the energy output of the Sun, without our atmosphere the average temperature on Earth would be -5 degrees F. At that temperature, the world's oceans and everything else on the planet would be frozen solid. With our present atmosphere all is well.

Photons of electromagnetic radiation leave the Sun in a wide range of wavelengths. The shorter the wavelength the higher the frequency and the higher the frequency the stronger (and more harmful) the radiation. As dermatologists continually caution, **ultraviolet** (UV) radiation from the Sun can damage the skin and cause cancers. Fortunately, most of the incoming UV radiation is absorbed by the **ozone layer** of our atmosphere. (Note: Ozone consists of 3 atoms of oxygen combined into a single molecule.) Also, heavy water vapor held in clouds serves to protect us from some UV radiation that gets past the ozone layer by radiating it back into space.

Infrared radiation warms the planet. Without our atmosphere, when these rays strike the planet they would be reflected back toward space and the heat would thereby be lost. **Greenhouse gasses** in the atmosphere serve to prevent the infrared radiation from simply bouncing back into space. Because of the greenhouse gasses our planet is warm and we have liquid water, essential for life.

The greenhouse gasses include carbon dioxide, water vapor, methane and ozone. They 'trap' the infrared radiation needed to warm our planet. Water vapor provides the largest percentage of the greenhouse effect. But the relative quantity of carbon dioxide in the atmosphere seems to be the most important variable for the heating effect of greenhouse gasses. Indeed, **carbon dioxide levels are the principle short-term regulators of planet temperature.**

Carbon dioxide is only a very small part of our atmosphere (about 0.03%). But, without carbon dioxide we would have no photosynthesis, no plant or animal life, and an altogether lifeless and frozen, cold world. Again, many times small things produce big and unexpected results.

Earth - the Water Planet

Gasses from the burning planet rose to the surface to form the early atmosphere. Hydrogen and oxygen were among those gasses and merged together to form water. Water, however, because of the tremendous heat, remained in a gaseous state for a long period of early Earth. More water was added to the planet by collisions with comets. Not until the Earth had cooled over a period of millions of years did water appear in its liquid form. The Earth then became fully covered by water as a world ocean formed. In time dry land appeared. Water collected in the lowest places on Earth to form the oceans. Since the basalt was heavier than silica, the oceans naturally formed over basalt and the continents formed over silica. The sum of the entire process results in the surface of the Earth today being composed 30% of land and 70% of water. (Note: 97% is seawater and 3% is freshwater.) And, on average, the oceans are five times deeper than the continents are high.

The consensus of the scientific community posits that for complex life-forms to exist on any planet, liquid water is required. In our solar system, in order for water to remain liquid on the surface of a planet holding an atmosphere, the planet must be located between 88 million and 127 million miles from the Sun, known as the 'Goldilocks' zone. The Earth orbits the Sun at an average distance of 93 million miles. As Goldilocks would observe…just right!

Liquid water needs not only the right temperature, but also the right pressure so that it may remain liquid over a fairly wide temperature range. As the Earth's atmosphere matured, the proper pressure was achieved. Water remains a liquid over a wide range of pressures and temperatures at which compounds of even heavier weight are gases. Because of the unique configuration of the oxygen and hydrogen atoms in a water molecule, the hydrogen side is positively-charged and the oxygen side is negatively-charged. This hydrogen bond results in the amazing fact that even where there is ice on the Earth, the crystalline structure of frozen water allows ice to weigh less than cold liquid water. This property allows oceans and lakes to freeze from the top down, which has proved to be most useful for the development of various forms of organic life.

Water acts as a virtually universal solvent that allows the formation of both inorganic and organic compounds. Water allows for electrical flow within. Water both absorbs and stores heat energy. And, water is essential for organic life.

All organic life requires water. Living cells contain between 70% and 90% liquid water, and all basic cellular processes that we know of occur only within liquid water.

The Stage is Set

Our Sun and our solar system, including our beautiful planet Earth, had been created out of swirling stardust from a supernova, the death of a stable star. Over a very long period of time, gravity and accretion had worked together to condense matter into the discrete heavenly body we call the Sun and the planets of our solar system orbiting that star. During the first one billion years of our planet's existence, we were blessed with good fortune. We were blessed with an abundance of metal elements and water. We were blessed by a 'big whack' with a renegade planet that served to create our large Moon, add heavy metals to our Earth and cleanse our atmosphere. We were blessed with just enough comet impacts to transfer abundant water to us while being protected by our big gas outer planets from too many cosmic and asteroid impacts. The Earth had cooled over a long period. Then, about 3½ billion years ago, something very strange began to develop on Earth. Life.

CHAPTER 4
How Life on Earth Began and Evolved

The idea of creating living things from dead things boggles the mind. But, most scientists believe that is just what happened. As some inanimate elements are placed in proximity to certain other elements their molecular components will engage in chemical reactions. The chemical reactions then lead to networks of continuing chemical and electrical reactions using the reactions of other elements and chemicals. This theory, called 'emergent evolution', postulates that life could emerge from numerous increases in chemical complexity of dead matter. I guess life could have started in that way. The fact is that no one knows.

The Beginning of Life

We do know that life did begin, somehow, on this planet some 3½+ billion years ago. And, that first life was ... bacteria.

It is quite unpoetic that the first life that appeared on our beautiful planet was in the form of bacteria. But, it is clear that bacteria has had staying power. Today bacteria exist in many forms. Some help our digestion and others kill us. Despite their microscopic size, their volume is huge, perhaps larger than any other living thing on the planet today.

When bacteria first appeared on Earth, the atmosphere of the planet was likely devoid of oxygen. Water on Earth had condensed from vapor and then froze throughout the world ocean due to the low temperatures of an infant Sun. But, occasionally the world ocean would vaporize after an asteroid impact. And eventually the Sun's temperature increased. At the time that life first appeared our planet was a pretty hostile place. And, the first life may not have been born on the planet's surface, but, rather, at a fissure on the ocean floor, far beneath the ocean surface, where pitch blackness prevailed and hydrothermal eruptions from the planet's core provided the hot energy necessary to support first life. Though this is but one theory, the consensus of the scientific community concurs that the first single-celled bacterial life most likely emerged from an environment of a hot watery ooze.

Evidence for the earliest form of life on Earth cannot be found directly by fossil research, because it would have been too small and uncompounded to leave a fossil record. However, chemical micro-analysis of ancient carbon deposits found in fossils has 'implied' a biological origin of bacteria over 3½ billion years ago. **The earliest fossil record of evidence of single cell bacterial life dates to about 3 billion years ago and is found in stromatolites, large masses consisting of microscopic single cell bacteria that had developed the process of photosynthesis.**

In perspective, that 500 million year period during which photosynthesis developed in single-celled bacteria is basically the same length of time (post Cambrian) that it took for all life forms to evolve from worms and crustaceans to racehorses and rocket scientists. But, the development of photosynthesis may be the most important evolutionary innovation on Earth to date, so it was probably worth the time. **Photosynthesis transforms the energy of sunlight into food calories which provide a reliable internal energy source for all of us.** Additionally, as bacteria developed photosynthesis, they began to give off oxygen that enriched the oceans and the sky.

A Cell Nucleus and How it Works

Bacteria cells did not contain a nucleus. In time they became more and more complex and produced a thicker algae, but they never 'learned' to develop a nucleus. **For two billion years non-nucleated single-celled bacteria remained the only living thing on the planet.** But, the two-billion-year 'learning process' did produce a new winner - yeast, a single-celled microscopic member of the fungal kingdom. **Yeast contained the first cell with a nucleus.** And that was a really big deal. The cell nucleus provided the means necessary to build more and more elaborate forms of not only fungi, but plants and animals as well. Indeed, all members of the plant kingdom and the animal kingdom (including us) have nucleated cells. The nucleus of the cell confines the chromosomes which house the DNA.

Chromosomes house all the genetic information for plants and animals and they are protected and contained in the cell's nucleus. The chromosomes are made up of proteins and strands of **DNA** (deoxyribonucleic acid). DNA strands for complex beings like us extend to an incredible length of several feet. Such a length is necessary to accommodate the vast amount of information that they carry, but the length is tightly coiled to neatly fit within the small nucleus. **DNA itself is composed of sub-units called nucleotides**

that chain together to form the DNA molecule. We are talking about really small stuff. Each nucleotide is but ten-millionths of a centimeter wide.

The DNA compound of nucleotides consists of three parts:

- a sugar;
- a phosphate; and
- a nitrogen base.

The sugar molecules of each DNA strand are connected to each other by a polymer created by the phosphate. A nitrogen base is attached to each sugar molecule perpendicular to the axis of the strand. The nitrogen bases form bonds with bases on adjoining parallel DNA strands (base pairs). **DNA is the universal carrier of hereditary information for all plants and animals.**

Segments on the strands of DNA, called **genes**, store information and **encode protein sequences** in the following manner. DNA passes on the information from a base-pair-code, contained in its nucleic acid to a specific combination selected from some of twenty specific amino acids. The selected combination of specified amino acids, in turn, is transformed into complex proteins that, in turn, become the structural elements of the cell and that perform the cell's specified work. All living things function only through the work of living cells. Each protein in a living cell is so small that it must be magnified a million times before we can see it.

The genes in DNA complete their work when they pass on the information. The genes and only the genes **carry the information code.** They are the only ones that do. And, **that's all that they do.** We will examine the DNA and protein 'stuff' of life in more detail in Chapter 10.

It took two billion years for complex cells to develop with a nucleus containing DNA. But, as we have seen, that was no small feat. The exquisite process of complex cell development had allowed for hereditary information to be passed on from generation to generation in elaborate detail and with great precision. That became the building block for the **genome** of each creature on earth. It was the method by which each fungi, each plant, and each animal would define who he was. However, cellular development was not complete at that point by any means.

Each cell is encased by protein-lipid membranes that allow some things to pass through it and to reject other things. The purpose of this primary function is to keep the interior of the cell in a state of chemical equilibrium or 'stasis'. The

evolution of a double cell membrane, composed of two layers of proteins and two layers of lipids, is found in more advanced, complex organisms. It **took yet another one billion years** for the nucleated yeast cell to develop this more complex structure and to develop multiple-celled offshoot organisms whose cells could communicate with each other and thereby work together. We will more closely examine cellular communication and cooperation in Chapter 11.

Evolution of Multi-Celled Organisms

The first multi-celled organisms were invertebrate (had no backbone) and seem to have evolved through an aggregation of single-celled organisms (called **flagellates)**. Scientists believe that the origin of virtually all the members of both the plant kingdom and the animal kingdom may be traced to such flagellates assembled into large groups. By this manner, cells comprising both plants and animals developed an interdependence and an elemental form of **symbiosis** (mutual dependence).

So, after about three billion years of development, life had become 'advanced' in terms of DNA, nucleated cells, and multiple-cell organism beginnings. But, the nature of such life still remained very small, co-existed with the much more numerous bacteria, and remained, itself, in very simple forms. Up to about 540 million years ago the most advanced life on Earth still lacked distinctive features, had no mouth or anus, no head or tail, and was not capable of self-locomotion. All life still remained in the seas, and the most advanced form of life, the jellyfish, simply drifted along at the whim of ocean currents.

And then about 540 million years ago complex life burst forth on the planet in great profusion and diversity in a very short period of time (in terms of geological time that we have been using). That began what is termed the **Cambrian Period explosion.** But, before we delve into that era, lets take a look at what has been happening to mother Earth during that first three billion years of life, because the evolution of life on Earth had dramatically changed the atmosphere and climate of the planet itself during that time. For, after all, mother Earth herself is dynamic and the actions of living organisms greatly affect her.

How Life Changed Mother Earth

At the time that the first bacteria began to engage in the process of photosynthesis, about three billion years ago, the Earth's atmosphere was devoid of free oxygen. The atmosphere was filled with heat-trapping carbon dioxide, ammonia, and methane. By the start of the Cambrian Period, that situation had changed dramatically.

Through the process of photosynthesis, living organisms (initially bacteria, and today green plants as well) capture the energy of sunlight. They take in carbon dioxide from the atmosphere, combine it with hydrogen which is extracted from water, produce organic molecules, and respire free oxygen back into the atmosphere (and the oceans). Gradually, as the levels of free oxygen increased, oxygen's reaction with ammonia and methane greatly reduced the concentration of those 'greenhouse' gasses in the atmosphere. The oxygen-rich (21%) atmosphere had in large part been created by the evolution of life itself on Earth.

The removal of heat-trapping gasses from the air, coupled with an increase of free oxygen through photosynthesis, thereby provided a fairly healthy atmosphere for the development of advanced life and was also propitious to our well being in another manner. It served to develop an ozone layer in the outer atmosphere which protects us from the harmful effects of the Sun's ultraviolet rays and other intense radiation from the Sun and other sources in the cosmos. The ozone layer allows useful visible light to pass through unaffected, but stops harmful ultraviolet radiation.

The compounded effect of photosynthesis was to create an atmosphere that was 'just right' for the emergence of a great variety of advanced life forms.

While the Earth's atmosphere of 500 million years ago contained about the same percentage of oxygen that our atmosphere contains today, the amount of carbon dioxide in the air was 18 times greater. And, many scientists feel that the level of carbon dioxide in the atmosphere is the greatest single determinate of our climate. If we had none at all we would freeze and have no food. If we have too much we would broil. In a manner similar to how the ozone protects us by absorbing harmful ultraviolet light, carbon dioxide absorbs infrared rays from the Sun. Carbon dioxide 'traps' energy leaving the earth, without stopping the incoming useful energy of visible light. The more carbon dioxide, the more energy is 'trapped' (just like in a greenhouse) and the hotter the climate becomes.

As an aside, the respiration of living things is not the great determinate of the amount of carbon dioxide in the atmosphere. As it turns out, in the long haul, the amount of carbon dioxide contained in the oceans and in the rocks of the planet determines the atmospheric quantities. End of atmospheric digression. Back to the Cambrian.

The Cambrian Period Explosion

With protection from harmful ultraviolet radiation, with a fairly healthy atmosphere, and with quite a warm climate, little organisms of complex life began to proliferate not only in the ocean but also in shallow pools of water inland.

Complex animal life began with a metaphorical bang at the start of the Cambrian Period about 540 million years ago. If you look at the three and one-half billion years of development from the beginning of the first life until the present as a 24-hour clock of geological time, the profusion of life that appeared over the thirty million years comprising that period took place in about 15 minutes of life's time line. And, we are not really sure why.

Evolution did not proceed along a straight line leading from bacteria to human beings. The evolutionary process was very much non-linear, with life developing in different forms and variations much like thorns and twigs and branches of an ever-growing and very thorny bush.

The first multicellular sponges had no nerve cell. Next, the jellyfish family developed two layers of cells, and, importantly, bilateral symmetry (a left and a right side and a head and a tail end) and locomotion. Flatworms developed three layers of cells (basic primary tissues) and a complex nervous system, and the wormlike animals that followed had it all: bilateral symmetry; a mouth and anus; a complex central nervous system; locomotion; and a defined body cavity. The increase in morphological complexity was immense over this brief period. And the variety of different bodily plans developed during this period is believed by some scientists to be greater than those living in the world's oceans today. Indeed, in all the time from the Cambrian to the present, very few new basic body plans have been added to the ones that apparently originated in the Cambrian Period (540-510 million years ago).

The 'Lay of the Land'

During the Cambrian Period, the 'lay of the land' of our planet was quite different than it is today, and it has continued to change radically over the course of geological time.

The geological measure of time from the start of the Cambrian Period up to the present is called the Phanerozoic Eon. The Eon is further divided into three Eras: (1) the Paleozoic Era (540-245 million years ago); (2) the Mesozoic Era (245-65 million years ago), and; (3) the Cenozoic Era (65 million years ago to present). Each Era is further subdivided into Periods (like the Cambrian Period), and the last Era (the Cenozoic) is even further divided into Epochs. So, the Cambrian Period is the beginning of the Phanerozoic Eon, which also starts the Paleozoic Era. Simple, huh? But, back to the point.

The 'lay of the land' during the Cambrian Period revealed most of the land on Earth to be under water. The land masses that existed were quite scattered and quite differently structured and located on the world globe, compared to today. Present day North America was under water. Siberia was a tropical island in the Southern Hemisphere, Europe was mostly under water and was located even further south than Siberia. Most of the existing land mass was fused into one continent called Gondwana.

The **Cambrian Period** did provide a great proliferation of complex life. But, it **was all sea life.** Before the Cambrian Period explosion the dominant organisms on Earth were microscopic blue-green algae residing in ocean water. These single-celled bacteria clumped together into huge masses of photosynthesizing plankton in the sea and released oxygen into the ocean and the air. That made further changes in 'the lay of the land' and in the Earth's climate necessary for life to arrive on dry land.

After the Cambrian Period, as evolution continued into the later part of the Paleozoic Era, the world's land masses drifted northward and eventually joined together into one great continent called Pangaea, or 'All Earth'. And, the world's climate had become periodically drier.

Life Moves Onto Land

About 470 million years ago plant life first began to appear on the dry land of the Earth. Land plants continued to draw-down carbon dioxide from the air, further cooling the planet's surface temperature and adding abundant oxygen to the atmosphere. By the time that **animal life first appeared on dry land, about 100 million years later**, the first tiny vascular plant life had evolved into great forests with 100-foot trees in the tropical swamps.

As the **Earth's climate became drier**, not only did the land mass of the world increase, but the inland waters began to dry up on a seasonal basis. As this occurred, the fish that found themselves trapped away from water struggled to push themselves to a nearby water hole or creek in order to survive. These fish evolved four-lobed structural fins to assist them in their trek over land. As poor aquatic conditions persisted over time, these fish developed lungs for breathing air. Their four-lobed fins evolved into four legs, and, thereby, the first amphibians appear in the fossil record about 365 million years ago. The amphibians had evolved a four-legged body structure that is common to all vertebrate animals that have followed after them, including us. Such is the process of biological evolution, a scientific theory overwhelmingly supported by scientists.

Ecomorphs

As long as life remained in the water, animal life was free to take on virtually any form that it wanted, as pointedly illustrated by the variety of the free-floating creatures of the plankton. As creatures began to swim, that locomotion required a body form of basically cigar-shape to accommodate that skill. As mammals moved on to land, they encountered the heavy on-set of gravity and the effects of heat. So, the laws of thermodynamics and gravity prescribed that land creatures must develop form and shape within certain parameters defined by the laws of physics. And, the larger they became, the more precise those parameters came to be.

Thereby, a land creature evolves in accord with the basic form or **morphology** that its forebears had developed to meet the challenges of the environment for its species. This form or morphology is termed an **ecomorph**, and all creatures of a lineage must use the basic form prescribed. This is known as **a phylogenic constraint.** Once a basic body plan has evolved, it is nearly impossible to change it. So, we are limited in our development to changes *within* the basic body plan. **For example, all of us mammals have a basic four-legged structure and we all have seven vertebrae in our necks.**

The forms are varied and become compounded: warm blooded versus cold blooded; vertebrate (with a backbone) versus invertebrate (without a backbone); reptile versus mammal; big versus small; herbivore (plant eater) versus carnivore (meat eater); and so on. But, creatures also adapt as need requires. For example, as vegetation dries up, herbivores may become carnivores or omnivores (eat both meat and plants). And, as prey becomes scarce, carnivores may become omnivores.

In this manner creatures evolved according to selected ecomorphs that they seem to have 'chosen'. They seem to have positioned themselves so that 'natural selection' could work on them. But, what really is natural selection?

Natural Selection

Natural selection is the basis for evolutionary change. So let's take a moment to review it. After all, Charles Darwin's seminal work on the subject of evolution was even entitled *On the Origin of Species by Means of Natural Selection*.

Evolution by means of natural selection, then, is simply this. Organisms having the genes that are best suited to meet the needs of environmental challenges and survival are those creatures that are also most likely to successfully reproduce and pass on the favorable genes to their off-spring, and so on and so on and so on, from generation to generation to generation. That's it.

Charles Darwin proposed, and the Modern Synthesis of Evolutionary Theory concurs, that the exclusive mechanism for the natural selection of *adaptive traits* in living organisms is random mutation.

As we previously discussed, discrete segments of DNA are called genes. Genes are the sole carriers of hereditary information in living organisms. They are the only ones that do and that is all that they do.

Natural selection through the mechanism of 'random mutation' starts when, during the life of an individual organism, a reproductive gene (or genes) of that organism becomes mutated in one of two ways:

- during DNA replication in a sex cell, the enzyme DNA polymerase makes a mistake (we will take a closer look at the DNA replication process in Chapter 11); or
- genetic DNA is altered in response to something that occurs in the organism's natural environment (chemicals, radioactivity or ultraviolet light from the Sun or cosmic rays).

These mistakes or accidents of nature, cause genetic change by bringing on chromosomal nucleotide rearrangements in the DNA through altering the order of genes along a chromosome (e.g., reverses a gene order along a chromosome; deletes the chromosome segment and the gene; or crosses over and fuses chromosome segments). That is how genetic mutations occur. Then through the reproductive process these genetic mutations are passed on to offspring in the next generation. As elegantly phrased by Carl Sagan in his book *Cosmos*:

> "A mutation is a change in a nucleotide, copied in the next generation, which breeds true."

These mutations are not purpose-directed. They occur at random. And, they occur without regard to the needs of the organism. Indeed, most mutations do not benefit the offspring of the organism by serving to better meet its needs. The requirement for reproductive success screens out most mutations that are not useful. However, **the off-spring of the individual organism** *whose genes are randomly mutated in a fashion that helps in adjustment to an environmental challenge and survival* **is benefited by the mutation and is viewed as being more 'fit'.** This is the idea behind the badly misused term in the common vernacular concerning evolution: 'Survival of the fittest'.

The increase in complexity and improvement over time that results from this reproductive cycle is the crux of the evolutionary process. **Natural selection thus changes the composition of the gene pool for all of the population of a species to include the favorable gene traits.** Species modification thereby derives from the modified gene pool.

Progressive Evolution and the First Mass Extinction

Evolution seems to be a collaborative effort. It is believed that complex life began when two simple organisms 'merged' together into one. One was subsumed by the other. This is a process called **endosymbiosis**, wherein each is benefited by the merger. Scientists posit that endosymbiosis first began as non-nucleated cells merged to construct more complex nucleated cells. Then, over time, complex nucleated cells formed even more cooperative interactions and networks that served to evolve the various species of the plant and animal kingdoms. We will more closely examine the wide expanse of plant and animal species in Part IV.

There seems to be a definite trend in evolution toward increased complexity and diversity and structural variety over time. Each level of complexity seems to lead to yet a more elaborate form of organizational complexity and interactions. And, species within a lineage seem to get bigger and better over time.

What in fact evolves is not the individual, but the genetic information of a species as the gene pool changes and as the gene frequencies change over time from generation to generation.

This simple but unbelievably complex evolutionary process really got rolling with the Cambrian Period explosion of life in the world ocean, occurring toward the start of the Paleozoic Era a little more than 500 million years ago. It continued apace as both plant and animal life gained a foothold on dry land about the middle of the Paleozoic Era around 350 million years ago. Marine animals and terrestrial vertebrates continued to proliferate until catastrophe occurred toward the end of the Era.

The predominant theory is that widespread volcanic actions occurred throughout the planet over the course of tens of thousands of years. This volcanism may have had a dramatic effect on the Earth's atmosphere. Although no one is sure why, the fossil record indicates that by the end of the Era (around 250 million years ago) 90% of the species of marine animals and 75% of terrestrial vertebrate species had been obliterated from the face of the Earth - extinct. This was most likely the largest extinction event in the history of complex life on planet Earth.

The Age of Reptiles and the Second Mass Extinction

Following the first mass extinction, the Mesozoic Era (245-65 million years ago) began. The Mesozoic was the Age of the Reptiles. During this Era the great dinosaurs of massive size and great diversity became and endured as the masters of the Earth for over 150 million years.

The reptiles derived from the amphibians. You will recall that the oxygen-breathing fishes developed their four-lobbed fins into forelegs and became the first land vertebrates, the amphibians. However, amphibians still had to lay their eggs in the water, requiring their offspring to be born and to develop in their early lives at great risk. Reptiles brought about a significant survival improvement. Reptiles developed a shelled egg to protect the developing embryo of their offspring. Thereby, the species was freed from reproduction in water, which gave them a significant advantage over amphibians.

The reptiles grew in number, variety, and size. Their most famous members were the dinosaurs, who became the largest land creatures that the world has ever known. Some dinosaurs were carnivores, some were herbivores, and others were omnivores. They were great brutes who enjoyed a bountiful food supply and who remained at the very top of the food chain for millions and millions of years.

While the reptiles were most fearsome in both size and appearance, they were not very bright. Reptiles made virtually no advance at all in brain size from that of their amphibian forebears.

Their great body size required that their food intake be enormous, and that may, in part, have led to their demise.

The Age of Reptiles was brought to a close by a second great mass extinction event that occurred toward the end of the Mesozoic Era. The majority scientific view is that a large asteroid struck the Earth at a point that is located in what is now called the Yucatan Peninsula in Mexico. The impact resulted in an incredible explosion that spewed a dust cloud into the atmosphere. The dust cloud encircled the Earth and blocked out the Sun's rays for many years. That resulted in greatly reduced photosynthesis, which naturally led to wiping out most of the world's food supply. Those animals needing huge quantities of food in order to live simply perished. Those who were small and tended to need relatively little food, survived the catastrophe.

The record is not clear how long the dinosaurs survived after the catastrophic event. Some scientists believe only a short while, while others believe they lasted for several thousand years following the asteroid. In any case, the result was the same. By 65 million years ago *all* dinosaurs on the planet were gone - extinct. The Age of Reptiles was over.

Along with the dinosaurs, the second mass extinction event eliminated about one-third of the families of terrestrial vertebrates. But, it also served to chart a new course for vertebrate evolution. Mammals.

The Age of Mammals

The earliest mammals appeared on Earth about 200 million years ago. They were little-bitty creatures about the size of field mice and coexisted with the dinosaurs for over 100 million years. They literally remained in the shadows as nocturnal foragers until the dinosaurs were no more. Mammals were smart enough to keep out of the way of the world's greatest brutes.

Mammal metabolism was greatly increased compared to the reptiles. Mammals are warm-blooded. They nurse their offspring. They have hair. They have a significantly more acute sense of smell and of hearing. The increased senses required the development of a bigger brain. And, their physiology changed in a manner that allowed them to chew food at the same time as they were breathing.

That was about the extent of mammal development until the dinosaur extinction occurred. Once the dinosaurs were gone the tiny mammals could venture forth from the night, grow, and expand with great variety and innovation. The Age of Mammals had begun. Within ten million years the orders of mammals had increased from five to twenty-five.

By about 100 million years ago, certain mammals developed an innovation that would prove to be of great significance to species survival - placental development of their offspring. By being developed inside the mother's body (in her womb) the fetal development milieu was changed in a manner that allowed for the building of even bigger brains.

One very special branch of mammals first appeared on the Earth about 50 to 60 million years ago. The primates. Our branch of mammals.

Primates differ from other mammals. Primates have five digits on each limb. Primates have a thumb on each hand that is opposed to the other four fingers, allowing for great improvements in grasping and dexterity. And, last, but not by any means least, primates have a much bigger brain.

Our fellow members of the primate branch include the apes and the monkeys. The monkeys have tails, while the apes do not. Of the apes, we are more closely related to the gorillas and the chimpanzees than to all the others. And of those two, we are closest yet to the chimps. But, how did we get to be so different? Let's turn to the next chapter to examine that question.

CHAPTER 5
Our Human Beginnings and Early Development

In 1871 Charles Darwin published *The Descent of Man and Selection in Relation to Sex*. In that seminal work he postulated that human beings are, in essence, descended from apes. That pronouncement, of course, set the civilized world of the 19th century on its ear, and began a controversy that continues into the 21st century. What exactly did Darwin pronounce?

Pongids and Hominids

Primates, those special mammals having four fingers and an opposable thumb, were subdivided by Darwin into two families: Pongidae **(pongids)** and Hominidae **(hominids)**. Orangutans, gorillas, and chimpanzees are pongids. Modern human beings and their extinct relatives are hominids. Darwin's evolution theory tells us that before pongids and hominids differentiated into distinct families they shared a common ancestor. Who was that common ancestor? That common ancestor is the infamous 'missing link' that has yet to be found. No one knows the answer.

From the fossil record and DNA analysis, the best that science has determined to date is that the split into the two families of primates occurred sometime between four and six million years ago (so, we will say five million years ago).

The earliest hominid ancestors in the fossil record date to about four million years ago and are classified as genus Australopithecus. The most famous fossil remains of that genus are known as 'Lucy.' Now aged 3 million years, Lucy was discovered in Africa, the fatherland and motherland of the human race. Hominids of the Lucy genus differed remarkably between males and females of the species: average male weight was 120 pounds while the female was about 60 pounds; average height ranged from 3 ½ feet to 5 feet. The average cranial capacity was about 400 cubic centimeters (cc), very similar to that of a chimpanzee. And all of that genus had the one significant feature that made them hominids - they walked upright as their predominant method of locomotion. That was a really big deal. Why did that happen, and why is that such a big deal?

Walking Upright

It is necessary to always keep in mind that momentous events are usually the result of an interrelation of causes. Such seems to have been the case in the phenomenon of hominids beginning to walk upright.

The primate apes were marvelously suited to the dense forests of Africa in the Miocene epoch. They could swing through the trees, using their highly developed hands to assist in grasping and developing wonderfully articulated shoulder joints that would allow a virtual 360-degree range of motion. Both skills would help in the adaptation of the hominids to bipedalism. As these skills developed, the climate changed.

Tectonic plate shifts pushed India into the Asian continent, thereby raising the Himalayan Mountains. As the mountains rose up and up, they cleansed more and more of the warming carbon dioxide blanket from the atmosphere. The cooling effect also dried things out. As things dry out, thirsty trees in dense forests thin out. Another aspect of dynamic plate tectonics created elevated highlands along much of eastern Africa, thereby cutting off moist winds and serving to further dry out the climate.

Less rain meant less forest and less fruits, nuts and other edible vegetation. Survival for the African primates got tougher. The fossil record shows that during the two-million-year period occupied by the hominid genus Australopithecus, between four million and two million years ago, as many as five species of that genus existed at the same time. As cooling and drying proceeded over this period in a non-uniform, yo-yo like manner, competition got intense. Adaptations would be necessary if species were to survive.

The dense-canopied forests contracted in size and thinned into lesser woodland areas, during the late Miocene-early Pliocene epochs. The earlier hominids divided their time between these less-dense woodland forests and the expanding grasslands. They likely would retreat to the safety of the treetops for nighttime safety and venture into the savannah grasslands to forage and hunt during the day. The chimps and gorillas stuck to the dense forests.

In dense forests there is no advantage to walking upright. The ecomorph of the ape is much better adapted to living in dense forests. But ape knuckle-dragging can be a real disadvantage in open terrain. Hominids began to stand up to walk as they ventured forth from the woodlands to the savannah grasslands by day and back to the safety of the forest treetops by night, in Africa. A primate can cover a lot more ground a lot faster by walking or running upright rather than dragging its knuckles on the ground.

The hominid ecomorph evolved as bipedalism evolved. Walking upright provided many advantages. In addition to covering more open ground quicker, walking upright provided the framework for a closer type of bonding - a marital type of bonding - based on shared responsibilities between male and female. Survival responsibilities began to be shared with the male venturing forth to search for meat while the female foraged for edible vegetation. Walking upright allowed the female to carry an infant in the crook of one arm while using the other to gather food.

Being cradled in one arm, instead of having to ride on mama's back while holding on for dear life, allowed the hominid infant to enjoy the advantages provided by a longer childhood development period. That in turn led to hominid development of tools and weapons and joint societal advancements. Groups of male stick-wavers began to protect the family. Female and infant no longer had to retreat to the forest and climb to safety when endangered. These developments led to larger brain development. A modern comparison may prove useful.

The brains of chimps and gorillas develop at the same rate as humans. But, chimp and ape infants are born with a closed skull and are somewhat capable of fending for themselves. By contrast, a human infant is completely helpless and is born with an open skullcap. That feature allows the human brain to continue to grow for about a year following birth. That then allows the adult human brain to develop to a far greater size than the adult chimp or gorilla brain. The fossil record seems to bear out the viewpoint that hominids became bipedal before they became brainy.

Walking upright was in fact a very big deal. It provided the basis for further hominid development. As further cooling near the end of the Pliocene, about 2½ to 2 million years ago, resulted in an even-greater differentiated topography of alternating woodland and open savannah terrain, the setting was right for the appearance of the second genus of the hominid family: the genus *Homo*.

The Arrival of Mankind

The final genus of the Hominid family is ours. The genus **Homo - man.** Our specific species is sapiens - wise. **Homo sapiens - wise man**. Our species was not the first of the genus. The fossil record provides evidence that our species certainly had forebears.

The first species of Homo appeared in Africa about 2.5 million years ago, as definite creatures of the grasslands. **Homo habilis**. Genus - Homo - man. Species - habilis - handy. **Handyman.**

Homo habilis was about the same size as the earlier members of 'Lucy's' genus. A female, long-armed skeleton, dated to about 1.8 million years ago, indicates a height of about 3½ feet. The 'handyman' had a larger brain, with an average brain size of about 550 cc.

The fossil record indicates that Homo habilis was the **first species to make stone tools**. These were merely hand-held simple chipped stones that were sharpened along the narrow edge for cutting, and most likely came into use about **2.5 million years ago.**

While there was likely much overlap of species populations and primitive cultures at this time, it is believed that Homo habilis became extinct about 1½ million years ago.

Homo erectus was the next species to arrive on the world scene, and **first appeared about 1.8 million years ago**. Genus - Homo - man. Species - erectus - upright. The name derived from the fact that Homo erectus bones were the first discovered that confirmed bipedalism. Later skeletal findings revealed that Homo erectus was, in fact, far from the first species to walk upright. But, the name stuck.

Homo erectus was larger than Homo habilis. They likely stood on average over five feet and had arms that were proportionally shorter than Homo habilis. Brain size had also increased, with an average brain size of about 900 cc. A further advancement was made in the use of stone tools. They now used 'hand axes' regularly.

Importantly, **Homo erectus was the first species to venture out of homeland Africa, somewhat more than one million years ago.** This diaspora proceeded apace, with fossil evidence showing that by 800,000 years ago these forebears had appeared in the Middle East, Europe, China, and Indonesia, and, indeed, were proceeding to use rafts to cross small sea-faring distances to off-shore islands.

With their dispersal to other locales around the world, they took with them the use of fire. They were **the first species to control fire**, with the important advancements that such control portended. Controlled fire not only provided warmth and cooking, it also could be used to create more grasslands by controlled burning, which ultimately would lead to the beginning of agriculture.

And then, the big event occurred. Homo sapiens arrived on the world scene. Genus - Homo - man. Species - sapiens - wise. Wise man. But, in reality, **the arrival of Homo sapiens was really a non-event. No big deal. Nothing changed quickly or exponentially. The fossil record simply shows that about 500,000 years ago there was a gradual transition from Homo erectus to 'Archaic' Homo sapiens.** Archaic Homo sapiens seems to have been an intermediate stage between Homo erectus and Homo sapiens Neanderthalensis. Brain size had further increased to an average volume of about 1,200 cc. And, brow ridges and teeth had grown smaller while skulls had grown thinner.

The first Homo sapiens Neanderthalensis are dated to about 250,000 years ago. They were indigenous to the Neander Valley in Germany, hence the name Neanderthals. The Neanderthals are commonly considered brutish, and may, or may not, be included in the ancestral line of modern man. Whether they contributed to our gene pool is still debated.

Neanderthals were stocky and very muscular. They were well-adapted for the European ice age in which they lived. They had prominent brow ridges, small chins, large jaws, and a forward-angled face. If you saw a Neanderthal on the street today you would be startled by the difference between his and your appearance.

Neanderthals may have appeared backward, but they certainly were not stupid. They **had a large brain, in fact a larger brain than the modern humans** that out-survived them. A Neanderthal's average brain volume was about 1450 cc, while a modern human's is around 1400 cc. So, we must look to reasons other than just a large brain size to try to figure out why modern humans prevailed.

The Arrival of Modern Man

Archaic Homo sapiens apparently evolved into two strains of Homo sapiens in the Middle East and Africa. The Neanderthal strain is dated to about 250,000 years ago. Our strain - fully modern Homo sapiens (Homo sapiens sapiens) - is dated to about 100,000 years ago. These fully modern ancestors were anatomically the same as us. The contrast between them and the Neanderthals was astounding.

We now know that the Neanderthals probably started out in the Middle East and later expanded into Europe. The fossil record shows that **Neanderthals were contemporaries of fully modern man in the Middle East about 100,000 years ago.** As Neanderthals migrated to the northern European climes they became even heavier and squatter. But, they retained a flat-face with a sloping forehead, long jaw, receding chin, wide protruding nose, and a heavy brow ridge. They were immensely strong. And, they likely had a larger brain than fully moderns.

By contrast, **fully modern Homo sapiens were anatomically just like us.** They had become 'gracile', or less densely boned. Their facial features had become as modern man. **Cleaned-up and dressed in business attire, the fully modern of 100,000 years ago would fail to turn a head while walking down Wall Street or Main Street today.**

The world of 100,000 years ago was one of complexity in evolution. It is believed that at least four species of Homo lived at the same time, including the last of the Homo erectus, Archaic Homo sapiens, Neanderthals and fully modern humans. It made for a rather bushy and disorderly family tree. **By 30,000 years ago only fully modern humans remained.** All other species had become extinct. Our species had prevailed. But, before we take a look at the likely reasons why that happened, let's explore the prerequisite question: 'How and why did we get brainy?'

How and Why Did We Get Brainy?

Our Hominidae family originated about five million years ago with the genus Australopithecus. Those ancestors had a brain about 1/3 of the size that we have. They were the first of us to walk upright, and they begat our genus, Homo-man- about 2.5 million years ago. They had become omnivores, eating both meat (carnivores) and vegetation (herbivores).

Meat eaters, carnivores, are by nature aggressive. And, aggressiveness is a tactical device for bringing focus to the task at hand - the attack. For modern man it is the bob and weave of the boxer, the thrust and parry of the swordsman. The heightened focus of chimpanzees participating in the group hunt is evidenced by an increasingly frenzied state of the hunting pack - shrieking, thumping, shaking branches - inducing terror and associated disorientation in their prey and culminating in the group tearing the prey animal literally into pieces. A similar frenzied focus in our distant ancestors may have assisted in their survival and led to an increasing percentage of meat in their diet. A hungry lion would likely engage a placid antelope instead of confronting those crazy hominids. From Australopithecus to Homo **meat intake increased** from 10% to 20% or more of the diet. And it is speculated that gang-hunting tactics increased in sophistication as such occurred.

The first species of the Homo genus, Homo habilis, had a brain about one-half the size that we have and was the first to use crude stone tools. A gradual increase in brain size continued at a fairly steady pace for the next 2.4 million years, until the fully developed brain of modern Homo sapiens was achieved about 100,000 years ago. From that point forward we possessed a finely tuned thinking machine unlike any the world had known. **The evolution of a larger human brain, for all practical purposes, stopped at that time, about 100,000 years ago.**

There are many theories as to what caused the human brain to continue to grow at a fairly steady rate over the course of millions of years, and then to stop growing 100,000 years ago. They are interesting theories, but we in fact will likely never know for sure how or why that happened. Looking at the fossil record will not shed any light on that issue. But, speculation by experts can prove enlightening.

As the ice ages began two million years ago, and the climate dried, and the trees of East Africa began to thin, our diet changed to include more meat. The brain could thereby grow through the advantage of a richer and easier-to-digest protein source. But, why would that result in a continuing selection toward a bigger brain? The whole point of Darwin's evolution theory is that natural selection only works to improve the next generation. Natural selection is not intended to help out in the long haul. And, increased protein can be used for a lot of anatomical improvements, not just increased brain size.

Some believe the increase in brain size is tied to bipedalism and the corresponding increase in use of the hands. As we began to walk upright, our hands were freed-up to do a lot of things. Since we no longer used them for walking, the hands could be used for many other useful things. The anatomical advantages presented by a very dexterous hand with four fingers and an opposable thumb were enormous. Things could be manipulated, seized, thrown, and held up to be examined. And, as you throw things and examine things that you hold in your hand, you ponder them, you think about them, you think about what you can do with them. This could have resulted in a hand-brain feedback loop that was unparalleled in evolution. That feedback loop may have enabled the brain to continue to grow from generation to generation. However, tool making itself did not seem to provide a feedback loop for brain growth. Indeed, once invented, crude stone tools were not improved upon at a very rapid pace. As their brains grew and grew, our ancestors did not put a handle on a stone 'hand axe' for more than a million years. So much for early innovation.

Over the generations, the human anatomical structure, our ecomorph, became more and more changed from that of the apes. Some of the greatest changes were the anatomical development of the tongue, the larynx, and the face muscles required for human speech. Why did natural selection provide for that? Why were those improvements selected by a force that is concerned only with survival of the next generation, and not for the long haul? No one knows.

Some conjectures, however, are both enlightening and fun. As our ancestors became creatures of the grasslands and not the forest, there was more and more of an advantage to verbal communication. It would help with survival. The group had to learn both how to work together to help each other survive and how to work together to ward-off or attack territorial invaders. The ability to successfully address those two dissimilar challenges would have been greatly enhanced by verbal communications skills.

Cultures of Hominids naturally divide themselves into discrete groups. Outsiders are inferior and are often considered 'non-human'. For four million years humans have lived in 'tribes' of less than 100 individuals gathered into a small community area. Even today our tendency to think of ourselves first as members of our smallest geographic unit says legions about our psychic roots. We are groupish and cliquish. It's simply in our nature.

Darwin, himself, did suggest in *Descent of Man* that intergroup warfare is a realistic theory for human brain expansion occurring continuously for millions of years. The need to balance between internal cooperation and aggression toward outsiders requires some quick thinking, and that requires brainpower. Aggression is needed to expand and control territory. Cooperation is needed to successfully meet the demands of group living. A successful group leader would have to be skilled in both aggression and cooperation. And, when worse came to worst, as the leader sacrifices himself to protect his family and community, that very sacrificial act helps ensure that such cooperative, aggressive and altruistic genes are passed on to the next generation. Altruistic, male group leaders, who were skilled in both aggression and cooperation, would pass along their genes to more and more offspring, as a consequence of being afforded favors by the appreciative group female members. Natural selection would take it from there.

Inter-group warfare is also a plausible explanation for the exodus out of Africa that took place about a million years ago. Such a territorial expansion would require even more of a combination of aggression and cooperation.

In the last stages of brain expansion, as weapons improved, brains improved. As brains improved, weapons improved. As spears replaced clubs, skulls and bodies grew thinner and more 'gracile'. Killing from a distance requires less brute strength than killing up close. Consequently, by 100,000 years ago, our fully modern anatomical structure had evolved and included an on-board, fully modern brain.

The upside of human brain expansion is firmly off-set by a downside that natural selection would most certainly recognize. The human brain is a very expensive organ to maintain. It requires a great deal of energy and, consequently, more rich food to keep it properly nourished. It causes bodily problems, like overheating. And, it increases the risk of death at childbirth. It further results in virtually helpless off-spring who, once they live through childbirth, need constant care. They are helpless babes for an inordinately long time compared to other creatures. So, the downside to human brain development is quite steep. But, we are living proof that it happened, whether by chance, or otherwise.

Regardless of how or why it happened, it happened. By 100,000 years ago we had become brainy, just like we are today. But, the completion of the anatomical development of the modern human brain did not result in an instantaneous leap-frog advancement of modern man. Quite the contrary. **With a fully modern human brain 100,000 years ago, fully modern Homo sapiens (and Neanderthals) continued to use the same crude stone tools that had been employed by their ancestors for over a million years.** Big brains had yet to result in significant innovation. Innovation inhabits the story of the great leap forward.

The Great Leap Forward

It is a fact that in human evolution a Great Leap Forward occurred. No one really knows why or how. But it did happen. The archaeological record bears that out.

Some scientists believe that the Great Leap was the result of modern man's anatomical vocal capabilities and the subsequent cultural development of language. Jared Diamond in his book, *The Third Chimpanzee*, observes that the small change of about 0.1% of DNA was the anatomical basis for complex spoken language. And, that small anatomical difference provided the necessary lay of the land for the cultural development of language by modern Homo sapiens. Language thereby enabled mankind to become the preeminent creatures on Earth.

Language, per se, does not make us think. Animals think. We think. Thought does not depend on words. Being conscious means knowing that we think. Most folks are aware that animals think, but if animals are conscious at all, they are not deeply conscious. They do not 'reflect'. Our higher degree of consciousness indeed seems to be linked to the use of words and language syntax, resulting in a feedback loop similar to the hand-brain feedback loop that may have helped to make us brainy.

Rudimentary human speech may be as old as the genus Homo itself. But, the development of a verbal syntax necessary for language - the ability to place words into categories and hierarchies -likely took a very, very long time. However, once that development occurred, we were positioned to take the Great Leap.

Other scientists contend that the Great Leap Forward occurred when Homo sapiens became endowed with the trait we call **'reflection'**. Reflection is the unique ability of human beings to be aware of the world around them while, at the same time, to be aware of themselves viewing and pondering the world around them. **We no longer just know things. Now we know that we know**. Reflection results not from intelligence or consciousness itself. Other creatures are certainly intelligent and may be somewhat conscious. But, we alone have our **consciousness 'turned in upon itself'**.

Evolution did not provide reflection gradually. **Reflection was a 'zero to everything' occurrence.** Up to the threshold of reflection, the ability to see oneself seeing the world is completely absent. Upon crossing the threshold, the ability is complete and constant. We can never not be aware of our self viewing the world.

After crossing the threshold of reflection, mankind was qualitatively different than any other creatures on Earth. We had been uniquely transformed. We had experienced a unique 'change of state'. We had achieved another order of complexity entirely.

Consciousness increases with advancements in the animal order. There is no doubt in my mind that a dog is more conscious than a fish. All one has to do is observe. My dog, Jake, is very intelligent and he knows a lot of stuff. But, he can never really know that he knows. The entire world that opens up with reflection is forever beyond his grasp. Jake cannot enter my thinking world and I cannot enter his. There is a fundamental difference in our brains and in our abilities to solve problems. My brain is bigger and more complex. I perceive more things in the world, more relationships and patterns. Because of my ability to reflect, my knowledge of the world is more ordered than his. I discern the rules that govern things which gives me an advantage. On the other hand, because Jake's biological senses are so much more keen than mine, his thinking world will always be more physically intense. But, his thinking world also will always remain more confused.

The Great Leap Forward, or 'change of state' caused by reflection, occurred in mankind without any apparent organic or anatomical change. Our modern anatomy and brain size was unchanged. But, the 'change in being' was enormous. We suddenly began to not only know, but to know that we know. For the first time we saw ourselves seeing the world.

Such a change of state is far from a unique occurrence in either biology or physics. Somehow primitive atoms undergo a change of state to become discrete molecules. And then further chemical processes result in cellular order where previous disorder existed. Distinct organs and living organisms result from further changes of state. In all these processes the initial components are completely transformed. What they were at the outset of the process had been completely abandoned to produce the new state. An entire state of being is completely changed by these natural processes. The complete change of state evidenced by the emergence of the butterfly from the caterpillar is one of beauty and grace.

Science evidences that many such change-of-states happen abruptly. Placid liquid water begins to boil at 212 degrees Fahrenheit and the state of placid water is quickly changed to roiling and swirling about almost instantaneously. Placid liquid water freezes at 32 degrees Fahrenheit and a change of state quickly occurs. The state of the same substance that you could swim in last summer is changed in a manner that allows you to skate on top of it this winter. One of the most awesome changes of state in nature is seen in the phenomenon of fire called 'flash over point'. Once the right temperature is achieved and the 'flash over point' is thereby reached, in the wink of an eye absolutely everything in sight, seemingly the air itself, is engulfed in a fiery furnace.

With the change of state resulting from reflection, modern man began a new kind of life. The world became centered around the person thinking about the world and marvelous new vistas were opened to human thought and achievement. In his book, *The Phenomenon of Man*, Pere Teilhard de Chardin summarizes that epiphany with these words:

"In reality, another world is born. Abstraction, logic, reasoned choice and inventions, mathematics, art, calculation of space and time, anxieties and dreams of love - all these activities of *inner life* are nothing else than the effervescence of the newly-formed centre as it explodes onto itself."

Natural selection, remember, is concerned only with the next generation. It could care less about the long haul. So, in order for the quality of reflection to become the preeminent characteristic of modern man, the 'flash over point gene' must have been passed on to the next generation by one couple, or two, or ten, or a hundred, or...we simply do not know. Whatever the biological method, the human metamorphosis caused by reflection seems to have occurred in modern Homo sapiens sometime between 100,000 years ago and 30,000 years ago. It simply failed to occur in Neanderthals.

Neanderthals and fully moderns coexisted for thousands of years using the same tools and weapons. Then they seem to have gone down entirely different paths. The Neanderthals were certainly as brainy as moderns. The Neanderthals were immensely stronger and more powerful than moderns. A modern would fare quite poorly in a one-on-one wrestling match with a Neanderthal. It would not be pretty. Yet, by 30,000 years ago, the moderns (called Cro- Magnons) flourished in Europe while the Neanderthals had become extinct. The Cro-Magnons, modern man, had become endowed with reflection and had developed a more-than-rudimentary language. So armed, they had taken the Great Leap Forward.

Neanderthals continued to use the same tools and weapons they always had. They seem to have been incapable of innovation. But, with reflection and language, modern Homo sapiens quarried stones of high quality to make more and more refined tools and weapons. Their tool kit included an axe with a handle, sewing needles, knives, fishhooks and harpoons. And, they developed more sophisticated weapons, such as balanced spears, battleaxes, and bows and arrows. Though just as brainy as moderns, it seems that Neanderthals failed to develop a more-than-rudimentary language. And, it seems clear that human reflection had eluded them entirely.

Jared Diamond summarizes the lay of the land and the state of human evolution 30,000 years ago eloquently:

"Until the Great Leap Forward, human culture had developed at a snail's pace for millions of years. That pace was dictated by the slow pace of genetic change. After the leap, cultural development no longer depended on genetic change."

After the Great Leap Forward, we did not have to adapt to our environment any longer. We began to change our environment to meet our needs. With that evolutionary paradigm shift, rudimentary human civilization began.

Wrap-up of Part I and Preview of Things Ahead

At this point we have completed our overview of the story of creation and evolution without God. It is a tale told by today's scientific elite and retold by today's mainstream scientific community. It is a fascinating and compelling story. But, the story is far from tidy. It has a lot of loose ends.

'The devil is in the details'.

That is a common expression we are all familiar with. It connotes the simple fact that generalized conclusions are sometimes artfully stitched together into an attractive fabric of 'truth' that begins to fray and unravel when the stitches of the fabric are examined in greater depth. On the other hand, if the fabric does not unravel upon closer examination, the 'truth' that is supported by the devilish details takes on a stature of unassailability. It simply must be true.

Parts II and III of this primer will examine some of the loose ends of the scientific story of creation and evolution without God. Do the devilish details unravel the fabric or support an 'unassailable truth'?

PART II

ASSEMBLING A LIFE-FRIENDLY WORLD WITHOUT GOD
- A CLOSER LOOK AT 'ACCIDENTAL' NECESSITIES -

It just so happens that our home planet, Earth, turns out to be located in an ideal location for intelligent life to evolve and inhabit. And, it just so happens that Earth turns out to be comprised of just the right stuff for such to occur.

These are most improbable things. How these improbable things happened can be explained in one of two ways:
- it was the result of planning, or
- it was accidental.

The elite and mainstream scientific community today believes that the accidental explanation is the best explanation.

In Part II we will examine some of the features of the universe that led to Earth being a 'Goldilocks Planet', just right for human habitation. And, we will review the fundamentals of why the elite and mainstream scientific community today believes that the accidental explanation is the best explanation.

CHAPTER 6
A Clever Universe

Intelligence is defined as **the ability to learn or to deal with new or trying situations.**

As the universe developed over eons of time, the early off-spring of the 'Big Bang' were faced with many new and trying situations. The things of the universe that we observe today give evidence to the fact that the early off-spring dealt with those situations in a manner that allowed a more and more complex and rich world to evolve. **They very effectively dealt with both new and trying situations.**

Does this represent intelligent evolution that is built-into universal development? If it is, what is the source of the intelligence? Where does the intelligence come from?

In an attempt to determine whether or not intelligence is built-into the universe, let's proceed with a step-by-step examination of the discoveries of our most brilliant scientists regarding how the universe began, how it developed, and how things within the universe work. The examination will look afresh at the fundamentals. We should repeatedly ask ourselves the questions: **'How did it know to do that?'** and **'Why did it get more complex?'** For those questions are not trite or beside the point. Those questions are precisely the point.

Students of the Universe

One perspective that clearly is presented through an examination of the fundamental building blocks of the universe is that every thing in the universe seems to be a student. All things seem to learn how to better adapt themselves to their changing environment. Evolutionary theory governing living things calls this propensity **evolutionary development through adaptive response to environmental change**. And, when you look at the big picture, the same adaptive response to environmental change seems to be just as important for non-living things.

There seems to be an intelligence of the universe that somehow becomes infused in all matter and all energy. For all things in the universe a step-by-step learning process seems to result in more and more and more complexity of things. And, in turn, the more and more and more complex things learn how to adapt through yet more complexity. The pattern is remarkably uniform and universal. It's really simple. And, it's really eerie.

Let's start at the beginning and review the work of the 'students of the universe'.

When Energy and Matter Were One

Science tells us that the universe began when a singularity, a tiny spot of unimaginable tinyness and unimaginable density, exploded. Somehow, we don't know how, it just exploded.

Before the 'Big Bang' nothing that we now know of existed. Matter did not exist. Energy did not exist. Fundamental forces did not exist. And, the fundamental laws governing the fundamental forces, the matter and the energy did not exist. There was simply an unimaginable void. But, somehow, and we don't know how, the fundamental laws, the fundamental forces and matter and energy sprang into existence within a tiny, tiny fraction of a second following the 'Big Bang'.

At the instant of creation of the universe, science tells us that only photons of electromagnetic energy - light - and only light, should have forever existed. Matter was, in essence, completely integrated within energy. Each particle of matter had a companion anti-particle and each obliterated the other. When a particle of matter meets a particle of anti-matter they annihilate each other and their rest-energy mass is thereby converted into photons of light. Therefore, only photons of light should have forever existed in the universe. Somehow, and we don't know how, anti-matter quickly morphed into ordinary matter.

In a tiny fraction of a second the universe was now composed of more than just light energy. It was now composed of both energy and ordinary matter. Albert Einstein discovered that, at core, energy and matter are the same stuff, but now they appear in two different forms. Fundamental quantum particles of ordinary matter - electrons and quarks - now spewed forth at the speed of light along with photon quantum particles of light traveling in waves of electromagnetic energy at the same speed.

Photons have no mass or electric charge. But they are believed to ultimately be responsible for producing all electric and magnetic fields. The elementary particles - electrons and quarks - have both mass and intrinsic electric charge. No one knows where electric charge came from. All ordinary matter is made-up of **irreducible** electrons and **irreducible** quarks.

The Evolution of Ordinary Matter

In still less than a second following the 'Big Bang' the **irreducible quarks of matter began to stick together to form protons and neutrons**. A proton contains two 'up quarks', each spinning clockwise and possessing an intrinsic positive electric charge of +2/3, and one 'down quark', spinning counterclockwise and possessing an intrinsic negative electric charge of −1/3. This combination gives each proton a positive charge of +1, the exact opposite charge necessary to offset an electron's intrinsic negative charge of −1. A neutron contains one 'up quark' and two 'down quarks' thereby giving each neutron zero electric charge. From what science can determine to date, these quarks are fundamental particles. They cannot be further sub-divided. The fundamental quarks had quite quickly learned how to stick together and compose a 'higher' form of matter, namely protons and neutrons.

The next evolutionary step for ordinary matter took about two minutes following the 'Big Bang'. At that early point in the development of the universe things had cooled enough so that the **strong nuclear force** was successful in **adhering protons and neutrons together** to form the nuclei of what would later become hydrogen and helium atoms. **Matter had evolved to form atomic nuclei.** Complexity of ordinary matter had increased greatly. Fundamental quarks had learned how to combine to form protons and neutrons, and then protons and neutrons had combined to form the nucleus for the first elements of the universe - hydrogen and helium.

About half-a-million years following the 'Big Bang' the universe had cooled enough for a most compelling complexity in the evolution of matter to occur. **Electrons, having a negative electric charge of -1, fell into orbit around the hydrogen and helium nuclei composed of protons and neutrons**. The negative charge of the electrons orbiting the nucleus exactly offset the positive charge of the protons in the nucleus, making each atom of matter electrically neutral. Individual **atoms** of matter were thereby formed. The newly formed atoms comprised a new complexity of matter. **Ordinary matter in atoms has evolved to have no electrical charge at all.** That development, in turn, will lead to more and more complexity, as atoms learn how to interact with other atoms in the neighborhood.

Hydrogen is the lightest and most basic of the 92 naturally-occurring chemical elements. It has one proton and one electron. Each of the 91 other natural elements has a unique and more complex atomic structure consisting of numerous electrons, protons, and neutrons. The number of protons in the nucleus determines the atomic number of the element. The heaviest naturally-occurring chemical element is uranium, containing 92 protons, 146 neutrons and 92 electrons.

The number of protons for each element is unique and constant, and determines the atomic number for the element. But the number of neutrons may vary, resulting in a heavier mass of a hybrid form of an element, called an 'isotope'. For example, oxygen always has eight protons, but it can have eight, nine or ten neutrons.

After atoms of hydrogen and helium were formed the force of gravity proceeded to collapse the atoms in upon themselves to form a dense gas core.

As the force of gravity caused the hydrogen gas to get denser and denser, the hydrogen atoms collided with each other at a faster and faster rate as gas pressure increased. This, in turn, caused intense heat. The heat intensity increased until it reached a point where the atoms no longer bounce-off each other. At that point the hydrogen atoms fuse together. Four protons of hydrogen fuse together to create a helium atom and release photons of electromagnetic energy in the process. When this thermonuclear reaction occurs, a star is born.

In this manner, one star, and then another and another began to form. Atoms had learned how to work together to form a huge mass. Those atoms then proceeded to work together to produce even heavier elements and, at the same time, to release enormous amounts of energy in the form of sunshine (or, if you prefer, starlight). The fundamental rule of the universe whereby matter and energy are, in fact, equivalent provided for this wonder.

This **thermonuclear reaction of atomic fusion** in the core of a star continues for millions or billions of years until all the hydrogen has been extracted. If a star is large enough, the core reactor then fuses together helium atoms to make each next-heavier element: lithium, beryllium, boron, carbon, nitrogen, and oxygen. The thermonuclear process continues through the successively heavier elements until finally manganese is fused to make iron.

Once iron has been created, further energy cannot be extracted by means of the star's thermonuclear process. The elements heavier than iron are formed when a massive star explodes as a supernova, thereby creating the remainder of the 92 naturally-occurring elements in the universe. **A lot certainly has been learned and accomplished by non-living things at this point. The atoms have followed the rules and learned to work together to develop all of the material elements and to produce abundant light energy in the process.**

After about nine or ten billion years of universal expansion and evolution, our Sun and its solar system were born out of the remnants of a supernova explosion. As the third closest planet to the Sun, our planet Earth began as a firestorm. But, as evolutionary time proceeded, the Earth cooled and became infused with just the right elements, including hydrogen, carbon and oxygen, in just the right proportions to support organic development. The natural elements then proceeded to discretely combine to form molecular compounds through the attraction and repulsion of fundamental particles of matter.

Molecular bonds are formed by the interaction of electrons in the outermost shell of an atom's structure. Weak molecular bonds are formed when atoms give-up or gain electrons and thereby become electrically charged (called **ionic bonds**). Strong molecular bonds are formed when atoms share electrons (called **covalent bonds**).

If atoms do not chemically bond they remain electrically neutral. But, all of the electrons in the outermost shell of each atom are negatively-charged. If like charges repel how does any ordinary natural matter stick together? The secret is called **residual electromagnetic interaction**. The electrons of one atom are residually-attracted to the protons in the nucleus of another atom in the neighborhood. As two atoms approach each other their electrons are attracted by the positively-charged nucleus of each of the atoms. The repulsive forces of each of the nuclei establish the closest distance that they can approach each other. Homeostasis of attraction and repulsion thereby establishes a molecular configuration that is even more stable than the two atoms in isolation.

Residual electromagnetic interactions between electrically-neutral atoms are responsible for the binding of atoms to form common molecules and most of the forces we experience everyday (except gravity). The rigidity of the floor beneath your feet, the friction between your feet and the floor that allows you to walk, and the wind resistance providing a cooling breeze on your face are all the result of this residual electromagnetic interaction. The repositioning of electrons or atoms, as matter is deformed by contact with other matter, results in these energy changes that allow us to experience the world in which we live.

Ultimately the electromagnetic force is what allows atoms to bond and to form molecules. It allows our world as we know it to exist. **At core, all the structures of the world exist simply because electrons and protons have equal and opposite electric charges.** Without these electrical force fields, the *things* of the universe would not exist. All that would exist would be simply diffuse clouds of protons and neutrons and electrons.

Two of the elements, hydrogen and oxygen, combined to form water, a necessity for all life.

Most fortunate for us, water molecules covalently bond two hydrogen molecules with an oxygen molecule and then add a double-bond (called a **hydrogen bond**) causing the now-slightly-positive hydrogen atom to have an affinity for the now-slightly-negative oxygen atoms of other molecules of the same compound. This fairly weak hydrogen bonding does not form new molecules. Rather, it plays a vital role in establishing the unique properties of water. This nearly-unique bond is responsible for both the high boiling point of water and the amazing fact that solid ice floats.

As the elements of matter 'learned' how to combine into more and more complex forms with more and more specificity, the stage became set for life on Earth. Carbon is the simplest element that can share the maximum number of electrons in its outermost shell. By its ability to share those four electrons, carbon can form strong bonds with hydrogen, nitrogen, oxygen, sulfur and itself. It thereby became the central elemental building block for the organic molecules that make up all living organisms. Most scientists believe that life began on Earth as molecules of more and more of these complex elemental compounds, proximate to each other, began to engage in chemical reactions. First the inorganic elements 'learned' how to become inorganic compounds, and then the inorganic compounds 'learned' how to become organic compounds, and then those organic compounds 'learned' how to become infused with life. They 'learned' how to do all these things by following the invariant rules of the universe. That is the mainstream scientific explanation. It is called emergent evolution.

The Evolution of Energy

Science tells us that from the instant of the 'Big Bang' until now, and until tomorrow and tomorrow and all other tomorrows, the total amount of energy in the universe has always been and will always be the same. The 'Big Bang' instantly created all of the energy in the universe. That energy was in the form of light. Light is composed of massless particles containing no electric charge. The particles are called photons and they travel through the void of spacetime in electromagnetic waves at the speed of light. Since all of the ordinary matter of the universe evolved out of that initial light, all the matter of the universe and all of the energy of the universe must necessarily have been initially contained within that light energy. So, the amount of that energy and matter has always been constant. It is the same amount now as it was then.

Science cannot tell us what energy really is or how it happened to result from that initial explosion. But, brilliant scientists have discovered an awful lot about it.

Electric charge is an intrinsic property of all of the fundamental particles of matter - all electrons and quarks. As quarks combine to make a proton, they infuse the proton with a discrete amount of positive electric charge (specifically $+1.60 \times 10^{-19}$ Coulomb). That exactly offsets an electron's negative electric charge (specifically -1.60×10^{-19} Coulomb). (Note: a Coulomb is the amount of electric charge transported by a current of 1 ampere in 1 second.)

As pure light energy quickly morphed into ordinary matter in the form of electrons and quarks, ordinary matter then proceeded to evolve. Quite wondrously, **as ordinary matter proceeded to evolve, it seems that energy proceeded to evolve as well.** All energy began as photons traveling at the speed of light in electromagnetic waves. Light is pure **energy of motion** that we call **kinetic energy**. The energy resides in the wave motion of the intrinsic electromagnetic field of light. The frequency of wave motion back and forth (vibration) determines the type of light. Some of that light energy must have been transferred to ordinary matter (electrons and quarks) as they came into existence in the infant universe.

All ordinary elemental particles of matter have three intrinsic properties: mass, momentum and electric charge. Momentum and electric charge provide the basis for energy use and transfer between ordinary matter.

In the early twentieth century, brilliant scientists discovered that each elemental particle has a unique energy state that is based on the frequency of the particle's vibration. The genius of Max Planck discovered that **energy comes in small chunks, called quanta**. And he found the constant that is used to define each unique energy state. In honor of his discovery the **Planck constant** is now used to determine most specifically the energy state of sub-atomic particles, atoms, and molecules. In calculations the Planck constant is denoted by the letter 'h'. The constant is quite specific: 6.626×10^{-34} Joules per second. (Note: the metric unit of energy that results from a force applied over a specific distance is expressed in Joules.) Energy (E) equals frequency of vibration (f) times the Planck constant (h). And amazingly, the energy content of ordinary matter must be a whole number integer of the frequency (f) of vibration times the Planck constant. You can have 2 hf or 23 hf or 1,987 hf, but you can't have 2.5 or 23.2 hf.

The energy of motion began to evolve into many different forms that would prove to be useful for the evolution of ordinary matter. Energy and matter began to work together as complexity begat more complexity. It seems that the ethereal substance that we call energy may be a student of the universe as well.

Textbooks on thermodynamics tell us that in order for all of the energy in the universe to remain constant we must include another fundamental state of energy in addition to the energy of motion. The two fundamental states are:
- **kinetic** (energy of motion), and
- **potential** (stored or rest energy).

As energy began to work together with ordinary matter to evolve the natural elements and to make chemical molecular compounds, energy began to be stored in a resting state in the electrical fields that hold molecules of ordinary matter together. **Stored rest energy increases the mass of matter. The release of that stored energy then reduces the mass of matter.** When any body is at rest, its energy is equal to its rest mass times the speed of light squared ($E = mc^2$). As Einstein discovered, matter and energy are fundamentally equivalent and can morph into each other. When photons are absorbed into matter the matter transitions to a higher energy level. When photons are emitted from matter the matter transitions to a lower energy level. And, it seems that the fundamental energy of motion learned how to morph into discrete energy types as it worked with ordinary matter.

Energy is transferred or transformed from a particle or mass of matter by either **heat or work**. Work and heat are actually methods of energy transfer or transformation.

Energy changes things and makes things happen. **In order for anything in the universe to happen, there has to be an energy transfer or conversion.** And, remember, the amount of total energy in the universe has not and will not change. It simply moves into different places and into different forms.

Energy has evolved to take many forms and to provide for the performance of all the functions of the universe.

Intelligence in the Universe

A true wonder to behold is how the very simple forms of matter and energy proceeded to become more and more complex as shaped by the fundamental forces, in conformance with the universal 'rules' that govern the development and evolution of all matter and all energy. How did matter and energy know how to do that? Why did they get more complex? **Where did the intelligence come from? The truthful answer is 'NO ONE KNOWS'.**

The rules of the universe (the laws of nature) are written in the language of the universe. We call the **language of the universe - mathematics**.

As we will review in the next chapter, brilliant scientists have discovered through mathematics that the rules of the universe are so finely-tuned and intricately orchestrated that non-living, inanimate matter and energy, in combination with the fundamental forces, are compelled to evolve over the course of time into more and more complex forms of matter and energy in accordance with the 'rules'. The 'rules' thereby direct the ever-more-complex interactions and outcomes. **Non-living matter and energy thereby evolve by adaptive response to environmental change. They learn to deal with new situations. That is the very definition of intelligence.**

Upon reviewing the discoveries of science to date, it seems clear that **the universe abounds with intelligence. What is the source of the intelligence that provides the information necessary for the evolution of matter and energy?**

Laws of Nature Supply Information in a Clever Universe

The 'information' that governs the evolution of non-living matter is quite limited. Non-living matter changes strictly in accordance with the laws of physics acting upon it. Each of the 92 elements that naturally occur in the universe is composed of atoms comprised of electrons and protons and neutrons unique to each element. They are assembled in strict accordance with the laws of physics governing the forces acting upon them. The protons and neutrons are, in turn, assembled in accordance with the information contained within the laws of physics governing the forces acting upon the quarks that comprise them.

Inorganic molecules are likewise the product of 'information' that is quite limited. They are created when atoms proximate to each other combine by bonding with other atoms in the neighborhood. The bonding, by either sharing electrons (covalent bonding) or donating or gaining electrons (ionic bonding) occurs in strict accordance with the laws of physics governing the electrical interaction of neighboring atoms.

Of the 92 elements of the natural universe only six (carbon, hydrogen, oxygen, nitrogen, phosphorous, and sulfur) have 'evolved' to make-up the 'stuff' of all life. Those six elements learned how to combine to form the four organic molecular compounds known as lipids, carbohydrates, nucleic acids and amino acids. The limited information governing physical chemical bonding requires that when atoms of one of these six elements comes into contact with atoms of some of the other five elements, under the proper conditions, physical chemical bonding occurs in the outer electron shells to form complex molecules of organic matter.

At this point it should become clear that 'intelligence' contained in non-living matter is really a misnomer. Non-living matter simply expresses itself in different forms strictly in accord with the information contained within the laws of physics acting on matter and energy. Non-living matter does not contain intelligence that governs its development. **The only intelligence that governs the evolution of non-living matter is the information contained within the very laws of the physical universe itself - the laws of nature.** Non-living matter develops strictly in accord with the laws of nature acting upon it. It contains no intelligence within.

But, the universe itself abounds in intelligence. We live within a very clever universe. **The intelligence of the universe resides within the laws of nature themselves that orchestrate the evolutionary development of all of the wonders we observe.**

The fundamental question regarding purpose and meaning remains: Where did that intelligence come from?

CHAPTER 7
The Invariant Rules of An Anthropic Universe

The 'Big Bang' produced all of the matter and energy in the universe that has evolved to detail all of the wonders that we observe each day of our lives. Matter and energy have evolved incessantly ever since that creation event.

The 'Big Bang' also produced the fundamental forces of the universe and the laws of nature that govern the fundamental forces as they mediate the interactions of matter and energy. The **fundamental forces and the laws of nature, unlike matter and energy, do not evolve**. They are now, and have always been absolutely the same. They are invariant.

The laws of nature establish the **rules of the universe**, and include within them the precise dimensions of the constants and fundamental parameters of the universe. We can never know what the initial conditions of the universe were. But, those initial conditions had to have included the methodology for producing the exactness of the universal constants and fundamental parameters that exist in the universe today. The present structure of the universe is a unique and continuous function of those initial conditions.

The universal constants and fundamental parameters, that were established at the beginning of time, have molded the features of the universe to include our Sun, our Moon, the sister planets of our solar system, our home planet Earth, and all living things on Earth, including us. And, amazingly, the rules of the universe have resulted in a universe that allows us to be its students.

The Invariant Rules

Science has discovered a lot about the constant, immutable, and eternal **rules of the universe**, that are called the **laws of nature**.

Textbook after textbook is filled with the complex rules of physics and chemistry and biology and astronomy and physiology and anatomy and on and on. Most of us struggle mightily to barely comprehend just some of the rules that science has discovered.

The rules seem to be described by mathematics. Scientific textbooks are replete with mathematical formulas and equations. The intricate and exquisite forms of mathematics go on page after page to describe and explain the complex interactions of forces with matter and energy as they manifest themselves in all phenomena.

Many scientists refer to **mathematics** as the '**language of the Universe**'. As in all languages, the language of mathematics can, at best, only attempt to provide a small glimmer of underlying meaning. The mathematical equations discovered by brilliant scientists describe more and more about the work of matter and energy spanning a scale from the tiny world within the atom to the unimaginable vastness of the universe itself. But, the language of the universe remains shrouded in mystery. In the words of wonder from the renowned physicist, Steven Hawking:

> **"What is it that breathes fire into the equations and makes a universe for them to describe? The usual approach of constructing a mathematical model cannot answer the question of why there should be a universe for the model to describe. Why does the universe go to all the bother of existing?"**

There will never be a scientific answer to those questions. Science deals in observable phenomena and facts. Answers to those questions are incapable of being derived from observable phenomena and facts.

As scientists have discovered more and more about the mathematics of the universe they have discovered some amazingly precise properties of the universe that cannot be predicted from mathematical theory. These precise properties can only be derived by experiments. Let's look at some of the discoveries made through experiments.

The Universe is Finely-Tuned

Brilliant scientists have discovered that the universe is quite finely adjusted. In order for the universe to develop as it has and to contain life within it, more than twenty separate physical parameters are precisely calibrated as fundamental constants of the universe. The universe appears to be 'fine-tuned' to allow the existence of life as we know it.

The Standard Model of Particle Physics includes some twenty-five arbitrary and precise fundamental constants of the universe that have been discovered by science. These non-derivative fundamental numbers cannot be predicted from theory. Science has discovered them by experiment and has found them to always be invariant. They preserve the same value at any place and at any moment in the universe (e.g., each proton is identical to every other proton anywhere and each electron has the same mass and the same charge as every other electron everywhere in the universe).

The constants of the universe are used by scientists to determine the cause and effect relationships of matter, energy and forces. **The constants are necessary in order for the laws of physics to work. But, the laws of physics cannot themselves explain the universal constants**.

Each of the twenty-five constants is very, very precise. If the precise value of any of the 25 were changed only very, very slightly, the world we live in would simply not exist. Some examples may prove useful.

The strength of the force of gravity and the strength of the other three fundamental forces in the universe are very precisely calibrated:

- If the force of gravity had been 10^{33} times weaker than the electromagnetic force, stars would be a billion times smaller than they are and would burn at a rate of a million times faster. Fortunately, gravity is precisely 10^{39} times weaker than electromagnetism.

- If the weak nuclear force had been only slightly weaker than 10^{28} times the strength of the force of gravity, all of the hydrogen in the universe would have been converted to helium. Water could not exist.

- If the strong nuclear force had been only 2% stronger no protons would have formed and thereby no atoms would exist. If the strong nuclear force had been only 5% weaker, there would be no stars in the sky.

The first constant of the universe to be discovered by science was Newton's gravitational constant (the inverse square law). The gravitational constant is universally used to determine the strength of the force of gravity between two bodies of known mass and separated by a known distance. But, that does not explain where the gravitational constant came from or why it is precisely what it is.

Max Planck discovered the quantum of action that is now called Planck's constant. It is used in physics to determine the precise dimensional properties of fundamental matter and energy in the universe and to calculate:

- the energy of light in any form;
- the energy in a vibrating electron, atom or molecule;

- the angular momentum of an electron; and
- the minimum uncertainty for simultaneous measurement of both the position and the velocity of an electron.

The Planck constant is precisely 6.626069×10^{-34} Joules-seconds. The constant is routinely used in quantum physics, but its routine use does not explain where it came from or why it is precisely what it is.

If the density of the universe at one second after the 'Big Bang' had been one part in a thousand billion greater, the universe would have collapsed after only ten years.

If the electromagnetic coupling constant (which binds electrons to protons in atoms) were only slightly different or if the ratio of electron mass to proton mass were only slightly different, then atoms and molecules would not have formed in the universe.

The charge-to-mass ratio of the electron is precisely 1.76×10^{11} Coulombs per kilogram. And the ratio of the masses of the electrons to the protons in the nucleus of an atom is very precise. Cambridge physicist Steven Hawking provides this provocative insight:

> "The laws of science as we know them at present, contain many fundamental numbers, like the size of the electric charge of the electron and the ratio of the masses of the proton and the electron. We cannot, at the moment at least, predict the values of these numbers from theory - we have to find them by observation....The remarkable fact is that the values of these numbers seem to have been very finely adjusted to make possible the development of life....In fact, a universe like ours with galaxies and stars is actually quite unlikely. If one considers the possible constants and laws that could have emerged, the odds against a universe that has produced life like ours are immense."

The Universe is Anthropic

Nicolaus Copernicus, a Polish amateur astronomer, proposed in the 16th century that the Earth was not the center of the universe. That, of course, is now an established scientific fact. The Earth and the other planets in our solar system orbit around the Sun. Scientists following in the long shadow cast by Copernicus have established a scientific principle in his honor.

The **Copernican Principle** posits that there is nothing special about the Earth's position in the universe. Modern scientists have expanded on that principle to include the idea that not only is there nothing special about Earth's position in the universe, there is nothing special about the status of mankind as well. The Copernican Principle denies that the status of human beings in the universe is in any way privileged. Most scientists support this principle.

At a 1973 conference to celebrate the 500th birthday of Copernicus, a Cambridge astrophysicist, Brandon Carter, proposed another principle thought necessary to explain the discoveries of scientists that seemed to be at odds with the Copernican Principle. Carter posited that:

"Our location in the Universe is necessarily privileged to the extent of being compatible with our existence as observers".

He called this the **Anthropic Principle**.

Carter's Anthropic Principle may have been intended to deal with bias that may skew experimental findings due to what scientists call a 'selection effect'. In their book entitled *The Anthropic Cosmological Principle* authors John Barrow and Frank Tipler explain the basics of a 'selection effect':

"The basic features of the Universe, including such properties as its shape, size, age and the laws of change, must be *observed* to be of a type that allows the evolution of observers, for if intelligent life did not evolve in an otherwise possible universe, it is obvious that no one would be asking the reason for the observed shape, size, age and so forth of the Universe. At first sight such an observation might appear true but trivial. However, it has far-reaching implications for physics. It is a restatement of the fact that any observed properties of the Universe that may initially appear astonishingly improbable, can only be seen in their true perspective after we have accounted for the fact that certain properties of the Universe are necessary prerequisites for the evolution and existence of any observers at all. The measured values of many cosmological and physical quantities that define our Universe are circumscribed by the necessity that we observe from a site where conditions are appropriate for the occurrence of biological evolution and a cosmic epoch exceeding the astrophysical and biological timescales required for the development of life-supporting environments and biochemistry."

Barrow and Tipler observe that in a sense Carter's Anthropic Principle

" ...may be regarded as the culmination of the Copernican Principle, because the former shows how to separate those features of the Universe whose appearance depends on anthropocentric selection, from those features which are genuinely determined by the action of physical laws.... The outstanding problem of ancient astronomy was explaining the motion of the planets, particularly their retrograde motion. Ptolemy and his followers explained the retrograde motion by invoking an epicycle, the ancient astronomical version of a new physical law. Copernicus showed that the epicycle was unnecessary; the retrograde motion was due to an anthropocentric selection effect: we were observing the planetary motions from the vantage point of the moving Earth."

Carter 's intention may not have been to insert a teleological idea into scientific discussions. The effort was likely intended to simply reconcile the Copernican Principle with the compounding facts discovered by science that were apparently at odds with it. But the use of the term 'anthropic' put the issue of teleology again on the scientific discussion table.

The 'Weak' Version of the Anthropic Principle

Teleology is the study of evidences of design in nature. On the face of it, the Anthropic Principle is a teleological idea. The finely-adjusted physical parameters of the universe have resulted in intelligent beings who can observe and write about them. That is a fact. Yet, most scientists who do not believe in teleology do not deny the Anthropic Principle. Here is how they deal with the issue.

Even the most staunch supporters of the Copernican Principle admit that the evidence in support of fine-tuning in the universe is most compelling.

Most scientists support the so-called **Weak Anthropic Principle**. The Weak Anthropic Principle recognizes the scientific fact that in order for intelligent life to exist on Earth it was necessary for the universe to be of a sufficient age and to contain certain specific properties and parameters that allow intelligent observers to evolve (a crude summary statement of Barrow and Tipler's 'selection effect'). These scientists simply contend that such a finding is no big deal. That is just the way things are. If the fundamental constants of the universe did not exist as they do then we human beings would not be here to observe them. The universal constants and the laws of nature themselves require no explanation. They are simply there. Just accept them as givens.

These scientists contend that the Weak Anthropic Principle actually predicts the existence of a 'multiverse' (an infinite number of possible universes) each possessing different specific properties and constants. **These scientists contend that there may be an infinite number of REAL universes possessing infinite sets of laws and parameters and constants. So they maintain that the fact that our universe has the specific universal laws and the twenty-odd specific parameters that it does is no big deal.** These scientists contend that **the fact that our universe contains so many specific universal laws and constants that are necessary for our being here is, in fact, presumptive evidence that many other REAL universes actually exist.** As cosmologist Sir Martin Rees explains:

> "If this multiverse contained every possible set of laws and conditions, then the existence of our own world with its particular characteristics would be inevitable."

This 'scientific' prediction is made in spite of the fact that **there is no scientific evidence whatsoever that any universe, other than our own, actually exists.**

The 'Strong' Version of the Anthropic Principle

A minority of 'contrarian' scientists support the so-called **Strong Anthropic Principle**. The strong version ascribes the existence of the specific universal laws and the finely-adjusted parameters of our universe to **actual teleology** (end reason or purpose). The Strong Anthropic Principle simply contends that the laws of nature, and the exquisitely-precise parameters and constants of the universe, are **designed in a manner necessary to create and evolve intelligent life.**

To these scientists the precision of the universal laws and parameters and constants are, indeed, circumstantial evidence that the universe was actually designed by a creative intelligent agent. Some call the creative intelligent agent God.

The strong version explains that the formation of our Milky Way galaxy, our Sun and solar system, and our beautiful planet, Earth, was the evolutionary result of those specific universal laws and finely-adjusted parameters of the universe working together and producing the wonders of our living world.

The wonders of our living world, including the wonder of ourselves, required the formation of a friendly planet in a nice neighborhood. Let's turn to the next chapter and take a closer look at that development.

CHAPTER 8
A Friendly Planet in a Nice Neighborhood

As astronomical discoveries are made and as NASA space probes continue to visit our closest celestial neighbors, news reports often give the impression that the discovery of life on another planet is right around the corner. Eminent scientists remark that the key to such a discovery is liquid water. If water is found then life will certainly follow. The impression is widely conveyed that science has found out that making life is an easy thing. Just put the right conditions together with the right chemicals and voila, life emerges.

Prominent scientists contend that life, including advanced intelligent living beings, is quite common in our own galaxy. In his popular book, *Cosmos*, eminent astrophysicist Carl Sagan writes:

"There may be a million worlds in the Milky Way Galaxy alone that at this moment are inhabited by beings who are very different from us, and far more advanced."

In this chapter we will review some scientific discoveries that seem to be at odds with such a conclusion. Scientific discoveries reveal that our galaxy is not an ordinary galaxy. Our Sun is not an ordinary star. And our planet is not an ordinary planet. Far from it.

In the Right Location

The discoveries of science have revealed that our planet Earth is located at just the right place in the universe to support the evolution of intelligent life. If Earth had formed just a little further from the Sun it would be an ice planet. If a bit closer, greenhouse gasses would run rampant.

As the universe evolved it took a long time for the natural elements to emerge, first from the thermonuclear reactors within the stars above, and then from the supernova explosions of very large stars at the end of their thermonuclear lives. That evolutionary process produced a lot of variety in the heavens. Scientific discoveries have revealed that the variation in our small part of the local universe turned out to be just about perfect for us.

Ninety-eight percent of nearby galaxies are metal poor and are, therefore, quite likely to be devoid of Earth-like planets. As part of the Milky Way galaxy, our Sun is a third generation star that happens to be located in an outer spiral arm far away from the dangers of deadly radiation in the galactic center.

The Sun and the planets of our solar system were born of a supernova explosion. They are thereby rich in metals. When our Sun and Earth were born the timing was just right so that they inherited an abundance of the chemical elements necessary to sustain life.

Science defines a 'habitable zone' for a planet to be an orbit located between 88 and 127 million miles distance from the Sun. The Earth travels around the Sun once each year in an elliptical orbit that averages a distance of 93 million miles. That is just the right location necessary for water to neither freeze nor boil. While proper distance from the Sun provides a fundamental basis for a habitable planet, the temperate conditions on Earth are the combined result of several seemingly serendipitous events and planetary circumstances.

As Earth formed through the gradual process of gravity and accretion, the planet was bombarded by asteroids and comets that added rich elements and an abundant water supply. Then an impact with a Mars-sized celestial body greatly changed the planet. The impact imbedded large quantities of the metal elements into our planet and the exploding remnants fell into orbit around the Earth creating our single large Moon. The creation of such a large Moon is unique in our solar system and it provides us with many conditions favorable for life.

Much like a spinning top, as the Earth spins on its axis, it tilts. Without a stabilizing influence the tilt would vary between 0 and 85 degrees. Because of the Moon's gravity, the Earth's angle of tilt relative to the plane of its orbit is very stable. On Earth the angle of tilt varies only by two and one-half degrees over a period of 40,000 years. This constancy of tilt angle provides for Earth's regular seasons and temperate climate zones. Such occurs because the Moon is about 2,000 miles in diameter and is located about a quarter of a million miles away. If the Moon was smaller or more distant, Earth's climate would be vastly different.

The gravitational effects of the Moon also slows the Earth's rotation, allowing for the 24-hour day we now enjoy and provides us with a hospitable climate. If the spin of the Earth were much faster, rainfall would be restricted to narrow bands around the planet and hurricane-force winds would be commonplace.

Big neighbors also help to make Earth a life-friendly planet. As the planets of the solar system matured, the gravity of the giant gas planets, Jupiter to a large degree and Saturn to a smaller, served to protect the Earth from frequent collisions by heavenly bodies. These gas giants either swallow or divert rogue comets and asteroids away from the Earth.

All these fortuitous events and circumstances served to provide Earth with just the right location to allow intelligent life to evolve.

Made of the Right Stuff

The discoveries of science have revealed that our planet Earth is made of just the right stuff to support the evolution of intelligent life.

The force of gravity crushed the elements within the sphere into a denser and denser core of iron and nickel. The heat caused by intense gravitational pressure and the radioactive decay of thorium and potassium deep within the Earth creates a liquid iron outer core. The Earth's spin causes a convective roiling movement in the liquid iron core. That convective movement creates a magnetic field that surrounds the planet. This magnetosphere protects us from the solar wind and deadly cosmic radiation.

The liquid outer core of the Earth also provides for planetary plate tectonics. Without plate tectonics relentless erosion would wear down all land above sea level. Plate tectonics is the only known method that can outpace erosion and thereby provide continuing dry land on the planet's surface. Without plate tectonics Earth would be a water world. As far as science can determine to date, Earth is the only planet with plate tectonics.

As heavenly bodies collided with an infant Earth the planet emerged as just the right size for a life-friendly world. If Earth were much smaller the interior would cool too much to generate a strong magnetic field. If the planet were much larger, the higher surface pressure of gravity would result in smaller mountains and shallower oceans. The planet would be covered with oceans of water too salty to support life. A planet twice as large as Earth would have 3½ times more gravity, would be covered with water, and would have a very dense atmosphere filled with methane and carbon dioxide.

Carbon dioxide is key to a healthy atmosphere. As a most-important greenhouse gas, carbon dioxide seals in heat by absorbing re-radiated infrared light from the Sun, while remaining transparent to visible light. That provides for an atmosphere with great visibility.

Too much carbon dioxide is deadly. So is too little. The carbon cycle process and the amount of carbon dioxide that has evolved on Earth is just about perfect for advanced life. Quite wondrously, plate tectonics provides a global thermostat by recycling the chemical elements and keeping carbon dioxide levels on a fairly even keel. That has not always been the case. Without life our planet would have an atmosphere laden with carbon dioxide. Today carbon dioxide is a trace gas in our atmosphere, far less than 1%.

The atmosphere of early-Earth was filled with the greenhouse gasses of methane, carbon dioxide and water vapor. The atmosphere evolved into an advanced-life-friendly one of mainly nitrogen and oxygen only through the presence and work of primitive single-celled microbes. As bacteria learned how to engage in the process of photosynthesis the atmosphere on Earth changed dramatically.

Photosynthesis is the process whereby microbes and green plants extract carbon dioxide and water from the atmosphere and combine them with the energy of sunlight. Through this wondrous process a form of usable energy supply is acquired for all higher living organisms, from plants to ferns, from worms to fish, from reptiles to birds, from mice to men. All higher living organisms ultimately are able to live because primitive microbes learned how to engage in photosynthesis some three billion years ago.

Because of photosynthesis, first learned by bacteria and much later by plants, the atmosphere on Earth changed dramatically. Today our atmosphere consists mostly of nitrogen (78%) and oxygen (21%). The remaining 1% consists primarily of the greenhouse gasses (carbon dioxide, methane, and water vapor) and ozone.

Atmosphere is important. If Earth was a smaller planet its gravity would be too weak to form an atmosphere. And, even if located within the 'habitable zone', without an atmosphere there would be no liquid water on Earth. Our nearby Moon is certainly within the 'habitable zone', but because it is only one-third the size of Earth, its gravity is too weak to form an atmosphere. Consequently, the temperature on the Moon is too cold to allow liquid water to form.

The supernova parent of our solar system provided us with a Sun containing much more carbon than the other stars in the galactic neighborhood and a home planet literally teaming with the stuff. Fortunate indeed for us, for between 10 to 20% of the weight of all living things is made of the element carbon. Because carbon can covalently bond with itself and as many as four partners (hydrogen, nitrogen, sulfur, and oxygen) it is the central building block of life. And, all of the carbon that is contained in all living things is initially taken out of the carbon dioxide in the air by microbes and plants engaging in the process of photosynthesis.

Made for Discovery

Not only is planet Earth located in just the right location and made of just the right stuff to support the evolution of intelligent life, the planet also provides us with an ideal laboratory setting for intelligent beings to make scientific discoveries.

The nitrogen atmosphere of planet Earth provides a transparency whereby we can clearly view the heavens and thereby learn about our solar system and our universe. Without an observable universe the scientific genius of Kepler and Newton would not have discovered the universal laws of motion and gravitation.

Our Moon is just the right size and located at just the right distance from Earth to allow us to view nearly perfect solar eclipses. Without nearly perfect solar eclipses science would not have progressed to be able to understand the nature of the Sun and the stars, the Earth's rotation or evidence of the bending of light to confirm Einstein's theory of General Relativity.

Our clear atmosphere and the location of our Moon are essential components for an environment that is conducive to the evolution of intelligent life. They also serve to provide an environment that is most conducive to making scientific discoveries. There is no logically necessary connection between the two, but that is precisely what we have on this planet.

Most wondrously, brilliant scientists have been able to learn more complex realities after discovering more simple realities. Physical reality can be learned piecemeal. The nature of the atom could be understood before scientists discovered the components of the atom.

At the level of Newtonian physics the laws of nature are in simple linear order. Without such we would have chaos in our everyday world. While quantum particles within an atom reside in a world of required uncertainty, the irreducible quarks comprising the mass of an atom must remain constant in order to build atoms and molecules that then form the reality of our everyday ordered world. The universe contains a higher-order-complexity that exists right alongside a quantum-uncertainty that tends to destabilize ordered complexity. Yet it all works out just fine.

In *The Privileged Planet* scientist Guillermo Gonzalez and philosopher Jay Richards make the following observations:

"What this suggests is that for some reason, the ultimate laws of physics give rise to mathematically simple theoretical laws at each conceptual level, even for those later judged inadequate, such as Newtonian space-time. This odd truth allows each conceptual level to serve as a ladder to the next level. If the theoretical laws could not be simple and yet relatively precise at each conceptual level, we could probably not discover them for that level, and hence could not progress from level to level toward the fundamental laws of physics....Like an excellent tutor, the universe has not been so demanding as to ensure failure but rather has allowed us to succeed while still presenting us with worthy challenges."
"The correlation between habitability and measurability seems to be the result of more than mere chance."

PART III

ASSEMBLING LIFE WITHOUT GOD
- A CLOSER LOOK AT 'ACCIDENTAL' NECESSITIES -

Part I over-viewed the story of creation and evolution without God. That story is the product of both scientific evidence and scientific belief. Most scientists believe in the story.

Part II examined, with greater specificity, what was necessary in order for a life-friendly world to develop accidentally, without God.

Part III will now examine, with greater specificity, what was necessary for the creation and evolution of life accidentally, without God. We will first examine what science has discovered to be the basic requirements of life and review the scientific theories that attempt to explain how those basic requirements came together at the same time to produce life. That's the story of Chapter 9.

Chapters 10, 11, and 12 will then examine some of the nitty-gritty of all living organisms. All living things have common components that somehow bring them to life. All living things have the ability to coordinate their internal functions and reproduce themselves. And, all living things have a common ability that allows them to sustain life by acquiring and using energy. A closer look at what is actually required to do those amazing things is necessary in order to gain a perspective of what is probable.

As you proceed to read these chapters keep in mind that this stuff is really hard to understand. Condensed into four short chapters is an examination of some of the most fundamental discoveries about life that took the combined genius and diligence of brilliant scientists more than two thousand years to accomplish.

CHAPTER 9
The Mystery of Life

Every living thing is composed of living cells. The essential individual components of all living cells are the nucleotides that string together to make up RNA and DNA, and the amino acids that string together to make up proteins. None of these sub-components are alive by themselves. A living organism is a cooperative venture. Each living cell contains a complete set of genes encoded by DNA that constitutes the blueprint and recipe for constructing and maintaining the complete, fully-functioning living organism of which it is a part. That set of genes is called the **genome.**

How Did Life Begin?

The **smallest genome for a living creature** that has been discovered by science to date is that of a tiny single-celled bacteria living within the cells of a host insect. The genome of that bacteria **contains about 160,000 base pairs of DNA coding for 182 discrete proteins.** This bacteria cannot live independently. There are specific genes that the bacteria's genome lacks that are necessary for independent life. Those genes necessary for independent life are compensated for by the insect host.

Science has calculated that the simplest independent single-celled living organism must have at least 200 genes coding for at least that many proteins in order to survive. So, **the first life on Earth must have had at least 200 different proteins in that first single-celled bacterium.** What are the odds of that happening by chance?

Noted agnostic astronomer Carl Sagan stated that the odds are 1 in 10^{130} for the DNA coded sequence necessary for the construction of a single protein of an average length of 100 amino acids to develop by chance. The total number of atoms in the entire universe is 10^{79}. So, the odds of constructing only one protein by chance is astronomically greater than the odds of selecting just one atom from among all the atoms in the whole universe. And the odds simply get much, much worse when you consider the same is true for each of the other 199 proteins required to bring just a single-cell to life. When you deal with such fantastic probabilities most statisticians agree that there reaches a point where the highly-improbable becomes impossible. The eminent statistician, Emil Borel defined that point on a cosmic scale as anything exceeding a chance of 1 in 10^{50}.

The information for constructing each of those 200 different proteins had to be contained within the cell's nucleic acids, copied, transcribed, edited and decoded into the language of amino acids, and then transported to the proper location within the cell for each protein to be built, and then further transported to the site in the cell for proper function. Not once. At least 200 different times **before the cell could be considered 'alive'.** All that information processing and transfer and protein construction had to occur before 'life' occurred on Earth. And, that information transfer and construction project is really just the start of the state called 'being alive'.

In order to be alive a living cell must be able to do three discrete things:

- A living cell must be able to obtain, store, and process all of the information necessary to first bring the cell to life and to then provide for its every function (a simple restatement of the immediately preceding paragraph).

- A living cell must be able to extract energy from its environment and then convert that energy into a useful form that is required to power the processes of cellular metabolism that are essential for all life.

- A living cell must contain the information and possess the ability to replicate. It must be able to clone itself, and the living organism must be able to produce offspring.

All of this exquisite orchestration of information and usable energy and function has to be explained somehow. How did life begin? What is the best scientific answer? **The best scientific answer is — NO ONE KNOWS.**

No one knows for certain if **life is the result of design _or_ happenstance**. But, we do know for certain that it had to result from either one or the other. And, that is not beside the point. That is precisely the point. It is a very important point to bear in mind as you explore the rest of Part III.

Emergent Evolution

The mainstream scientific community today believes that the theory of **emergent evolution** best provides the explanation for life. It **just happened**. The theory maintains that life was not created by an intelligent agent. Rather, life was the result of happenstance - **chance and necessity**.

The product of chance events operating in accord with the laws of nature.

The theory of emergent evolution had its origin in the writings of Charles Darwin himself. In one of Darwin's letters, following the publication of *On the Origin of Species*, he put forth his idea of how life may have evolved from non-life, as follows:

> "It is often said that all the conditions for the first production of a living organism are now present, which could ever have been present. But if (and oh! what a big if!) we could conceive in some warm little pond, with all sorts of ammonia and phosphoric salts, lights, heat, electricity, etc. present, that a protein compound was chemically formed ready to undergo still more complex changes, at the present day such matter would be instantly devoured or absorbed, which would not have been the case before living creatures were formed."

The idea that life so emerged from the primordial ooze was thus born. That has now become the mantra of the mainstream scientific community. Scores of scientists have since sought to scientifically explain the mystery of the origin of life through experiments aimed at proving Darwin's hypothesis of life being assembled from non-life. Let's look at a few.

Creation of Life Experiments

Thus far **all scientific experiments** that have been conducted in the laboratory in an attempt **to create life have failed**.

In the 1950's Stanley Miller assembled in the laboratory the conditions believed to have existed on the early Earth. He set out to see what chemicals could be produced in an early-Earth atmosphere containing ammonia, methane, water vapor and hydrogen. Into this mixture he sparked electrodes to simulate lightning on early Earth. His experiment was successful in producing several amino acids, the building blocks of those workhorses of life called proteins. The scientific community became energized with the idea that creation of life in the laboratory was almost at hand.

Scientists following Miller's lead varied the experiments and used ultraviolet radiation to simulate sunlight and pulses of pressure to simulate explosions.

Through their efforts nearly all of the twenty amino acids naturally-occurring in living things have been observed in these origin-of-life laboratory experiments. But, **all attempts to develop proteins naturally from amino acids have failed.**

Experiments to link amino acids under a variety of laboratory conditions have successfully joined amino acids. But, such joinder is nowhere near to the creation of a protein from amino acids. Undeterred, researchers claimed victory and pronounced that they had created '**proteinoids**' which must have been an ancestor of actual proteins. Such an ancestor is absolutely necessary in order to conform with the tenets of Darwinian evolution by natural selection of random mutations.

Since the nucleic acids are the other essential ingredient necessary for all life, many scientists investigating the question of origin of life have sought to create RNA or DNA to no avail. Theories abound, like the hypothesis that RNA existed first and was an evolutionary ancestor to DNA, and that these nucleic acids evolved from ancestral building blocks called purines and pyrimidines that have been observed to self-assemble in the laboratory. That would comport very well with Darwinian evolution by natural selection. **The problem with this scientific theory is that evidence is entirely lacking.**

Other research scientists have hypothesized that life simply learned to self-organize from non-living mineral inorganic crystals. Crystals are an orderly array of atoms or molecules like the silicates found in clay and presumably in the primeval muds. Again, no evidence at all in support of this theory.

If none of these experimental paths have proved fruitful in explaining the origin of life in a scientific manner, not to worry. Most preeminent scientists insist that the only way that life could have originated and progressed is through evolution by natural selection of random mutations. Indeed, some in the mainstream scientific community have chosen to ignore the scientific laws of probability altogether and simply assert that, contrary to the laws of probability, chance alone must have brought the nucleic acids and proteins together in the manner necessary to create life. Since they are wedded to the proposition that life must have evolved from non-life, they believe that a single instance of spectacular luck, that would have been necessary to create the first life by pure chance, is rightly hypothesized. No evidence is necessary. And the laws of probability can be ignored.

Oxford Professor Richard Dawkins in his book *The Blind Watchmaker* imagines that if life only originated but once in the entire universe then we can assume a very, very large amount of chance luck to create it. He contends that **just because we cannot imagine the amount of luck that it would take in order for life to have originated de novo by chance does not mean that such did not occur. As long as the chance is not zero, it may have happened**. In his book, he analogizes this to the improbability of a marble statue waving its hand at us in the following passage:

> "In the case of the marble statue, molecules in solid marble are continuously jostling against one another in random directions. The jostlings of the different molecules cancel one another out, so the whole hand of the statue stays still. But if, by sheer coincidence, all the molecules just happened to move in the same direction at the same moment, the hand would move. If they then all reversed direction at the same moment the hand would move back. In this way it is *possible* for a marble statue to wave at us. It could happen. The odds against such a coincidence are unimaginably great but they are not incalculably great."

That is Professor Dawkins' scientific explanation for his belief that the beginning of life was a chance event. Not all eminent scientists concur with Professor Dawkins.

Francis Crick, who won the Nobel prize for his discovery of the nature of DNA structure, views the level of chance required for the origin of life by chance to have been too great to even calculate, let alone happen. In his book, *Life Itself: Its Origin and Nature*, he provides this contemplation:

> "An honest man, armed with all the knowledge available to us now, could only state that in some sense, the origin of life appears at the moment to be almost a miracle, so many are the conditions which would have had to have been satisfied to get it going."

He then offered a solution in what he termed **panspermia**. He proposed that the first life on Earth could have occurred when bacterial spores were delivered to our planet, complete with the nucleic acids and proteins necessary for life. But that still begs the question: Where did the life contained in those bacterial spores come from?

Hubert P. Yockey in his book, *Information Theory, Evolution and Molecular Biology,* pronounces that the information in the DNA and RNA nucleic acids necessary to begin sustainable life is simply too complex to have developed by chance. He then suggests that **life be viewed as simply a given**, like energy and matter - just givens.

The problem is not an easy one to solve. **All cellular life in order to live needs proteins constructed in chains of specific amino acids. And all cellular life needs the information-bearing molecule DNA working in conjunction with RNA as a decoding and translating molecule to tell the amino acids how to chain together in very specific ways to make very different proteins to perform very specific functions. And, the real conundrum is that some of those proteins have to already be built in order to assist DNA and RNA in performing their informational-delivery functions necessary to make proteins in the first place.**

That conundrum is just the start of the mystery. Those molecules of DNA and RNA and amino acids and proteins had to cooperate in a coordinated fashion to build, maintain and replicate themselves. And, they had to establish a sustainable energy source in order to come alive and to sustain life.

Let's now turn to the next three chapters to examine the scientific discoveries that provide some insight into how these simple molecules manage to do these amazing things.

CHAPTER 10
The 'Stuff' of Life - What All Living Cells Have in Common

All living things, all bacteria, all plants and all animals, including us - all living things - are made-up of one or more living cells. Cells are the building blocks of life. Because living cells are ubiquitous to all life, this chapter will examine the nature and structure of living cells. All living cells have these things in common.

The Basics and Building Blocks

Let's first look again at the big picture. It may provide a clearer perspective of the place of living cells in the overall scheme of things.

There are 92 naturally-occurring elements on planet Earth. Each of those elements is comprised of discrete atoms.

The atoms of each element are unique to that element. The unique atomic structure of each atom is determined by the number of protons and neutrons residing in its nucleus, and the number of electrons orbiting around the nucleus of the atom. For example, each atom of the element **carbon** is comprised of 6 protons, 6 neutrons, and 6 electrons. The neutrons have no electric charge, each proton has a positive electric charge ($+1$), and each electron has a negative electric charge (-1), making the atom electrically neutral. Each atom of each of the 92 naturally-occurring elements is, by itself, electrically neutral.

The atoms of the elements combine into compound **molecules** by the **bonding** of the **electrons** in the outer orbital shell of an atom of one element with the electrons in the outer orbital shell of another atom in the neighborhood. That atomic bonding of atoms to form molecules is how chemistry works. Chemistry is ultimately based on the electrical interaction of atoms.

Organic molecules are naturally produced only in living things. Everyone has heard the term **'carbon-based life form'**. Well, that term reflects the fact that all of the macromolecules that comprise all organic compounds and which are present in the cells of all living organisms are based on the element carbon. **All organic compounds are of just four classes, and all have a carbon base.**

The four classes of macromolecules comprising organic compounds are:
- Carbohydrates (e.g., sugars, starches, cellulose),
- Lipids (e.g., fats and hormones),
- Nucleic acids (e.g., RNA and DNA), and
- Amino acids (20 specific kinds that serve to make up proteins).

All organic compounds are held together by a strong **covalent bond** whereby the atoms of the different elements making up the molecule **share electrons**. All of the organic compounds are large bio-molecules that are formed by connecting small building blocks together into long chains. And, all of the organic compounds exist in the cells of all living organisms.

- **Carbohydrates** are bio-molecules made up of chains of carbon, hydrogen and oxygen which are the building blocks for the sugars. Carbohydrates serve as the principal **source of energy** for a living organism.

- **Lipids** are bio-molecules made up of long chains of carbon atoms bonded to each other and to hydrogen atoms. Lipids serve as the fatty acid building blocks for the fats and oils. One very important function of lipids is to help in the construction of the **membranes that enclose each cell**.

- **Nucleic acids** are bio-molecules made up of long chains of nucleotides. Each nucleotide is made up of a nitrogenous base, a phosphate group and a five-carbon sugar. Nucleotides chain together to convey genetic information through the nucleic acid molecules DNA and RNA. **DNA** serves as the information library containing both the **blueprint and the recipe necessary for** the construction of all of the **proteins** in a living organism. The nucleic acid **RNA** serves first to copy (**transcribe**) the information contained in DNA into RNA format, next as a **messenger** to deliver the information contained in DNA to the proper location in the cell for building proteins, and finally to **translate** the information contained in DNA into a language that the amino acids can understand in order to construct the proteins.

- **Amino acids** are bio-molecules with a backbone consisting of a carbon atom in the middle and surrounded by an amino group, a carboxyl group and a hydrogen atom. The backbone of each amino acid has a specific side group of atoms. The twenty **amino acids** found in living things **differ only by their side group of atoms**. The amino acids bond together into long chains called **polymers** (between 50 and 5,000 amino acids) to form a **protein**.

These four classes of organic compounds are all carbon-based. They naturally occur in the cells of all living organisms. They serve as the basic building blocks of all cell development, maintenance and reproduction. They are the basis of life. But, knowing that these four types of organic compounds repose in all living things is a far cry from explaining how life began. Combining all of these organic compounds in all sorts of combinations outside of a living organism has never resulted in the creation of a living organism. All living things are made-up of living cells. **No scientist has ever developed a living cell. No scientist has ever created life.**

The First Life on Earth

How did life on Earth begin? The scientific answer is 'We don't know.'

Some scientists believe that life began as an evolution of non-living matter. Somehow the organic compounds that existed on the planet in its early years coalesced in the presence of an energy source, like lightning strikes or lava eruptions from fissures in the planet's surface far under the sea, to produce the first living organism. This theory is called **emergent evolution**.

Other scientists believe that the origin of life was the product of design, that an intelligent agent created life. Some call the intelligent agent God.

The first living organism was a single-celled bacteria, specifically an archaebacteria. It came to life in the oceans of the Earth about 3½+ billion years ago. Most scientists believe that all life on Earth has evolved from that single ancestor.

The bacteria had but a single cell, and that cell did not have a nucleus. An organism containing a cell without a nucleus is called a **prokaryote.** For about two billion years the single-celled prokaryote remained the only life on the planet.

About 1½ billion years ago the first organisms who's cell had a nucleus appeared on the Earth. An organism containing a cell with a nucleus is called a **eukaryote**.

It took almost another billion years for the first organisms to appear on Earth that had more than a single cell. That was a really big deal that has resulted in the diversity of plants and animals that exist in numerous forms and sizes on the planet. We will pick-up that story line in the next chapter when we examine how living cells communicate with each other. But, first, let's examine the structure and life processes of a living eukaryotic cell. After all, living eukaryotic cells are ubiquitous to all plants and all animals, including us.

The Structure of a Eukaryotic Cell

All plants and all animals, including us, are composed of cells that have a nucleus. Cells that have a nucleus are called eukaryotic cells.

Each eukaryotic cell is a highly complicated and sophisticated structure. It is beyond the scope of this primer to delve into all of its discrete parts and functions. So, only the most basic will be reviewed.

The cell is enclosed with a **membrane** comprised primarily of lipids as a support structure that is interlaced with various proteins to assist in transfers between the outer cell and the inner cell's milieu. The membrane provides for protection for the inner cell's working structures and climate.

The heart of the cell is the **nucleus.** The nucleus contains the **chromosomes** for the organism which holds the library of genetic information about the organism. The chromosomes hold the **DNA** which contains the entire blueprint and recipe information necessary for the construction and functioning of the organism itself. The **genes** are specific information units that serve to make-up a strand of DNA. The information necessary for the construction and functioning of the single cell itself is contained in the genes that reside on a single strand of DNA that is a part of only one of the chromosomes that specify the uniqueness of the organism itself. For example, the **human genome** resides in 46 chromosomes consisting of 23 **pairs**. The information necessary for the construction and operation of a particular cell in our body will most likely be contained in the genes of a tiny, tiny portion of a single strand of our DNA residing on a particular chromosome.

The **cytoplasm** of the cell is the watery milieu outside of the nucleus that contains the various working parts of the cell including the **organelles** and the **endoplasmic reticulum** (ER).

The **organelles** of the cell are each surrounded by their own additional membrane and perform specialized functions necessary for life. For example, in both plants and animals the organelles called **mitochondria** produce ATP, the molecule that synthesizes energy for plants and animals. In plants, the organelles called **chloroplasts** capture the Sun's energy and store it in the form of high-energy organic molecules of glucose sugar. The organelles called **ribosomes** are the cell's **factory** where the information contained in the genes of the DNA particular to that cell is first translated and then used to construct proteins. We will take a closer look at the whole process of protein construction in due course. And, the organelles called **lysosomes** serve to rid the cells of the waste products of cellular metabolism.

The **endoplasmic reticulum** (ER) forms a membranous network attached to the nucleus and to the ribosomes, and further synthesizes proteins for use within and without the cell. It also provides for intercellular transportation of proteins and acids and performs a detoxification function.

Basics of Cell Biology

The simplest living creature known to science is the single-celled bacteria. Remarkably, the single-celled bacteria contains an intact set of the same ingredients contained in the cells of all living creatures, including all plants and animals, and including us. But, while the single-celled bacteria is the simplest of living creatures it is, indeed, a highly complex organism. **The simplest living organism – a single-celled bacteria - is far more complex than the most complicated machine that mankind has ever been able to produce.**

All living cells are governed by a basic but exquisite chemical plan that is the same for all living organisms. The single-cell of a bacteria uses the same genetic code and code translation mechanisms that are used in the cells of all living creatures, including all plants and animals, and including us.

A living cell is indeed a wonder. Cells are the basic units of function and structure of all living things. All of the activities of all living organisms are done by cells. Breathing, moving, fighting, loving, laughing, body repair and thinking - all bodily functions are performed through cells.

For all living creatures on Earth the basic design of the living cell system is the same. The two principal components are the same for each living cell - proteins and nucleic acids. And the roles of the nucleic acids (RNA and DNA) and the proteins are the same for each living cell.

Proteins

Protein molecules are the **work horses** of life. Each protein molecule is a micro-mini-machine that is constructed from simple amino acids. Each amino acid is a small organic compound consisting of ten to twenty atoms. **Of the hundreds of amino acids that science has discovered to date, only a specific few - precisely twenty - are used by all living organisms in constructing proteins.**

While the number of amino acids - twenty - used by all living organisms is the same, **the sequencing of those same amino acids is highly disparate.** In most proteins the amino acids selected form a linear chain between 100 and 500 amino acids long but some can number as great as 5,000. **For each disparate protein a unique amino acid sequence is followed to form the protein structure.** The sequence of amino acids in the chain determines the size, shape, and function of the protein that it makes. **The side chain of the molecular structure of each amino acid specifies the chemical interactions.** For example, the side chain of some amino acids are attracted to water (polar or hydrophillic) and twist outward to become the protein's outer layer while others are repelled by water (non-polar or hydrophobic) and twist inward to avoid water by becoming the internal milieu of the protein. The amino acid sequence thereby follows the laws of physics and chemistry to **fold each protein into an intricate three-dimensional structure.**

DNA directs the amino acid sequence that is specified by nature for the protein. And, **the sequence of amino acids then determines the protein's specific 3-D shape.** To be biologically active, proteins are required to adopt these folded 3-D shapes. And, **the function of any particular protein is largely determined by its shape.**

Proteins perform all sorts of different biological functions. Proteins serve as skeletons for our cells and scaffolds for the chromosomes housing our genetic DNA. Proteins bind to protein receptors on the outer surface of cell membranes to send messages between cells. Proteins form the structural connections between cells. Some proteins serve to build structures and organs in living organisms while other proteins (known as enzymes) perform catalytic (chemical reaction) functions. There are an enormous number of protein functions and a corresponding enormous number of types and shapes and sizes of proteins.

However, while proteins are capable of performing all these miraculous functions, they ultimately require the instructions contained in and provided by the nucleic acids in order to come into existence in the first place. The nucleic acids are analogous to the blueprints and recipes and messengers and translators required to build the protein machines.

Nucleic Acids

There are **two types of genetic nucleic acids, RNA and DNA**. The **DNA molecules reside in the cell's nucleus and supply the blueprint and recipe for cell development.** The RNA molecules are the messengers (**mRNA**) that convey the proper blueprint information to the specific location in the cell where the protein machine is being assembled (ribosomes) and then translate and transfer (**tRNA**) the information into the language of amino acids necessary to build the protein in accordance with the specific sequence of amino acids originally dictated by the DNA in the first place. Simple, huh?

Like proteins, the RNA and DNA molecules are linear chain molecules called **polymers.** While the **chains of proteins are sequences of amino acids** that bond to form highly complex 3-D shapes, **the chains of the RNA and DNA molecules are sequences of nucleotides.**

The DNA molecule has two strands of nucleotides in the shape of a double-helix, formed by the complimentary base pairing of the nitrogenous bases of the nucleotides. **Each turn of the helix contains ten nucleotides.** The mRNA molecule has a single linear strand of nucleotides. The tRNA molecule has a complex 3-D structure that arises from complimentary base pairing. The unique structures of these nucleic acids enable them to perform the elegant and orchestrated ballet of chemicals required to form each protein of each and every living organism.

In this way the synthesis of proteins by a cell can be seen as the result of the miraculous partnership of one group of bio-molecules (amino acids) with another group of bio-molecules (RNA and DNA nucleic acids). The nucleic acids supply the blueprint and recipe and extract the information from the blueprint at the just-right location in the cell for protein construction. Then, through a series of exquisite transformations, the amino acid building blocks fulfill the construction project by synthesizing the protein needed by the cell in order for it to properly function.

That's it in a nutshell. But, how does all that stuff really happen? Let's take a little closer look.

The Blueprint and Recipe for Life - DNA

The blueprint for all information about each and every part of each living organism on this planet is contained in the DNA of that organism. **DNA is the repository of all life information and contains both the blueprint for the construction of each part of every living organism, and the recipe for how to bring the blueprint to life.**

DNA (deoxyribonucleic acid) is comprised of **building blocks** called **nucleotides**. Each nucleotide contains three specific parts:

- a 5-carbon sugar group,
- a phosphate group, and
- a nitrogenous base.

The sugar group and the phosphate group chain together with a linear chemical bonding that forms the nucleotide backbone of the DNA. The nature of their chemical bonding results in a 3-carbon atom at one end of the nucleotide chain and a 5-carbon atom at the other end of the chain, thereby making the nucleotide chain electrically charged in order to electrically and chemically bond with the next nucleotide sequence in the DNA chain.

Attached to the **side** of each alternating sugar group and phosphate group is a distinctive **nitrogenous base**. The nitrogenous base is the only functional unit of the nucleotide and is, therefore, the defining feature of each segment of DNA constructed by the nucleotides. There are only four nitrogenous bases and they are configured along opposite sides of DNA strands as **complimentary base pairs**. The four bases are: T (thymine); A (adenine); C (cytosine); and G (guanine). A always bonds with T and C always bonds with G.

The alternating base pairs are strung out in linear sequences along long strands of DNA. The nucleotides making-up the DNA are wound in a right-hand direction into complimentary strands to form the famous **double-helix**. And, the DNA strands are always anti-parallel running in opposite directions, thereby becoming polarized.

DNA is housed in **chromosomes**. The chromosomes are made-up of both DNA and various proteins. A chromosome contains 5 to 10 times more protein than DNA.

However, only the DNA portion of the chromosome contains the blueprint and recipe information. The specific segment of a DNA sequence that holds the information required to express a specific protein is called a **gene**.

The blueprint and recipe information contained in the genes along the strands of DNA is in a code. The code is a simple yet elaborate structure. Once the code is translated, the message simply specifies the **sequence of amino acids** necessary for the construction of a single protein to be used by the cell. Again, the code information contained in DNA is simply the sequence of amino acids to be used in the construction of proteins. But, what is this cryptic DNA code, and how is it decoded?

Deciphering the DNA Code

The historical central dogma of genetics informs us that **a gene is simply a specific segment of DNA that contains the information required to construct a specific protein.** That information directs the specific sequence of amino acids necessary to construct a particular protein. But, the nucleotides defining the genes of DNA cannot talk directly to the amino acids that make-up proteins. A code is needed. **Each gene codes for a protein.**

Nucleotides that chain together to make-up DNA use a language containing but **four** base pairs of nitrogenous bases (remember A always pairs with T and G always pairs with C). Living proteins contain amino acid sequences always comprised of some of **twenty** specific amino acids. **A code is needed to translate the language of DNA nucleotides (four units) into the language of amino acids (twenty units).**

It took a brilliant scientist, Francis Crick, and his colleagues many years to first discover the amazing properties of the DNA double-helix and then to discover the three-letter genetic code necessary to translate the language of DNA and RNA nucleotides into the language of amino acids in order to construct proteins.

The genetic code is a **three-letter code** using **specific combinations** of **three nucleotide bases** called **codons**. Crick found that the three-letter code needed to uniquely identify sequences of the four possible bases ($4 \times 4 \times 4 = 64$) provided 64 combinations.

Each unique combination of three nucleotide bases is called a codon. **Each codon codes only for a single amino acid.** Sixty of the possible 64 codons code for a specific amino acid. **Three** of the codons code only for a **stop signal. One** of the codons codes as a **start signal** and thus provides an unambiguous marker for the first nucleotide base in a sequence of codons in mRNA.

Now that we know the code that is used to translate the genetic information contained in the genes of DNA into the language of amino acids necessary for the construction of proteins, let's look at how we actually get those amazing proteins and how we get them to the right locations in order for a healthy cell to properly function.

How We Get Those Amazing Proteins

Let's review. We know that the nucleus of each and every cell in an organism contains an information library of DNA molecules that spell out the blueprint and recipe for the construction and operation of all of the proteins throughout each cell of the body. We know that the information is in code, and we know how the code is deciphered. We know that the 'just right' information for the construction of a specific protein needed by an individual cell resides on a specific segment of a DNA strand called a gene. And, we know that such information is absolutely useless by itself. The cell must now extract and use that information in order to construct the protein that it needs to perform its specific function in the organism.

Remember the role of proteins in the body. **Proteins are the work horses of life**. All of the living functions of an organism involve proteins in one way or another. Proteins perform all sorts of biological functions. Proteins form the girders to support the cell's structure. Proteins serve as transport vessels for other substances (like hemoglobin proteins that transport oxygen in the bloodstream). Proteins are the cell's mechanics, acting as small motors and transporting materials within cells. Enzyme proteins serve as catalysts for chemical reactions making the chemical processes of life work in the cell's environment. Proteins regulate cell activity by switching on and switching off cellular processes. There are an enormous number of protein functions and a corresponding enormous number of types and shapes and sizes of proteins. And, during their life the interaction of proteins with each other and with other organic compounds will change the shape and thus the function of the proteins themselves. Amazing!

DNA contains all of the library information holding the blueprint and recipe for the construction of all of the proteins in the body. But, that information is absolutely useless by itself. The DNA information must: first be transcribed into RNA format; transported by messenger RNA (mRNA) out of the nucleus to the location in the cell containing the protein factory (ribosome); translated into the language of amino acids by transfer RNA (tRNA) at the ribosome site; and finally used at the ribosome site to construct the proper protein in accordance with the exact sequence of amino acids as originally specified by the DNA. Now that is thoroughly confusing. So, let's go a little slower and proceed step by step.

First Step - Transcription

The DNA library contains all of the information for constructing each and every protein in each and every cell in the body. Protein construction leads to functional operation for each and every part of the body and all the connections therein. The DNA library reposes intact in the nucleus of each and every cell of the body.

For the **DNA information** to be useful in constructing the protein needed by a cell it **must be extracted from the DNA**. Each living cell will construct only those proteins whose functions are **expressed** by the segments of the DNA particular to that cell's proteins. Each particular segment of DNA is called a **gene**. You have heard the term **gene expression**. Well, that's what it means.

Let's look at the enormity of this. A single human cell may contain thousands of different proteins. So, thousands of DNA segments (**genes**) will need to be **expressed** in order to complete that particular cell's development. Thousands of genes need to be expressed to make thousands of different proteins in a single cell.

Merge that thought with the fact that the process of **transcription** can make thousands and thousands of copies of those expressed genes for a single cell, and with the further fact that the human body contains trillions of cells. The numbers become incomprehensible. But, let's get back to the transcription process.

OK. The chromosomes reside in the cell's nucleus. The DNA for the organism resides on the chromosomes. There is a specific segment of the DNA, called a gene, that contains the specific information necessary to construct a specific protein needed by the cell. How does the particular cell know which proteins it needs?

The DNA itself tells the cell which proteins it needs. The beginning sequence of nucleotides on a gene is a 'start' signal. So, the DNA gives instructions to the cell for not only which genes to copy but where to begin copying.

At the start signal on the gene a protein enzyme called **RNA polymerase** attaches at the just-right location on the DNA and begins to unwind the two strands of DNA. **RNA polymerase then begins the transcription process.** Transcription is correctly understood as DNA-directed RNA synthesis. Remember, the functional role of mRNA is to simply act as a messenger after it transcribes DNA's language into a form that can be accurately handed-off to tRNA at the protein building site in the cell (the ribosomes).

RNA polymerase then proceeds to copy the entire gene segment of one of the DNA strands and thereby transcribe it into RNA format. Only one strand of the DNA gene segment is copied into RNA form because the exact sequencing nature of base-pairing requires only one strand.

After a 'rough copy' has been transcribed, a number of specific protein enzymes attach to the 'rough copy' and proceed to edit out a large sequence of the information that has been transcribed. Amazingly, as much as 90% of the transcription is at this point removed and discarded. The 10% remainder is spliced back together to make a much-smaller mRNA strand. Science does not know the purpose of the 90% that is edited out.

Once copied, edited-out, and spliced back together the new strand of RNA is snipped-off at a 'stop' signal which is designated by DNA as the last series of nucleotides on a gene. The DNA strand then rewinds to its original double-helix form once the RNA strand is snipped-off.

While still in the nucleus, the final editing process for the mRNA strand takes place. The ends of the strand are chemically modified by other enzymes. A guanine (G) cap is added to one polarized end and an adenine (A) tail is added to the other end. We now have a 'mature' strand of mRNA.

At this point the mature mRNA strand, containing the encoded information necessary to construct the protein, physically departs the cell's nucleus. The mature mRNA strand travels to the correct protein construction site in the cell's cytoplasm (the ribosome) and hands off its work to its teammate, tRNA.

Second Step - Translation

At this point we have the necessary information for building the protein, which is now positioned at the just-right location in the cell, to begin construction. But that **information is still coded in the language of RNA nucleic acids. That information must now be accurately translated into an entirely different biochemical language of amino acids in order to be used to construct the protein. That job is done by tRNA**. Here's how.

Remember that specific sequences of bases (triplets of nucleotides called **codons**) are needed to code for the amino acid sequences to be used in constructing proteins. What is needed is a mechanism that can understand the language of both nucleotides and amino acids and then accurately translate the triplets of the four nucleotide bases (the codons) into the twenty different amino acid sequences. Here's how that happens.

The mRNA arrives at the ribosome site located in the cytoplasm of the cell. Here the mRNA docks on the ribosome and begins an interaction with another nucleic acid known as tRNA.

Each specific strand of tRNA has on one end a three-base sequence called an **anticodon.** The anticodon binds to a specific codon on mRNA. The other end of the tRNA binds to the correct amino acid. Translation takes place at this point. A specific enzyme called **aminoacyl-tRNA synthetase** performs this amazing translation. That enzyme catalyzes a chemical bond between the appropriate amino acid and tRNA that thereby electrically charges and activates the tRNA.

Once this complicated 'hook-up' is completed between the appropriate amino acid, the appropriate tRNA, and the appropriate codon on the mRNA, the next codon of mRNA and then the next and the next moves into place as on a conveyor belt and is translated in this same fashion until the 'stop' codon on the mRNA is reached. At this point the translation process is completed.

Third Step - Protein Construction

As the ribosome factory thus receives the order for a protein now written in the language of amino acids it proceeds to build the protein as ordered. The factory employs a large aggregate of protein enzymes to assist in the task all along the amino acid chains. And an additional type of RNA, ribosomal RNA (**rRNA**) is also employed to help in construction. In fact, about 60 % of the ribosome structure itself is comprised of rRNA.

When the 'stop' signal is reached on the amino acid chain, construction is completed. The bond between tRNA and the amino acid chain is broken, and the 'newly-minted' protein emerges. But, where does it go from there? How does a protein get to its proper location in the cell in order to properly function?

Fourth (and final) Step - Getting to the Right Location

In order for a cell to properly function it needs more than just the proper set of proteins. The cell needs to have those proteins located at the proper place in the cell (the nucleus, the cytoplasm, or in one of the organelles).

During the process wherein the protein is being synthesized through the proper sequencing of amino acids during construction at the ribosome site, further amazing things happen.

We have seen that **a eukaryotic cell has specialized work sites called organelles. Each of those organelles has its own biological membrane to separate itself from the remainder of the cell.** Without the membrane the organelle could not properly function. Those organelles need specialized proteins in order to perform their proper function. **But, proteins are electrically-charged molecules and, as such, cannot cross cell membranes by themselves. They need to be transported by other organic compounds, called vesicles,** that can pass through the membranes. It turns out that such a feat is nowhere near a simple process. Here's what happens.

During protein construction all protein synthesis begins in the cytoplasm of the cell located on 'free-floating' ribosomes. You will recall that translation from the language of nucleic acids into the language of amino acids is happening at the site on the ribosome where mRNA has linked-up with tRNA and tRNA is docked with the translation protein enzyme to begin the amino acid sequencing process necessary to build the protein. As soon as an 'address tag' sequence of amino acids is reached, the synthesis process stops. **The 'address tag' sequence calls into play a 'signal-recognition' particle which then pulls on the ribosome to attach the ribosome to the membrane of the endoplasmic reticulum (ER).** After this docking procedure, **synthesis of the protein restarts within the ER membrane**.

When synthesis is then completed within the ER, the protein is further processed by the endomembrane system. For example, if a protein is destined to function as part of the **lysosomes** (organelles filled with digestive protein enzymes that act like the 'stomach' of the cell), the endomembrane system attaches a 'sugar coating' to form the protein into a **glycoprotein.** The sugar coating itself serves as an additional 'address tag' labeled for transport to the lysosomes. The sugar-coated protein then bonds with compatible vesicles, which then transport the glycoprotein to and through the lysosome membrane so that the glycoprotein can now perform its useful function in the life of the cell.

Let's sum-up protein location. Many ribosomes in the cell's cytoplasm are transformed from their free-floating state. These ribosomes become attached to the endomembrane system, specifically at the ER. The proteins constructed at these ribosomes are then released into the ER. The ER then adds various 'finishing touches' to those proteins. For example, sugars may be added to the proteins to form glycoproteins, and 'chaperone proteins' may help the newly-minted proteins to form the correct shapes. Lastly, these newly-minted proteins may be transported from the ER in vessels, called vesicles, who take the proteins to their proper location in the cell or beyond.

Finally, to complete the picture, let's re-focus on where the 'address label' for each protein ultimately comes from. The 'address label' is a specific sequence of amino acids coded on mRNA. And, mRNA simply transcribed the coded sequence from the original gene located on a strand of DNA located in the cell's nucleus. So, **DNA not only tells the cell what proteins to construct. DNA also tells the cell where and how to place those proteins within the cell for proper cell function.** Is this stuff amazing or what?

A Perspective Thought

The human genome consists of about 30,000 different genes that are located on 46 chromosomes (23 pairs) that house the 3 billion base pairs of our DNA. Each of the genes codes for several different proteins and the DNA code is transcribed and translated via RNA in order to construct the specific sequence of amino acids necessary to build each specific protein.

This entire chapter provides but a rough capsulization of what happens each and every time a protein is needed by a human cell. Now consider the magnitude of this feat.

The average human cell contains about 50,000 different specific proteins. The average number of cells in the human body is approximated by science to be 100 trillion. So, this wondrous process of information transfer and cellular construction does not occur just once. It occurs 50,000 times for each living cell. Fifty thousand times for each of the 100 trillion cells in the human body. So multiply 100,000,000,000,000 times 50,000. Think about the magnitude of that for more than a moment.

Now consider that the prevailing elite and mainstream scientific belief is that all of that information transfer and cellular construction ultimately happened accidentally, without any intelligent input and without any planning.

This completes our very brief review of the stuff of life - what every single living cell has in common. What lies beyond the single cell is a tale to be told in the next chapter.

CHAPTER 11
The 'Bonds' of Life - How Living Cells Communicate and Work Together

We have now examined the structure of a living cell. We know about DNA and genes and how those amazing proteins are made. And, we know that the picture of life at the level of a single cell looks pretty-much the same for nearly every living organism on the planet that we can see. The cellular structure is basically the same for each and every living creature that consists of more than a single cell, including us.

But, there is a huge gap between a single cell and a multi-celled organism, like us, consisting of literally trillions of cells. The way that gap is filled is, in large part, the story of how living cells communicate and work together. Without inter-cellular communication and cooperation between those amazing proteins and other organic compounds, multi-celled organisms, like us, simply would not exist. That inter-cellular communication and cooperation, at the molecular level, in our bodies is a silent world of hidden wonder that forms the 'bonds' of life.

The story of inter-cellular communication and cooperation must, of necessity, begin with the creation of new cells from a single cell. Let's look at the two ways that new cells are formed. We will turn first to the creation of a new cell through sexual reproduction and then continue with the process of simple cell division.

Living Cell Creation Through Sexual Reproduction

For all multi-celled organisms, for all plants and all animals, including us, the creation of a new living cell occurs through sexual reproduction. Sexual reproduction occurs when one of the special sex cells (sex cells are called gametes) of the mother (the egg) is fertilized by one of the special sex cells of the father (the sperm). The methods of fertilization vary widely between plant and animal species (e.g., sexual intercourse for mammals and pollination for seed plants), but the result is always the same - a fertilized egg, called a **zygote**.

Each of the cells of a multi-cellular organism, except the sex cells, has tightly-wound within its nucleus the DNA of that organism contained within a number of **pairs of chromosomes**. In contrast, the sex cells, called **gametes**, contain only one-half of each of the chromosome pairs. The process whereby the new zygote cell receives a complete paired set of chromosomes (half from each parent) is called **meiosis**.

Through meiosis the sex cells join together to form a single new and unique living cell - a **zygote**. Only the sex cells of the father, gametes called sperms, and the sex cells of the mother, gametes called eggs, contain but a single copy of each of the DNA chromosomes. The sex cells are produced by an unwinding of the DNA into single strands. Only one strand of each of the chromosome pairs is supplied by the father. The other strand of each of the chromosome pairs is supplied by the mother. Through sexual reproduction the single strands are mated to produce a zygote. The result is a **recombination of chromosomes**, whereby the mother's gamete sex cell (egg) is fertilized by the father's gamete sex cell (sperm) to become a new living organism's first cell. The first single cell of new life, the zygote, thereby contains DNA chromosomes that are double-stranded, with one strand each coming from the father and one strand each coming from the mother. The DNA of the new living zygote single cell is thereby unique. **A new unique life has been created to begin as a single living cell.**

The story of how a complex multi-celled organism results from a single living cell, called a zygote, is the story of embryonic development, called **embryogenesis**. And, the very basis of embryonic development is **cell division**, a process called **mitosis**, whereby **new living cells are created that contain the same exact DNA contained in the original zygote single cell**. Here's how that happens.

DNA Replication

Just prior to cell division, whereby one living cell splits apart to become two living cells (a process called **mitosis) the DNA double-helix, wound-up in the cell nucleus, is copied exactly.**

DNA is copied in such a manner as to exactly replicate the order of nucleotides along the polymer DNA strands through the complimentary base pairing of the nucleotides with each other across the two strands of the double-helix of DNA. That process is, indeed, a wonder.

First, a protein enzyme called **helicase** attaches to a specific point on the double-helix, breaks the hydrogen bond between the two strands of DNA, and proceeds to unwind the double-helix, starting as a small bubble and proceeding outwards in both directions. Another protein enzyme holds the double-helix open. Yet another protein enzyme, called **DNA polymerase**, starts at the fork and simply continues to make a new and complimentary strand of DNA from the template of the old DNA strand. The DNA polymerase that has attached to each strand of the DNA proceeds to copy the mirror image of the DNA strand simply by lining up and bonding the nucleotides in a sequence that creates the complementary nucleotide base on the new strand. As DNA polymerase completes a segment and moves forward, another protein enzyme causes the two strands to rewind together into the double-helix. **Two identical DNA double-helix molecules have emerged from one.** Each now contains one old strand of DNA and one new strand of DNA that forms a double-helix for each of the two identical DNA molecules.

Remember, **all DNA is structured by a simple relationship of the matching of A with T nitrogenous bases and C with G nitrogenous bases in alternating pairs of nucleotides.** In this manner, the nucleotides line-up in the sequence necessary for accurate replication of the entire DNA structure. **The DNA replication process is not completed until DNA polymerase and yet other protein enzymes conduct tests for accuracy and correct mistakes.** The result is an error rate in DNA replication that is less than one mismatch in over 1 billion nucleotides.

Cell Division

After DNA replication has taken place, the nucleus of the cell now contains two identical complete sets or copies of the chromosomes containing the DNA for the organism.

In the life of a cell, once DNA replication has taken place, the cell is ready to divide in two. The division process begins with the membrane surrounding the nucleus starting to break down. As the nucleus membrane disintegrates, the chromosomes containing the DNA condense and then the two copies of the chromosomes (called **chromatids**) physically migrate to the center-line of the cell. A special protein attaches one of the chromatids to the top of the cell and the other chromatid to the bottom of the cell by a kind-of guy-wire arrangement.

The two sister chromatids are thereby pulled apart with one on each half of the cell. The cell then is cinched-in along the centerline until it physically splits in two. Two daughter cells have now been created where only one existed before. **Each of the daughter cells now has the exact sets of chromosomes containing the exact DNA that the original cell contained.**

Throughout the process of embryo genesis and then throughout the life of the new offspring child, when new cells are needed to perform specific functions this process of cell division, called mitosis, will be employed to make exact copies of the parent cell. Each new cell will thereby contain and be able to extract the specific information necessary for the proper functioning of that cell from the DNA library residing in the cell's nucleus. Amazing!

Embryo Genesis

We now know that the nucleus of the new single zygote cell contains the blueprint and recipe for its unique life, for the unique life of the organism that it is about to become. And we know the process of cell duplication used throughout the process of embryonic development. So, let's return now to continue the story of how a single zygote cell becomes a fully functioning plant or animal.

Embryonic development begins as the act of fertilization itself triggers a specific realignment of the contents of the zygote cell producing the first asymmetric distribution of proteins and mRNA within the cell, thus creating a north-south axis of the cell. Next, the single zygote cell begins to divide into daughter cells, wherein, each contains an exact copy of the unique DNA contained in the original zygote cell.

The first rapid series of cell divisions is called **cleavage**. Specific patterns of orientation occur through cleavage as thousands and thousands of cells get smaller and smaller and become, in mass, about the same size as the original zygote cell. **DNA is replicated, but not transcribed, during cleavage.** Thus all the cells at this stage of embryo genesis are **embryonic stem cells. No genes have yet been expressed.**

As embryonic development proceeds, through further and further cell divisions, gene expression begins to occur as RNA transcribes and translates DNA and those amazing proteins are born and become more and more profuse. Through protein interactions the developing embryo folds into more and more complex shapes and orientations. The further folding and bending results in cellular tissue layer formation that produces structural movement. As this process repeats and repeats, the morphology of the organism takes shape. Body organs are thus developed and placed in the proper positions, legs and arms and everything else takes shape and everything fits into its proper place. Horses thereby become horses, and roses become roses and fish become fish and humans become humans. That's an over-simplified overview, but that's what happens.

How all that happens is the story of how cells communicate and work together and, not surprisingly, that all has to do with those amazing proteins that are encoded on the genes on DNA. Let's take a look.

How Genes are Expressed During Embryo Genesis

We know that all of the cells in a living organism have the exact same genes encoded in the same DNA in each cell's nucleus. An easier way to say it is **'all cells have the same genetic material'**. We also know that the cells of an adult organism comprise a vast variety of different types and shapes and sizes. That vast differentiation is determined by which genes are **expressed** in which cells. Gene expression determines size, shape and function of body cells. And size, shape and function of body cells determines the size, shape and function of the adult organism. So, **gene expression** is pretty important.

Science does not know what causes gene expression, but some really brilliant scientists have discovered a lot about how gene expression works.

The original single zygote cell of an organism has the potential to become each and every part of the adult organism-to-be, including all cell types making up all body organs, structures and interrelationships. It has the potential to become a cell for each and every part of the organism. It is the original **stem cell.**

In the early stages of embryonic development a vast number of undifferentiated stem cells are formed. At each stage thereafter, the potential or **potency** of new cells to specify the entire organism diminishes. As potency reduces, the fate of the cell becomes more and more restricted until finally the body cells of the fully-developed organism become fully determined. Determination is a progressive process. It does not occur all at once. Once determination is complete, a cell is said to be **differentiated.**

Once differentiated, most of the cells of the organism have lost all their potency. However, some body cells that must be frequently replaced during the life of the organism (like our skin cells and blood cells) still retain some limited potency. That is why in stem cell research the blood and skin cells can only be useful in a most limited way, while the stem cells that are produced during the earliest cleavage stages of embryonic development are highly prized because their potency is so great.

The progressive process of cell differentiation in embryo genesis is controlled by **transcription factors.** Remember that the genetic information encoded in DNA and residing in the nucleus of the cell must first be transcribed into mRNA. Then the mRNA travels outside of the nucleus to the site of the protein factory (ribosome) in the cell's cytoplasm where it is translated through tRNA, docked on the ribosome and decoded into the proper sequences of amino acids. The amino acid chain is constructed into the specified protein by following the sequences of amino acids originally dictated by and encoded in the specific gene segment of the DNA. That's the process that happens each and every time that a gene is expressed. The historical central dogma of genetics informs us that each time a gene is expressed a protein is created.

A gene can only be expressed when it is 'freed' by a transcription factor. And the transcription factors are themselves proteins whose whole job is to enable the expression of genes. Here is how they work.

A specific protein transcription factor must contain the specific shape required to exactly fit into the portion of the DNA molecule that defines the start of the specific gene. **The insertion of a specific protein key into the DNA lock at the specific regulatory region on the DNA strand is what allows gene expression.** All gene transcriptions will begin with the protein enzyme RNA polymerase binding to the DNA at a specific site. The protein transcription factors enable RNA polymerase to bind at the proper site to begin transcription of the gene. Usually only one protein transcription factor will allow RNA polymerase to bind to a specific regulatory region. The presence or absence of the transcription factors determines whether or not a gene will be expressed.

As the embryo develops, more and more transcription factors trigger more and more complicated gene expressions. The transcription factors themselves get more and more compounded as the process of cell differentiation proceeds.

With the continuation of embryonic development, it seems that the cell's final destiny depends more and more on its location or position in the developing embryo. Signals from other cells become the dominant force for development. Cells stimulate each other to develop in certain ways. That is a process called **induction**. An example may be useful. Let's look at the mammalian eye.

The lens tissue of the eye forms out of the same basic ectodermal cells that make up other ectodermal tissues of the body, like our hair and skin. The lens begins to become a highly differentiated tissue when it comes in contact with the optic vesicle that has formed beside it. When the optic vesicle physically contacts the lens, the contacted area of the lens begins to thicken and proceeds to fully develop. Once formed, the lens' contact with the optic vesicle reciprocates the favor and the optic vesicle begins to differentiate into what will ultimately become the retina. As the lens contacts other surrounding cells it induces them to differentiate to become the cornea.

This fascinating example underscores the point that **induction often involves reciprocal effects on the organism's development triggered by inter-cellular contact**. But, it gets even more fascinating.

Some induction does not require direct cell contact. Indeed, as arms become arms and legs become legs, this pattern formation is induced by signals that are spread across great expanses of tissues. Specific organizing regions of the developing embryo secrete growth factors and positional molecular orders to developing cells that are then diffused throughout the limbs.

As pattern formation proceeds in a developing embryo there is literally **a regulatory cascade of interacting genes to guide development.** The transcription factors become so complex that in the development whereby arms become arms and legs become legs these **transcription factors allow the expression of segmentation genes that are themselves more finely classed as gap genes, pair-rule genes and segment polarity genes.**

In the last stages of embryo genesis a morphogenesis ('creation of form') occurs. Bending and folding changes cell shapes and cells literally get up and crawl during body development. This is accomplished by glycoproteins on the cell's surfaces reaching out and adhering to the right kind of cell and then dragging the rest of the cell along with it. **Some cells are even destined to die at the proper stage of embryo development**. For example, cell death accounts for the removal of the webbing between our fingers and our toes prior to our birth.

As more and more complex switching mechanisms turn on and off genes the ever-more-complex **transcription factors produce more and more refinement of genes and proteins and cell structure and function.** And, the cells work together in concert toward the ultimate end of embryonic development - the fully-formed organism.

Science is not sure exactly how the exquisite process of embryo genesis came to be as finely tuned and orchestrated as it is, except to say that it occurred as the result of adaptive change through natural selection of random mutations over evolutionary time. But, **some master controller genes must have evolved early-on for all this order and precision to have developed so perfectly.** The wonder of it all.

Growing Communication in Plants and Animals

Technically, the period of embryo genesis ends with the birth of the new plant or animal. However, birth as a functioning organism is not the end of the developmental process. Not by a long shot.

At birth the newly-functioning organism has all the parts necessary to support and sustain the organism throughout its life. The period of embryo genesis has supplied the new living creature with the systems and processes required to grow to adulthood. We are all personally, and many times painfully, aware of the growth process that takes us from infancy, through childhood and adolescence to adulthood. 'Growing pains' seem to be most acutely experienced during that stage of adolescence, those teenage years, where we experience a growth metamorphosis of sorts, that seems to actually propel us past adulthood in both body and mind. We only later learn that such was an illusion. In actuality, our physical growth during these developmental periods is caused mainly by **hormones.** And the same is true for all multi-cellular organisms. **For both plants and animals the most important system for regulating growth and development to adulthood is the hormonal system.**

Hormones cause changes in the metabolic activity of specific cells in the organism. Hormones are, in fact, chemical signalers that specialize in long-range communication between cell tissues. And, these physiological messengers are absolutely essential for complex multi-cellular organisms to exist in the first place.

Plant Hormones

Plants do not have specialized organs for the production of hormones. Rather, plant hormones are synthesized in the cells of the different parts of the plant organism - the roots, stems, leaves, and flowers.

The major **growth hormone** in plants is called **auxin**. At the tip of each plant stem and root, cells are contributed to elongate these organs at the **apical meristem**. Auxin is secreted from the apical meristem cells. **The amount that the cells in plant stems elongate and grow is directly proportional to the strength of the hormone secreted.** As auxin moves toward the roots it dilutes in concentration, but strangely, the lesser amount of auxin in the roots' upper cells seems to cause the root tips to grow downward. No one knows why. They just do.

Auxin also causes plant stems to grow toward sunlight (a growth tendency called **phototropism**). This happens as auxin is redistributed within the stem's cells, regulating the cellular growth in a manner that always results in **stems growing toward the light.** The higher concentration of the hormone on the shady side apparently causes the cells on the shady side to grow longer. When the cells on one side of a stem thereby become longer, the laws of physics require the stem to bend toward the side having the shorter cells - the side nearest the light. No one knows what causes more auxin to move from cells in the light to cells in the shade with the resultant growth of stems toward the light. They just do.

Auxin is most essential for plant growth because **plant cells cannot move in relation to each other.** That is because plants have a thick outer membrane, called a **cell wall,** that provides structural rigidity. Control of cell shape in plant development therefore relies on controlling the cell wall. And that is the job of auxin.

Other plant hormones cause seed germination, plant dormancy, opening and closing of flowers, twining of vines, and ripening of fruit. As fruits ripen, dramatic biochemical changes take place resulting in fundamental cellular change through the breakdown of membranes.

Animal Hormones

Animals have specialized organs, called **glands**, used for the **production of hormones**. The large number of hormones produced and stored in these organs is collectively called the **endocrine system**.

The major growth hormone in animals is aptly called the **growth hormone**. In humans, this hormone is produced and stored in one of the endocrine glands, called the **pituitary gland.** It is located at the base of the skull. The pituitary is called a 'master gland', but it is, in fact, controlled by the brain in the region called the hypothalamus. The hypothalamus releases a specialized signal hormone to the pituitary gland that, in turn, releases the growth hormone into the bloodstream.

Growth hormone affects all of the cells of the body. It is essential for linear growth. The hormone causes an increase in lean body mass, a decrease in body fat, an increase in metabolic rate, and the growth of bone and cartilage. Its release is greatest in children, but it retains a reduced function throughout life.

While growth hormone affects all of the cells of the body, most hormones do not. Most are very selective. Since they are released into the bloodstream they flow throughout the body. Whether a specific hormone affects a specific organ of the body depends on whether the cells of that organ have a chemical **receptor** for that specific hormone. An example may prove useful.

FIRE! FIRE! THE HOUSE IS ON FIRE!

Upon hearing those words a growing child, or an adult, will experience the infamous 'fight or flight' response. Your awareness is heightened. Your muscles tense. You are filled with the energy necessary to assist in resolving this emergency life-threatening situation. What actually happens at the molecular level is that you undergo an 'adrenaline rush'. How that occurs is, indeed, a wonder. Here's how.

The hormone **epinephrine** is produced and stored in an endocrine gland called the **adrenal medulla.** The brain's response to the verbal warning is to release that hormone quickly into the bloodstream whereby it flows throughout the body. However, not all organ tissues of the body will be affected by the epinephrine chemical messenger. **Only a very few organs have 'receptors' for the hormone.**

Epinephrine is a water-soluble molecule. As such, it is electrically charged, which makes it unable to cross the cell membrane of a receptor cell by itself. In order to affect the receptor cell an elegant system is used to affect the interior of the receptor cell by transducing across the receptor cell's membrane. As a molecule of epinephrine passes in the blood stream by a body tissue or organ that does have a receptor for the hormone, it **binds to the target cell's receptor on the exterior of the cell.** Binding occurs because the shape of the epinephrine molecule (figurative hand) exactly fits into the (figurative glove) pocket on the outer contour of the cell's receptor protein. The receptor is, in fact, a protein embedded in the membrane of the cell in such a fashion that it passes entirely through the membrane and is exposed to both the outside of the cell and the inside of the cell.

When the molecule of epinephrine binds with the receptor protein in the cell's membrane it **causes the shape of the receptor protein to change inside the cell** and that change triggers other protein reactions and shape changes. For instance, when epinephrine binds with **liver** protein receptors it initiates a sequence of changes that cause liver cells to **convert sugars** that are stored in the liver as **glycogen into a simpler form, called glucose**, that is then released into the bloodstream to provide a quick burst of energy. It seems to happen instantly.

But the conversion of glycogen to glucose in the liver is far from a simple process. **Epinephrine activates the shape change of the liver receptor protein** that, in turn, activates an enzyme called cAMP that, in turn, triggers a series of proteins to sequentially add a phosphate group to their molecules. At each step in the process the signal from the hormone messenger gets more and more amplified as **each step exponentially activates an ever-larger number of protein enzyme molecules.** The magnitude of this **signal transduction cascade** is staggering. The last step in the process is the activation of the enzyme glycogen phosphorylase which synthesizes glucose, which is then released into the bloodstream. Through this amazing process **a single molecule of epinephrine will cause 100 million molecules of glycogen phosphorylase to be activated, thereby producing abundant glucose.** Yet, checks and balances are also built into the process. If initial blood glucose levels are already high when 'fight or flight' triggers, the quick production of high concentrations of more glucose may be very harmful to the body. In this case the signaling cascade is modified to modulate glucose production to avoid damage. (To keep all this in perspective, remember that each of those 100 million molecules had to have been already produced by the complex [DNA to mRNA to tRNA to ribosome] protein building process in the liver cell in the first place, only to lie dormant in the liver cell awaiting a call to action.) But, we are not through.

Epinephrine affects not only the liver. As it passes in the bloodstream throughout the body, it also affects other specific body organs and tissues. The heart has a receptor protein that binds epinephrine. Upon binding, the shape change of the receptor protein transduces through the heart cell's membrane to trigger protein enzyme cascades within the heart cell that, in concert, causes the heart to beat faster thereby rushing blood throughout the body. And, blood vessels in the intestines are constricted through the same process in the intestines. **The point is that different kinds of cells in different body organs and tissues have receptors for epinephrine, and the specific response of each receptor cell depends on the specific type of protein metamorphosis that epinephrine triggers.**

Epinephrine's journey throughout the bloodstream, initiating specific effects on specific organs and tissues, is but one example of how cells communicate with each other over long distances in the body. The entire endocrine system performs that communication function with great regularity during growth and throughout the life of the organism. In totality, the goal of the entire system is to keep everything in balance.

Homeostasis

All parts of a single-celled organism are, of necessity, in direct contact with the natural environment surrounding them. Their cellular processes and biochemical reactions are directly dependent on the surrounding external environment. They have no buffer from the surrounding external environment.

Plants and animals, on the other hand, have an internal cellular environment that protects the internal cells from the vicissitudes of an ever-changing external environment. The maintenance of internal temperature and bio-chemical conditions within narrow parameters is especially important for animals. This allows biochemical processes to work in an optimal fashion. For example, **if**

proper internal temperatures are not maintained, protein enzymes may lose shape and become dysfunctional. Plants and animals have bodily systems that work together to regulate internal temperature and bio-chemical balances. The goal of these systems is to maintain a stable internal environment for the proper functioning of the whole organism. A great many of the functions and structures of multi-celled organisms, like us, are aimed at maintaining a stable internal environment that is optimal for the survival of the organism. This allows plants, to a lesser extent, and animals, to a greater, to occupy a broad range of external environments.

While all complex multi-cellular organisms have varying degrees of internal balance controls, this primer will briefly review only the system employed by mammals and will use our own body system as the pointed example.

The human endocrine system is an elegant system of checks and balances that regulates body growth and body systems' operations. Each endocrine gland has a specific function and helps to maintain an optimal internal environment for the organism, a balance called **homeostasis**. The body constantly strives through growth and adulthood to maintain **a stable internal environment.** And, **the key to homeostasis is communication between cells.**

Homeostasis is fascinating because it requires exquisite cooperation and teamwork among different cells in order to keep everything in balance. To achieve that, different cellular groups must each perform specific regulatory functions. **Three delicate body balances that must be regulated to stay within rather narrow parameters are: (1) blood pressure, (2) blood sugar, and (3) body temperature.** If the levels deviate too far from a 'normal' set point, various body systems become damaged, and we may even die. So, these delicate body balances, and the endocrine system employed to keep them 'just so', are quite important.

Let's return again to FIRE! FIRE! THE HOUSE IS ON FIRE!

We have already seen how inter-cellular communication quickly provided for a blast of glucose to enter the bloodstream in order to give us the energy boost necessary to effectively pursue 'fight or flight.' At the same time that we get a sugar energy boost our blood pressure skyrockets. That causes certain nerves which are sensitive to pressure in the blood (called baroreceptors) to signal this condition to the vasomotor control center of the brain which, in turn, sends nerve impulse instructions to the heart to reduce the heart rate and thereby reduce blood flow from the heart which, in turn, reduces blood pressure back to the normal range. Homeostasis is thereby achieved. If balance is not regained, we may suffer ruptured blood vessels, a stroke, or even die.

Blood sugar levels in the body also must remain within a fairly narrow range. The endocrine gland that provides the necessary regulation of blood sugar is the **pancreas.** The Alpha cells of the pancreas secrete glucagon. Glucagon stimulates the biochemical processes to maintain blood glucose (sugar). The Beta cells secrete insulin. Insulin stimulates movement of blood sugar across the cell membrane and thereby lowers blood sugar levels. **The physical condition that results when blood sugar levels are too high is called diabetes**. If excess blood sugars are not lowered through exercise, diet or insulin injections, debilitating bodily disease and even death may occur.

Each of the delicate body balances must be regulated within rather narrow parameters. That regulation requires three basic things: (1) a **receptor** to provide information about the present state of the parameter; (2) a **comparator** to compare the present state with the desired state and to give instructions to; (3) an **effector** that changes the value of the present state in the appropriate direction of the desired state. An example of temperature regulation may prove useful.

Our primary temperature control organ is the hypothalamus, located at the base of the brain. The hypothalamus serves as the primary temperature receptor and **integrates signals of hot or cold received by the skin**. The hypothalamus is the temperature comparator, comparing the present body temperature against the desired body temperature of 98.6 degrees F. If body temperature falls significantly below that desired temperature, the hypothalamus triggers nerve impulses that cause blood vessels to constrict and muscle tissues to shiver, both of which have the effect of increasing body temperature. If body temperature rises significantly above 98.6 degrees F, the hypothalamus triggers nerve impulses that cause blood vessels to dilate and the skin to sweat, thereby lowering body temperature. And, **the defined parameter for temperature in our bodies is most narrow - only about 15 degrees F.** The failure to maintain temperature within that narrow range for any significant period of time results in organ failure and ultimately death.

The Cellular Electrical System

One feature of life that is not frequently discussed in primers is that the failure to maintain proper electrical charge in a living organism will also result in organ failure and ultimately death. The living organism contains complex electrical circuitry that is used to communicate not only between cells but also between the interior of the cell and the environment outside of the cell's plasma membrane. All living cells run on electricity and the electricity is generated through organic chemistry.

All living cells have an excess of electrical charge on the inside of the cell relative to the outside. As a result there is an electrical voltage that runs across the plasma membrane of the cell. Elite and mainstream science maintains that this most-complicated system simply evolved by chance, yet it is pretty hard to understand, let alone to construct and maintain. Let's take a few moments to explore some intriguing questions about cellular electricity. Without cellular electricity we simply would not be here.

How Does a Living Cell Become Electrified?

Most all living cells have an excess of **negative** electrical charge on the inside relative to the outside. Scientists describe this condition of the cell as being **polarized**. As a result of being polarized there is an electrical voltage across the plasma membrane of the cell. Voltage is caused by the separation of charges by the membrane. The electrical voltage is the result of the difference in charged particles residing within the cell compared to the charged particles in the fluids directly surrounding the outside of the cell.

In order to establish and maintain a state of electrical charge a living cell constructs an elaborate system. It starts with the recognition of charged particles within the cell. A living cell contains all sorts of both negatively-charged particles (called **anions**) and positively-charged particles (called **cations**) within. The anions are generally organic molecules like amino acids and proteins that have lost electrons and thereby become negatively charged. The cations are generally potassium ions ($K+$) or sodium ions ($Na+$) that have gained an electron. It turns out that the potassium ions and sodium ions are the principle regulators of the cell's electrical charge. Here's how they do it.

Cells are constructed so that charged particles cannot pass directly through the plasma cell membrane. This feature lays the foundation for the electrical interactions.

In order to provide for the selected transport of charged particles through the cell membrane, cells construct **ion channels** that allow for sodium ions (Na+) and potassium ions (K+) to flow back and forth between the inside and the outside. These ion channels are actually specialized proteins that run through the lipid plasma membrane and allow these chemicals to pass through the cell wall.

Most of the ion channels are very selective, allowing only one particular chemical to travel through. And, the channels vary greatly in number. In general there are twenty times more ion channels for potassium than there are for sodium. Thereby, potassium is said to be more 'permeable' than sodium.

In general, potassium ions are the most important for establishing what is known as the **resting potential** of the cell. There is a huge excess of K+ inside the cell compared to the outside. As a result, **K+ passively flows** through its ion channels virtually all the time from the inside to the outside **down its concentration gradient**, as it strives to equalize concentrations.

Sodium ions (Na+) have a greater concentration outside than inside the cell. Most of their ion channels are gated and remain closed most of the time preventing sodium from flowing down its concentration gradient from the outside to the inside of the cell. When the concentration of sodium ions outside and inside are in equilibrium the electrical equilibrium potential for sodium is about +60mV (millivolts). When potassium ions are in equilibrium the electrical equilibrium potential for potassium is about −80mV.

If potassium was the only charged particle in the cell the resting potential for the cell would be the same as the equilibrium potential for potassium. But the charges of sodium and all the other charged particles in the cell serve to off-set the potassium equilibrium potential. So, collectively the negative and positive charges of all these particles establish the **resting potential** for the cell. As we have seen, the resting membrane potential of the cell is dominated by potassium so, the resting potential of the cell is usually a negative charge of about −60 or −70mV, somewhat less than the potassium ion's equilibrium potential (−80mV). Heart cells are an exception, being even more negative than −80mV.

So, a cell's membrane potential is the electrical potential difference (known as voltage) across the cell's plasma membrane. Voltage is caused by the separation of charges by the membrane. With cell activity the magnitude of membrane potentials will usually be in the range of 10 to 100 mV. For comparison a 1.5 volt flashlight battery has a 1,500 mV charge. The charge represents the stored energy in the flashlight battery or the cell membrane, and can be used to do work.

How Does a Cell Remain Electrified?

We have now reviewed how a living cell constructs itself in a manner that allows it to become electrified. But, the construction of the gated ion channels that run through the plasma membrane of the cell and thereby allow the cell to establish a resting potential of negative electric charge is not, by itself, sufficient to maintain a resting potential which allows for the electrical work required for life.

This resting state of electrical charge within a cell is absolutely essential in order for electrical transmission of information between cells and for the proper work functioning of the cells themselves. If the passive ion channels were the only cellular mechanism in play simple diffusion of chemicals down their concentration gradients would soon nullify the resting potential of the cell.

The maintenance of this state of cellular charge is so important that all cells have constructed **sodium/potassium pumps** in their membranes. The pump is itself a specialized protein that runs through the cell membrane. Each cell constructs literally thousands of these sodium/potassium pumps to actively transport sodium and potassium ions back and forth between the cell's interior and outside the cell's membrane. For each cycle of the pump three sodium ions are pumped out and two potassium ions are pumped in. It is indeed a pump, because the flow of ions is against the concentration gradient. In effect the ions are forced to run uphill. To do that requires energy. Indeed, the **energy used** to operate the pumps and, thereby, force ions to run against their concentration gradient is immense, about **30% of the body's metabolic energy.**

The pumps all act the same way. Each time a pump operates, three sodium ions are pumped out of the cell and two potassium ions are pumped in. The **rate at which the pump operates is**, however, highly variable. Pump activity is mainly **regulated by body hormones.** Thyroid hormones seem to be a major player in maintaining steady-state concentrations of pumps in most tissues. Also, insulin is a major regulator of potassium homeostasis and has multiple effects on sodium/potassium pump activity.

Dedicating almost 1/3 of the body's resting energy to run the sodium/ potassium pumps is a pretty good indication that the pumps are very important. But, why must the resting state of the cell remain polarized?

What Purpose Does Cellular Polarity Serve?

Simply stated, our bodily systems can't work, our bodily organs like our heart and kidneys can't work, and our brain can't work without the operation of these sodium/potassium pumps. They maintain our internal cellular polarity and thereby allow the electrical circuitry that provides for life to work.

The sodium/potassium pump that is used to maintain cellular polarity allows for many wondrous functions to occur within the body. To name but a few:

- the export of sodium from the cell provides the driving force for many facilitated transporters, which import glucose, amino acids and other nutrients into the cell;
- the export of sodium and import of potassium creates an osmotic gradient that drives absorption of water, which allows our small intestine and kidneys to work; and
- the rapid re-establishment of the resting potential of a neuron cell after it 'fires' enables a neuron to quickly 'fire' again and again and again in response to additional stimuli.

And, that provides a pretty good segue for wrapping-up our exploration of how living cells communicate and work together by mentioning neurons. They are the specialized nerve cells that provide for literally lightning-fast communication of information within living animals.

Communication by Nerve Cells (Neurons)

As we have seen, the hormonal cellular communication system operates effectively in both animals and plants. And, it works at a pretty fast clip, as we observed in the 'adrenaline rush' reaction needed for a 'fight or flight' response. However, a much, much faster cellular communication system is essential in order to effectively operate the basic sensory mechanisms for animals. **Plants**

lack the system entirely. It is termed the 'central nervous system' and it provides for virtually instantaneous cellular communication in animals.

The key to the central nervous system is the operation of neurons, specialized cells that transmit information by electrical impulses. Because the system in mammals is highly specific, we will examine it in some detail in Chapter 18, The Mammal Class. But, for now suffice it to acknowledge that **neurons, at core, function by changing the permeability of sodium ions and potassium ions across cell membranes**. The action of these amazing ion channels govern all the responses of the central nervous system, from hand-eye coordination, to falling in love, to remembering what you just read.

Onward to Energy

This concludes the brief examination of how information is transmitted by chemistry and electricity to all parts of the body, and how living cells communicate and work together. Cellular communication is, indeed, amazing stuff.

But, none of this stuff could happen if the cells of the body were not supplied with the proper energy necessary to perform their work. What energy do they use and where does it come from? Let's turn to the next chapter.

CHAPTER 12
The 'Power' of Life - How Living Cells
Acquire and Use Energy

Let's recap and review where we are in looking at the fascinating subject of living things.

All living things, all plants and animals, including us, are made up of living cells. Each and every one of the living cells comprising an organism contains, within its nucleus, the complete library detailing the blueprint and recipe for constructing each and every part of the organism.

We each start off as a single cell. Then the information contained in our DNA library instructs the cell to divide and those cells to further divide, and to differentiate and differentiate until a finished product emerges. A 'higher' organism, like ourselves, then contains literally trillions of cells. And those cells employ a highly-sophisticated communication system that constantly transmits the information necessary for the organism to function and prosper in both the internal and external environment.

Take more than a moment to let the wonder of all that sink in.

Energy is Needed to Power Life

All of that building and functioning and communicating, however, could not happen in the first place without **energy**. Each of the raw materials from which cellular life comes from are **organic molecules**. And, in order for organic molecules to form in the first place, energy is required. Without energy, no life. Period. So, we turn in this chapter to the story of energy on Earth and within living cells. Indeed, another wonder to behold.

And if you think the last couple of chapters have been obtuse, wait till you get a load of this one.

The prevailing scientific belief is that all this stuff that we have examined **just happened** over the course of evolutionary time. No creative intelligent agent was involved. All of this stuff within all living things grew, without direction, from organic molecules. Those molecules figured out how to construct various combinations of amino acids to make the vast array of those amazing proteins that make-up and run every living thing. They figured out how to encode the amino acid sequences necessary to turn those amino acid building blocks into protein substances to form living tissues. And, they figured out how to translate the nucleic acid code into the language of amino acids to accomplish that. And, they figured out how the various combinations would intricately interact with precision fit to create the protein shapes necessary to carry out all the functions of the organism. And, they figured out how to assemble and store that information, with fantastic precision, in the nucleus of each living cell, and how to pass on that information intact, with a minuscule error rate, from cell to cell. And, how to extract just the right information from the encoded DNA library at just the right location and at just the right time to form a living, breathing, fully-functioning organism that is itself blithely unaware of all these wondrous things that are taking place.

All of this stuff was accomplished, at core, by microscopic nucleic acids and amino acids that have no thoughts, no feelings, and no brains. And, not only did they figure out how to do all this stuff, **they did it!**

You and I represent the current zenith on the evolutionary chain. As human beings we are the uncontested most intelligent creatures that have ever lived on the face of the Earth. That is also the prevailing scientific viewpoint. Yet we struggle to simply understand, to barely grasp the intricate processes whereby these tiny unintelligent nucleic acids and amino acids accomplished all this amazing stuff.

Another wonder is that all this amazing stuff could never have occurred in the first place until even-smaller building blocks, the underlying elemental atoms that make up the nucleic acids and the amino acids, figured-out how to harness and use nuclear power.

For every living creature on this planet requires energy to run living bodily systems. Without energy life does not exist. And, all energy used by living organisms on this planet originates in the nuclear furnace of our Sun.

All living cells require energy. Without energy they die. This chapter first examines how the elemental atoms of inorganic molecules learned how to capture the Sun's energy and convert it into useful organic molecules that then formed those amazing proteins. Then we will examine how the organic molecules figured-out how to keep the cells alive by recycling energy first derived from the Sun's nuclear power.

The Sun That is the Source of All Our Power

You may recall that between four and five billion years ago our Sun and our solar system were born out of the remnants of an exploded star, called a supernova. Each of us and every other organic and inorganic thing on this planet are, quite literally, made of stardust.

As our Sun formed into a new-born star, gravity created immense pressures and temperatures at its core. Those immense pressures and temperatures served to fuse together the nuclei of hydrogen atoms. The process is called nuclear fusion. Hydrogen is burned with great intensity in this process and immense heat energy is poured outward into space in the form of electromagnetic radiation. We call it sunlight.

The process of nuclear fusion at the core of our Sun has been proceeding apace with great regularity for the past four to five billion years. Scientists tell us that it will continue to radiate a steady-state and reliable stream of light energy for yet another five billion years into the future.

Light energy is in fact electromagnetic radiation. It includes all spectrums of light wavelengths and frequencies, such as x-rays, gamma rays, ultra-violet rays, visible light, and radio waves. Types of light differ only by the rate at which their waves vibrate. The shorter the wavelength the higher the frequency.

All of the light types are composed of electromagnetic waves consisting of changing magnetic and electric fields. And, all of the light waves, regardless of wavelength or frequency, are traveling at exactly the same rate of speed, the speed of light, which is 186,000 miles per second, in a vacuum.

While light energy travels in electromagnetic waves it is not continuous. Light energy is discontinuous. While light is a wave it is, quite amazingly, also a particle. Light energy comes in discrete packages of energy, called **quanta,** that are of a specific size. These quanta of light are called **photons**. So, light has a dual nature. It is both a wave and a particle.

The immense heat and pressure in the thermonuclear reactor at the core of the Sun fuses together the nuclei of hydrogen atoms. This nuclear fusion serves to convert hydrogen into helium and yields free photons (particles of light energy) as a by-product. Photons of electromagnetic energy are released when hydrogen is fused into helium. That's it. That's the source of all our power.

After they reach the surface of the Sun, waves of light energy, consisting of discrete photons, are radiated into space at the speed of 186,000 miles per second. When that light energy reaches the Earth, about nine minutes after it leaves the surface of the Sun, it is useless to us unless it can be converted into an energy form that can be used to perform useful work. **Photosynthesis** is the process of capturing sunlight energy and transforming it into a form that can eventually be used by living organisms to perform useful work.

About 3 billion years ago single-celled bacteria developed the fundamental method of energy conversion called photosynthesis and thereafter green plants refined the system into the exquisite process that exists today. All multi-celled organisms on Earth, including us, are dependent on this solar energy transformation process called photosynthesis. Let's take a look at how it works.

Photosynthesis

Brilliant scientists have discovered the intricacies of photosynthesis. They know how photosynthesis works, but they have never been able to duplicate the process in the laboratory. Scientists at the Brookhaven National Laboratory are diligently working in the first decade of the 21st century to produce fuels like methanol and methane directly from water and carbon dioxide, using renewable solar energy in a project called 'artificial photosynthesis'. It is self-evident that our society would benefit greatly from being able to develop a process like photosynthesis whereby we could extract useful energy from just sunlight, air, and water. But, just as scientists have never been able to produce a living cell from non-living matter, scientists have never been able to replicate the process of photosynthesis that provides all of the energy needed by a living cell for life.

Photosynthesis is the process whereby green plants capture the energy of sunlight and then synthesize and store it in high-energy organic compounds. Only a small amount of sunlight energy is captured by green plants in this way for use in making organic compounds. The rest is dissipated back toward space. However, the small amount of light energy that is captured by plants and synthesized into high-energy organic compounds represents the total amount of energy available to power all living things on Earth. Without photosynthesis, life as we know it would not exist.

The element carbon is the basis of all known life. Living things are made of carbon-based organic compounds. These carbon-based organic compounds are highly ordered. To maintain this high level of order, energy must be constantly added to them. Without the addition of energy the 2nd law of thermodynamics dooms them to quickly increasing disorder through entropy, and living things die. (Note: We will review the laws of thermodynamics and entropy further in Chapter 19.)

All of the carbon that is contained in all living things is initially taken out of **carbon dioxide** (CO_2) by plants through the process of photosynthesis. Carbon dioxide is the compound of carbon that contains **the lowest form of energy** (the lowest form of energy is said to be the **most oxidized**). The process of photosynthesis increases the low-energy form of carbon in carbon dioxide into the high-energy form of carbon in the compound sugar called **glucose**.

At first glance, the process of photosynthesis looks like a pretty simple chemical process. Sunlight energy is added to the inorganic compound of carbon dioxide and the inorganic compound of water. The result is the inorganic compound oxygen and the organic compound of simple sugar, called glucose. At the end of the process the energy of sunlight has been captured and stored in the form of high-energy organic sugar that is used as the fuel to power the work of all living cells on the planet. The chemical formula for photosynthesis is $6CO_2$ + $6H_2O$ + energy >> $C_6H_{12}O_6$ + $6O_2$ (carbon dioxide plus water plus energy yields glucose sugar plus oxygen).

Photosynthesis is an energetically-uphill process. It requires energy to first break chemical bonds and then to store even more of the Sun's energy in high-energy organic compounds.

To accomplish this conversion of sunlight into useful energy is actually the story of how the electrons are excited and allowed to be captured and stored at a higher-energy level. It is an amazing task that requires two discrete steps. Let's take a look.

Step One - The Cell Captures Energy From Sunlight

Light from the Sun is electromagnetic radiation. It is both a wave and a particle. Electromagnetic radiation travels in waves and interacts with matter as a particle. Particles of light are called photons.

After its nine-minute-long trip from the Sun, photons of light encounter matter on Earth. When a photon encounters matter it will do one of three things:

- pass through the matter,
- reflect off the matter and dissipate toward space, or
- be absorbed by the matter.

When a photon is absorbed by matter it transfers its energy into that matter. However, photosynthesis is not as simple as plant matter just absorbing photons of sunlight. Far from it.

When I'm out fishing on the lake in the summertime, I absorb photons of sunlight quite readily. The absorbed photons briefly excite the electrons of my skin cells to a higher energy level. However, I do not convert the sunlight into useful energy. I quickly lose most of that energy as heat and light of a longer wavelength that is radiated back toward space (a process called fluorescence). Most of the energy of the Sun that I keep is not useful to me. We call it sunburn.

Green plants, on the other hand, have figured out a way to capture the energy in the photons of light that they absorb and convert it into useful energy without it being lost as heat. This light-energy-capture occurs within the cellular structure of the leaves of green plants.

The cells of plant leaves (plant cells, never animal cells) contain distinctive organelles called **chloroplasts.** The chloroplasts have their own membrane system, and, embedded in the membranes are lots and lots of chlorophyll and other pigment molecules. Their specific job is to capture sunlight energy. Here's how they do it.

When a photon of sunlight encounters and hits a molecule of chlorophyll it excites the electrons in that molecule to a higher energy state. In plant leaves the molecules of chlorophyll are tightly packed together into what is called a **photo system**. So, when a photon strikes a molecule of chlorophyll the electron that it excites does not quickly decay to a lower energy level and is not fluoresced back into space at a longer wavelength (like me and my sunburn). Rather, **the chlorophyll molecule passes the photon on intact to the next molecule in the photo system,** and so on and so on. When the photon reaches the chlorophyll molecule next to the primary **electron acceptor molecule** in the center of the photo system a very strange thing occurs. The excited electron in that adjacent chlorophyll molecule is transferred intact into the electron acceptor molecule. The chlorophyll molecule loses an electron (losing an electron is called **oxidation**). The electron acceptor molecule gains an electron (gaining an electron, while counter-intuitive, is called **reduction**). This intricate process is called oxidation-reduction, abbreviated as a **redox reaction.**

Once a redox reaction occurs, once the energy of a photon causes an excited electron to leave the chlorophyll molecule and be added to the electron acceptor molecule, the captured energy is stored in a more lasting state. Now it must be converted into a more usable form of energy and stored for future use. Let's look at how that happens.

Step Two - Increasing and Storing Energy in Plant Cells

Immediately after the electron acceptor molecule gains an extra electron (is reduced) it transfers that electron to an electron transport chain embedded in one of the chloroplast's membranes. The energy level of the added electron is gradually stepped down through a series of redox reactions as it proceeds along the electron transport chain. **The energy released from the electron in these reactions is used to pump hydrogen ions against their concentration gradient across the chloroplast's membranes.** The energy, stored in the concentration gradient of hydrogen ions, is used to drive the production of the ubiquitous energy molecule called ATP, used by all living organisms, through the help of an amazing protein enzyme called ATP synthase. (Note: I am not making this stuff up.)

Simple water (H_2O) is a necessary reactant in photosynthesis. Electrons from water fill the holes created through the redox reactions and donate electrons to build high-energy glucose molecules and produce oxygen as a waste product.

The chemical reactions through the electron transport chain result in the production of a small amount of ATP and energy retained in the electron transport molecules themselves. However, these forms of energy do not have a very long 'shelf life'. ATP has to be used fairly quickly once it is produced, or it decays into inorganic ADP. Therefore, **green plants had to figure out how to store this captured energy in a longer lasting form** (i.e., high-energy organic molecules) that can be used later. That wonder is performed in a three-phase process, called the Calvin cycle. The Calvin cycle is brought into play in order to use the initial organic energy source to now synthesize the high-energy sugar called glucose.

The intricacies of the Calvin cycle are far beyond the scope of this primer. Suffice it to say that for every three molecules of CO_2 that enters the cycle, the plant cell produces three organic high-energy G3P molecules. The plant can use the G3P molecules directly for its energy needs or it can combine two molecules of G3P to produce a higher-energy molecule of glucose sugar. The energy that is now stored in the high-energy molecule of glucose has a very long 'shelf life'. It can be used by the plant for its future energy needs. And, most important for us, it can be used as an energy source by animals. We simply eat it. **In order to live we have to eat other organic things.**

The food that we eat powers the processes of glycolysis and cellular respiration that make all the ATP that we need to maintain the high order state we require in order to live. (Note: Glycolysis and cellular respiration will be examined in some detail shortly.) It turns out that we acquire most of our ATP energy molecules by recycling them. But, that process requires energy from an outside source (e.g., glucose) to get things going in the first place. To get energy we first have to eat other living things or things that used to be alive.

We need to now take a look at the specific processes whereby animals acquire all the energy that they need in order to live and to meet all the challenges of living. But first we need to take a brief foray into the basics of cell biochemistry.

Basic Biochemistry of Cells

Life requires a high degree of order. Energy is required to maintain that high degree of order. If a living thing runs out of energy it dies. It then decomposes in accord with the fundamental second law of thermodynamics. Entropy decomposes dead things.

How living things create and maintain order is mostly a matter of the biochemical processes of chemistry.

The entirety of all of the biochemical processes occurring in the cells of a living organism is called **metabolism**. And all of the actions of metabolism, called **metabolic reactions**, either require or produce energy.

Metabolic reactions that produce energy are called **catabolic**. Those metabolic reactions that require energy are called **anabolic**.

A run-of-the-mill cell performs literally thousands of intricate and interconnected biochemical reactions, both catabolic and anabolic. It performs these reactions with great precision akin to a small chemical plant. It takes in raw materials and uses them to construct chemical compounds to perform useful functions. And, it also de-constructs other complex chemical compounds and uses the de-constructed products as a supply of materials and energy.

A run-of-the-mill cell's interconnected biochemical reactions traverse intricate biochemical metabolic pathways. All of those metabolic pathways are potentially interconnected. The product of one pathway is often the beginning (called a **substrate**) of another reaction. And the products and substrates of that reaction may then perform other critical linkages. And, remember that we are talking about thousands and thousands of reactions for each cell.

Catabolic pathways release energy. They do this by de-constructing the high-energy organic molecules we consume as fuel when we eat. That is basically the metabolic processes of glycolysis and cellular respiration.

Anabolic pathways require energy. They initially get the required energy from glycolysis and cellular respiration and then benefit from a wondrous energy recycling system.

Energy is the ability to do work. All energy is either moving (called **kinetic** energy) or it is stored (called **potential** energy). The chemical definition of energy is the ability to create or break stable chemical bonds in molecules by rearranging the atoms in those molecules.

In general, simple molecules, like carbon dioxide are low-energy molecules. They have only a small amount of stored energy. Complex molecules, like glucose sugars, are high-energy organic molecules. They have a great deal of stored energy. An energy source (ultimately the Sun) is required to produce them.

Both the anabolic reactions and the catabolic reactions in cells require some energy to get started. While that is self-evident for anabolic reactions, it is not intuitive for catabolic reactions.

Catabolic reactions run 'downhill' but they do not normally do so spontaneously. If the high-energy organic molecules like glucose sugars and proteins and fats broke their chemical bonds spontaneously at our normal body temperature, we would all simply spontaneously combust. Or, more likely, the glucose sugars and proteins and fats would never have formed in the first place.

Spontaneous combustion is avoided because an **activation energy is required in order to begin catabolic reactions**. This necessity for activation energy to begin catabolic reactions is a **necessity of life**. Once activated, the catabolic reactions run 'downhill' and release energy. We start out with high-energy organic compounds like glucose and end up with low-energy inorganic waste products like carbon dioxide.

One more basic of cellular biochemistry before we return to how we actually get energy from the food we eat. **Enzymes**. Remember enzymes? An enzyme is a special kind of protein that is employed to catalyze chemical reactions. An enzyme is a highly selective catalyst that increases the rate of a chemical reaction by **lowering the amount of activation energy required** to get the catabolic process rolling. An enzyme usually catalyzes only a single chemical reaction.

A reactant that binds to a catalyst is called the **substrate** of that enzyme. The location on the enzyme protein where the substrate binds is called the **active site**. The binding requires an exact fit. The substrate induces this exact fit by causing the enzyme to change its shape (those amazing proteins). Once the substrate tightly binds to the enzyme the substrate is catalyzed into the reactant's product and the enzyme drops off. The enzyme is then free to bind to another substrate, and so on, and so on. A run-of-the-mill enzyme may catalyze 1,000 substrate molecules a second. Others are really fast.

Because enzymes are not used up in these catalytic reactions, only a tiny amount of a particular enzyme may have a most dramatic effect on the speed of a metabolic reaction.

These basics of cellular biochemistry govern the catabolic reactions that release the energy contained in the high-energy organic molecules like glucose and other carbohydrates and fats and proteins. The energy that these catabolic reactions release is ultimately converted into the ubiquitous energy molecule called ATP. Now, let's return to the metabolic processes whereby this wonder occurs - glycolysis and cellular respiration.

How Cells Burn Fuel to Produce Energy

When we are exercising or doing hard work we aptly say that we are 'burning calories' (calories being a simple measure of heat release). The statement is very true because the combined action of the catabolic processes of glycolysis and cellular respiration is the biological equivalent of burning high-energy molecular fuel. While our body cells can burn all forms of organic high-energy molecules, the most common high-energy fuel is the sugar called glucose.

You will recall that plants 'construct' high-energy glucose sugar through the process called photosynthesis. And the chemical formula for photosynthesis is:

$6CO_2 + 6H_2O$ + energy $>>$ $C_6H_{12}O_6$ $+6O_2$

(**carbon dioxide plus water and energy yields glucose and oxygen**).

Well, it turns out that we extract the energy from glucose by kind-of reversing the process. The **chemical formula for 'burning' glucose to obtain energy** is:

$C_6H_{12}O_6$ $+6O_2$ $>>$ $6CO_2 + 6H_2O$ + energy

(**glucose plus oxygen yields carbon dioxide plus water and energy**).

The **chemical energy in glucose** is stored in the 6 carbon atoms of the molecule. Specifically, it **is stored as potential energy in the chemical bonds of the carbon atoms**. That **chemical energy is released through redox reactions**. Remember that in a redox reaction one of the elements loses an electron (**is oxidized**) and one of the elements gains an electron (**is reduced**).

In the redox reaction whereby glucose is burned as fuel and its energy is released, the carbon atoms and the oxygen atoms are of particular significance. The carbon atoms in glucose occur at a higher energy level where the carbon atoms have a weaker pull on their electrons. On the other hand, oxygen atoms that occur at a lower energy level hold on tightly to their electrons in a state of extreme electro-negativity. **Oxygen is the extreme of electro-negativity. Oxygen is an electron hog with an inexorable pull on carbon electrons. When electrons are ultimately transferred from carbon atoms to oxygen atoms through redox reactions, energy is released.**

Not just glucose sugar, but all of the high-energy organic compounds, including carbohydrates, fats and proteins, have an abundance of carbon-carbon and carbon-hydrogen bonds that can readily be moved to lower energy states when attracted to oxygen (hence, the vernacular terms **cellular combustion** and **burning calories**). But, as we have noted, unless the release of energy through oxidation occurs in a controlled fashion, living cells would face an insurmountable problem - they would simply burn up.

To overcome this fatal problem, living cells have learned how to avoid passing electrons directly to oxygen. They do this by passing electrons to organic compounds called **electron carriers** that function as specialized oxidizing agents. The most important electron carrier (and the only one that will be mentioned in this primer) is called **NAD**. NAD is responsible for the orderly transfer of electrons in the cell. It **removes electrons from glucose through enzymes called dehydrogenases, which remove hydrogen atoms from organic substrates. It then carries them to the locations within the cell where the processes of glycolysis and cellular respiration occur.**

Glycolysis

Glycolysis occurs in the cytoplasm of the cell. Glycolysis is the process whereby the cell breaks down one molecule of glucose (a six-carbon sugar) into two molecules of pyruvate (a three-carbon sugar).

The process of glycolysis involves 10 discrete reactions, with each of the reactions being catalyzed by a different enzyme.

Through this complex metabolic process the net result in energy harvested from the molecule of glucose is actually quite small. For each molecule of glucose the cell processes through glycolysis it obtains a net of two molecules of ATP. That represents a harvesting of about 3% of the potential energy stored in glucose. But, the cell also harvests through this process reduced electron carrier molecules of NAD (called NADH). However, while the energy in ATP is now available for direct use by the cell, the stored energy in NADH is not in a form that is directly useable.

Cellular Respiration

Following glycolysis, **if oxygen is present the cell proceeds to harvest additional energy from the product of glycolysis** (the two molecules of pyruvate). This harvesting is done not in the cytoplasm of the cell but, rather, in a special organelle called the **mitochondria**. The harvesting occurs through two distinct processes: the **Krebs cycle** and the **electron transport chain**.

STEP ONE – The Krebs cycle serves to harvest as much of the energy remaining in the carbon bonds as possible. It does so through **an 8-step cyclical process**. The 8 compounds are never used-up in the process. The **intermediate compounds regenerate themselves in a cyclic fashion.** The last compound in the cycle becomes a substrate for the first reaction, ergo, a cycle (named for the brilliant scientist who discovered it). **The Krebs cycle, in essence, squeezes all the energy that it can from the reduced electron carriers that enter the cycle and then yields carbon dioxide as a waste product.**

Glycolysis nets 2 ATPs per glucose molecule processed. The Krebs cycle produces an additional 2 ATPs. However, through the Krebs cycle, the cell has now captured 12 electron carrier molecules (e.g., NADH). Those electron carrier molecules now contain the vast majority of energy extracted from the glucose molecule. But, again, while the energy contained in the ATP molecules is available for immediate use by the cell, the energy contained in the electron carrier molecules is not. The release of that energy is the story of the **electron transport chain**.

STEP TWO – The electron transport chain is a series of redox reactions that **releases the energy contained in** the electron carrier molecules (e.g., **NADH) in a step-down fashion.** Most of the molecules in the electron transport chain are proteins that are specialists at accepting and passing on electrons in redox reactions. **Each compound in the chain is slightly more electro-negative than the previous compound. In this fashion, an electron that has entered the chain is 'energetically pulled downward' losing a portion of its energy at each step in the chain.**

At the bottom of the chain the electrons are passed on to molecules of oxygen. Remember, oxygen is an electron hog. Oxygen serves to 'pull them in' because of its extreme electro-negativity. It is the ultimate electron receptor. Oxygen seems to drive the movements of electrons down the electron transport chain.

Then, as oxygen receives electrons it combines with the hydrogen ions (naked protons after hydrogen has lost its electron through a redox reaction) to form water.

The energy stored in the captured electrons in the electron transport chain can now be released in a form that can do work in the cell. This is called **oxidative phosphorylation.** The process of oxidative phosphorylation links the electron transport chain to the production of the ubiquitous energy molecule called ATP. Here's how.

Oxidative Phosphorylation

The energy once stored in the electrons in the electron transport chain is now used to pump hydrogen ions (hydrogen atoms that have lost their electron - a naked proton) to create a strong concentration gradient. **The concentration gradient itself becomes a form of stored energy.** And, energy stored in the concentration gradient, like any stored energy, is energy that can be used to do work. **This energy is then used to add a phosphate group to ADP thereby producing ATP.** Where all those molecules of ADP come from in the first place is a recycling story that we will get to in just a while. But, first, here is how a phosphate group is added to ADP to produce an energy molecule of ATP.

As the concentration of hydrogen ions increases, the rate of ATP production increases. This is done through a highly complex enzyme, called ATP synthase, that acts as an ion channel across the inner membrane of the mitochondria that is selective for hydrogen ions. Remember, these ions carry a positive electric charge and therefore cannot cross a cell membrane.

The numerous protein sub-units of ATP synthase are grouped into 3 distinct parts:

- a rotor that traverses the inner membrane of the mitochondria,
- a knob on the other side of the inner-membrane, and
- a rod that connects the rotor to the knob.

Apparently the movement of hydrogen ions through the rotor makes both the rod and rotor spin. And, this spinning motion (estimated at 3,000 rpms) somehow transfers energy that is then used to form a new bond between ADP and an inorganic phosphate group to form a molecule of ATP.

Living cells have learned how to break down the energy in glucose, into useable energy molecules called ATP, through a controlled step-by-step process, with each step being catalyzed by a different enzyme. And, they have learned how to harvest that energy in a form that is usable by virtually all living organisms - ATP, the ubiquitous energy molecule.

Living cells have learned how to harvest useable energy in a highly productive fashion. While only 2 molecules of ATP are harvested through glycolysis, 36 ATPs are harvested through cellular respiration. **From one molecule of glucose we extract 38 usable molecules of ATP.** That is an efficiency rate of 40%. By comparison, the brilliant scientists employed by automobile manufacturers have been able to develop a car engine that is only about 25% efficient.

Lastly, the metabolic pathways of cellular respiration represent a kind-of crossroads in metabolism. Metabolism is a series of interconnecting pathways that connect all of the biochemical reactions of the cell - literally thousands. The pathways of cellular respiration are at the very center of the metabolic system in the cell. So, **cellular respiration does not need to use only glucose to power the process. Cellular respiration can accept a wide array of fuels that we eat. The other high-energy organic molecules of sugars, other carbohydrates, fats, and proteins simply enter the metabolic pathways through different portals and then proceed apace to provide us with the energy we need for life and living**. Amazing.

We have seen how all of the catabolic reactions (reactions running downhill to release energy from high-energy organic molecules) require a little activation energy, a little ATP, to get started and to then generate more ATP. But, **anabolic reactions** (reactions that run uphill) **require a great deal of energy**. Let's take a look at how that need is met by cellular respiration as a large-scale energy recycling factory.

The Cell's Energy Recycling Factory

Anabolic reactions go from simple to complex. They run uphill. They are the reactions needed to convert nucleotides into RNA and DNA and to convert amino acids into those amazing proteins. And, anabolic reactions always require energy. ATP is the ubiquitous energy molecule that powers the anabolic reactions of all living cells.

As we have seen, glucose (or other high-energy organic molecules) stores energy and serves as a cellular fuel for making ATP. ATP is then produced through glycolysis and cellular respiration.

Energy is used by a cell through a process called energy coupling. One way in which energy coupling occurs is when the energy stored in ATP is coupled (by an enzyme) to the reaction being powered. In fact, **anything in the cell that involves work requires energy that is released in this way.**

When a cell makes RNA or DNA out of a series of nucleotides, or when the cell combines long strings of amino acids to make proteins, these actions are anabolic reactions powered by ATP doing chemical work. **ATP supplies energy to do this chemical work by phosphorylating (donating a phosphate group to) one of the substrates produced by an enzyme that catalyzes the reaction.**

The movements produced by muscular reactions, like those that allow my fingers to type this sentence or that allow my feet to dance (kind-of), are powered by ATP again donating a phosphate group (phosphorylating) to the protein myosin. This phosphorylation does not catalyze an anabolic reaction. Rather, it changes the shape of the molecule thereby allowing for mechanical function.

To be clear, ATP powers all work that the body's cells perform. Whenever the ubiquitous energy molecule called ATP releases its energy to do work in a bodily cell, it does so by donating a phosphate group. And, **after ATP donates a phosphate group ATP is transformed into a molecule of ADP, containing no energy**.

Then, through the wondrous process of cellular respiration that we have just examined, those molecules of ADP are recycled and recharged through the addition of a phosphate group (called **oxidative phosphorylation**) to become an abundant source of new ATP energy molecules. And so on, and so on.

Without this elegant recycling process, we would never get the immense amount of energy that we need for living by simply eating food. We use a lot of ATP. A typical muscle cell may consume **ten million molecules of ATP each second**. And, on an average run-of-the-mill, non-athletic day, **we consume our body weight in ATP**. For me that's 190 pounds of ATP each and every day. Although I eat a lot, I certainly don't eat 190 pounds of anything each day. So, the body's energy needs are primarily met through the recycling factory of cellular respiration. The wonder of it all.

Wrap-up for Cellular Wonders

In the last three chapters we have taken a brief glimpse into the world of living cells. We have examined, on the cellular level, the basics of:

- the 'stuff' of life,
- the 'bonds' of life, and
- the 'power' of life.

Let's reflect a moment on the magnitude of the cellular world that we have been exploring.

Plants and animals, as complete living organisms, contain an almost incomprehensible number of living cells. And, each and every one of those cells is engaging in the wonders that we have covered in the last three chapters.

Consider this perspective. Your body and my body each contains about **100 trillion cells**. While the size of the many different kinds of human cells varies widely, an average size is estimated to be about 50 micrometers (0.002 inches) in length. If we travel around planet Earth at the equator we would cover about 25,000 miles. In terms of inches, that is about **1.6 billion inches** or roughly 800 billion cell-lengths. So, for a rough estimate, **if we laid all the cells in our body end-to-end they would circumnavigate the Earth over 100 times**. And, forget not that each of those trillions of cells is doing all of this intricate and exquisite stuff all the time.

Ultimately, living cells simply make up living organisms. Over the course of 3 and ½ billion years of life on this planet, living organisms have evolved to become absolutely amazing creatures doing absolutely amazing things.

Let's turn now to Part IV and begin to examine the diversity of life on planet Earth and the mechanisms whereby life became so diverse.

PART IV

DIVERSIFYING LIFE WITHOUT GOD
- A CLOSER LOOK AT 'ACCIDENTAL' NECESSITIES
AND THE AMAZING RESULTS -

At this point we have now examined the features of the universe that had to begin and evolve by accident in order to produce a life-friendly world, without God. And, we have examined the necessities of life that had to occur by accident in order for living things to exist at the most fundamental levels, without God.

In Part IV we will now examine the diversity of life. We will first review the process by which such diversity has occurred - evolution. We will then look at the scope of such diversity and view with amazement the wide array of living organisms that most scientists believe have occurred accidentally, without God.

CHAPTER 13
Evolution of Living Things By Natural Selection

For billions of years of time following the 'Big Bang' matter and energy 'learned' how to combine into more and more complex forms with more and more specificity until the stage became set for **life** on Earth. Molecules proximate to each other began to engage in complex chemical reactions. First the **inorganic compounds learned how to become organic compounds, like nucleotides and amino acids. Then** somehow, and we don't know how, those **organic compounds learned how to become infused with life**. This is the mainstream scientific explanation for the beginning of life. It is called **emergent evolution**.

The things of the universe 'learned' how to become more and more complex until they 'learned' how to become infused with life. We may not know how they did that. But we do know that life happened. And, with the advent of the first life on Earth, wondrous advancement followed.

The **first life on Earth** appeared about 3½ billion years ago. That first life took the form of a type of **single-celled bacteria**. Those first living organisms had no cell nucleus. They learned how to reproduce and multiply by, in essence, cloning themselves through the process of simple cell division that we call **mitosis**.

By about 3 billion years ago the single-celled bacteria, that is still found today in stromatolite formations, had learned how to capture the Sun's energy through the process of **photosynthesis**, whereby they convert the energy of sunlight into useful energy to sustain life.

By about 1½ billion years ago single-celled organisms learned how to **develop a cell nucleus**. Today we call those organisms **yeast**. The nucleus of the yeast cell, like the nucleus of all plant cells and all animal cells, confines the organism's **genes in DNA tightly-wound on chromosomes.**

During the next billion years, single-celled organisms with a nucleus learned **how to differentiate cells**. They became **multi-celled organisms**. In these organisms, different cells would now perform different functions and the **different cells**, performing different functions, had learned how to communicate with each other. Living organisms now had **multiple parts**. And, the parts now **communicated and worked together**.

That was pretty-much the state of living things on Earth until about 550 to 600 million years ago. Those simple organisms then proceeded apace to learn how to become more and more complex. Multi-celled organisms learned how to amass like-cells into tissues. Then they learned how to combine various tissues into organs. And, they learned how to tie all the organs together through vast inter-connected metabolic pathways and to thereby evolve more and more and more complex living creatures. They began to thrive on dry land. Living organisms learned how to reproduce sexually. 'Learning' occurred through the process called 'natural selection'. That is the mainstream scientific explanation for the advancement of living organisms. It is called the **theory of evolution**. According to the modern theory, that is how life begins, proceeds and advances.

As of today, at this point in the evolution of life, we humans seem to be the most advanced living organisms that have developed through this wondrous process of increasing complexity begetting ever more increasing complexity. And we, like all other matter and energy, remain students of the universe.

From Natural Theology to Natural Selection

As with many things modern, the roots of evolutionary theory are found in ancient Greece. Democritus and Epicurus, members of the philosophical school of atomists, pronounced that there was a **continuity in nature and that changes were random. Those are the core components of modern evolutionary theory.** But, in the march of time Aristotelian teleology became the prevailing viewpoint of Christian Europe.

In the Greek language, *telos* (end) combines with *logos* (reason) to make *teleology* (the study of end reason or purpose). Teleology is firmly based on the idea that the universe and the things of the universe are designed for a purpose.

For Aristotle, teleology was the study of final causes. The final cause is the reason for which a natural phenomenon was created. Aristotle believed that natural things developed in order to realize the end that is internal to their own nature.

Following in the footsteps of Isaac Newton, scientists of the 17th and 18th centuries devoted themselves to finding mechanistic explanations for natural phenomena that operated within a 'Clockwork Universe'. All natural phenomena, both physical and biological, were viewed as machines designed by an intelligent agent. The intelligent agent was called God and the prevailing scientific belief was **natural theology**.

In the late 18th and early 19th centuries an English naturalist and clergyman, William Paley, published his teleological arguments for the existence of God in such writings as *A View of the Evidence of Christianity* and *Natural Theology*. Paley's writings were required reading for undergraduate students at the University of Cambridge. Charles Darwin was such a student.

Natural theology proclaimed that the nature of God can be best understood by studying the natural world that he created. Paley maintained that God's design was evident in all living things of nature. In *Natural Theology* he used a watch as an analogue for a living organism. He argued that logic compelled the conclusion that both must have been designed, and design necessarily requires an intelligent agent. As he stated in this abridged quote:

"…when we come to inspect the watch, we perceive that its several parts are framed and put together for a purpose…the inference we think is inevitable, that the watch must have had a maker; that there must have existed at some time and at some place or other, an artificer or artificers who formed it for the purpose which we find it actually to answer; who comprehended its construction and designed its use."

Paley contended that the only way to account for the vast variations and adaptations of living organisms on the planet, each of which is far more complicated than a watch, is to recognize that each was created by an intelligent designer.

As Paley observed the world around him he saw harmony and happiness everywhere.

"It is a happy world after all. The air, the earth, the water, teem with delighted existence. In a spring noon, or a summer evening, on whatever side I turn my eyes, myriads of happy beings crowd upon my view."

Thomas Malthus, a British economist, saw things in sharp contrast to Paley. He observed the world of the British poor and saw them living in wretched conditions. Their lot was one of human despair. Malthus pronounced that human population would grow exponentially (2,4,8,16) while the food supply would grow only arithmetically (2,3,4,5). He predicted that the growth of population would quickly outpace the food supply. Population could only be kept in balance with food supply if human population is subjected to plagues, starvation, and infanticide. The weak will die. The strong will survive. Those people in the human population who can best adapt to their change in fortune will be those who survive to raise their children. Others will die of starvation and disease. Only the fittest will survive. (Note: World population in 1900 was 1.8 billion; in 1950 it was 2.1 billion; in 2000 it was 6.0 billion; in 2050 it is projected to be 9.2 billion.)

By the 19th century most scientists had abandoned the Aristotelian view that species of living organisms are immutable, never change, and exist forever. The new sciences of geology and paleontology were revealing a 'fossil record'. That record contained evidence of past extinctions of great beasts who obviously did not roam the Earth today. And, anatomists were amassing a great deal of data that revealed amazing similarities among many species, including the species Homo sapiens. The evidence had become compelling that living organisms had evolved. The question became: 'What was the **mechanism** that guided the evolution of life?'

Most scientists then believed in a vitalistic evolutionary theory proposed by a Frenchman, Jean-Baptiste Lamarck. He used the example of the evolution of a giraffe's long neck. He theorized that a giraffe really strived (worked and worked) to stretch his neck to reach the desired leaves residing on a higher branch. Because of this striving the giraffe's offspring would inherit a slightly longer neck. As this striving was passed on generation to generation the final result was the greatly elongated neck of the modern giraffe. The crux of such a theory of evolution was that evolutionary change over time was directed in accordance with the requirements of the organism evidenced by its striving. In Lamarckian evolution the mechanism for change over time is a **vital inner drive.**

In the mid-19th century two Englishmen concluded that the real mechanism that accounted for the evolution of living organisms over time was not some vital inner drive. As they independently developed their theory of an evolutionary mechanism, they were each much influenced by the writings on populations by Thomas Malthus. Malthus had developed his ideas by looking at humans not as individuals but as groups of individuals and introduced the concept of population dynamics. And he had introduced the concept that would become ever-after associated with evolution - survival of the fittest. This British economist had provided both Russel Wallace and Charles Darwin with the mechanism of evolution that has become the centered heart of modern biology - **natural selection**. Charles Darwin published his book on natural selection before Wallace. Thus, Darwin today is a household name while Wallace is little known.

Darwin's Theory of Evolution by Natural Selection

In 1859 Charles Darwin, an Englishman, published his famous book supporting modern evolutionary theory. It is called *On the Origin of Species by Means of Natural Selection or the Preservation of Favoured Races in the Struggle for Life*. Today he is widely recognized as the 'father of evolution'.

In his seminal work, Darwin began by taking great pains to describe his theory. At core **Darwin's theory of evolution** is this. Living organisms inherit traits from their parents that may or may not help them in their struggle for survival. Those traits that prove to be valuable to a species will be retained in the gene pool of the species because those individuals possessing the trait will be the most likely to reproduce and pass it on. In essence, **heritable traits that prove valuable in the struggle for the survival of the species will be retained.** Much better put, in Darwin's own words:

"Owing to this struggle for life, any variation, however slight and from whatever cause proceeding, if it be in any degree profitable to an individual of any species, in its infinitely complex relations to other organic beings and to external nature, will tend to the preservation of that individual, and will generally be inherited by its offspring. The offspring, also, will thus have a better chance of surviving, for, of the many individuals of any species which are periodically born, but a small number can survive. I have called this principle, by which each slight variation, if useful, is preserved, by the term of Natural Selection, in order to mark its relation to man's power of selection."

By Darwin's lifetime, man had been selectively breeding plants and domestic animals for literally thousands of years. Man was the intelligent agent who 'selected' the new and improved plant or animal by selective breeding. Darwin proposed that the same type of thing was going on with non-domesticated plants and animals all the time. Yet, no intelligent agent was required. The mechanism of 'natural' selection was one of random mutations.

Darwinian evolution is descent with modification by natural selection over evolutionary time. It is **natural selection working on random mutations** (i.e., genetic modifications due to mistakes or accidents). The modifications selected from random mutations are not selected by some vital inner-drive as suggested by Lamarck. **The selected modifications happen purely by chance and are retained in the gene pool only if they prove useful for survival of the species.**

With the unfolding of the science of genetics in the 20th century, evolutionary change mechanisms have been expanded to include such things as genetic drift and gene flow, as well as the natural selection of mutations. However, the only mechanism that can result in **adaptive modifications** over time is still believed to be **random mutation**. The modern mainstream evolutionary belief is succinctly stated by Harvard University Professor, Ernst Mayr, one of the founders of the modern synthesis of evolutionary theory:

"Ultimately, all variation is, of course, due to mutation."

Darwinian evolution, and the modern mainstream evolutionary belief, are all about adaptive traits. The contention is that the evolution of adaptive traits by random mutations enhances the survivability of individuals within a species population who inherit those traits. And, on a large-enough scale, the evolution of adaptive traits by random mutations may result in the development of a new species. For Darwinian evolution and the modern mainstream evolutionary belief, **random mutation is the key**.

Post-Darwin Discoveries in Genetics and Cellular Biology

Genetics, broadly speaking, is the branch of biology that deals with heredity and variation of living organisms. More narrowly, in common parlance today, it is the study of heredity linked to genetic makeup. Cellular biology is the study of the composition of living cells. When Charles Darwin shocked the world with his theory of evolution by natural selection he and his scientific colleagues were wholly ignorant of genetics and cellular biology. Darwin believed that 'pangenes' from each and every part of the body came together to form the egg and the semen that fertilized it. And, Darwin thought that living cells were filled with a kind-of mushy protoplasm.

Genetics

Variation in living organisms had been observed since first there were observers. And, since the beginning of recorded history people had been selectively breeding domesticated animals and edible plants. A speedy champion mare was mated with a speedy champion stallion to produce a new horse race champion three years later in an ancient analogue of today's Kentucky Derby. The 'trait' of swiftness was inherited by the colt. In early agricultural efforts, farmers cross-pollinated plants with the 'traits' of disease-resistance and hardiness and thereby produced more abundant and healthier crops for the following year's harvest. It was common knowledge that 'heritable traits' were passed on to offspring through breeding.

For thousands of years scientists believed that the factors responsible for trait inheritance simply blended together as a result of sexual reproduction. But, if such blending occurred then, over time, the distinct variations that we see in species of living things would vanish. After all, if you pour and stir two different buckets of paint together, eventually all of the blended paint will be identical. That clearly does not happen in living organisms. Quite a conundrum for science. How then is variation maintained in natural species? Answers needed to be found.

Gregor Mendel was an Austrian monk who lived in the mid-19th century. He was greatly interested in solving this conundrum. He wanted to find the secret of how variation is naturally maintained in living populations. So, he started to investigate and experiment. The subjects he chose for experimentation were pea plants. And, he experimented a lot, using more that 28,000 pea plants. He made 287 crosses between 70 different purebred pea plants. While in his lifetime no one knew what a gene was, his experimental results, records and theories describing 'Mendelian traits' accord him the niche in history as the 'father of genetics'.

In carefully controlled experiments, Mendel tested the 'blending hypothesis' in peas. He crossed true-breeding purple-flowered pea plants with true-breeding white-flowered pea plants. His results disproved the blending theory. The offspring plants did not have blended whitish-purple flowers. All offspring of the

1st generation had purple flowers. In the 2nd generation some offspring had white flowers and some had purple flowers, but none had flowers of a blended color. The white flower trait that had apparently disappeared in the 1st generation had somehow reappeared in the 2nd. How could that be?

Further experimentation and documentation of generation after generation of pea plants led Mendel to develop a scientific hypothesis that accurately described the working of genes and chromosomes long before genes were ever discovered. Mendel concluded that:

- each offspring must have two 'heritable factors' (known today as **alleles**) for each 'heritable trait' (known today as **a gene**); and
- one 'heritable factor' (allele) is dominant for each 'heritable trait' (gene) and the other is recessive. Today we refer to **dominant and recessive genes** for things like hair color and eye color in children.

As Mendel continued to experiment, he crossed plants with multiple differences, like both seed color and flower color. The results of his experiments led to his conclusion that, in today's language, says that **alleles of different genes segregate independently of each other during the formation of gametes (sex cells).** This is known as the **Law of Independent Assortment**, and it is a very important part of natural genetic variation. Independent assortment is, in essence, the **natural shuffling of alleles that leads to variation** '*within species*'.

Cellular Biology

In the 17th century Anton van Leeuwenhoek of Holland developed the first practical microscope. Through it he observed bacteria and the circulation of blood in capillaries. A primitive microscope first revealed to Robert Hooke, an English physicist, small cubicles, or rooms, in dead bark cork. He called them cells. Cellular biology thus began at that early date. By the later part of the 19th century improvements in the microscope instrument allowed biologists to observe the nuclei within living cells. Walter Sutton and Theodor Boveri observed that the nuclei of cells split in half during the formation of gamete sex cells. By the early part of the 20th century they published their theory that each gene has a specific location on a chromosome in the cell nucleus.

As the 20th century dawned, Thomas Hunt Morgan began to investigate how mutations occurred in living organisms and **discovered that most**

mutations make alleles dysfunctional. Morgan was an American zoologist who experimented a lot with fruit flies. Through his experiments he discovered the linkage of chromosomes in cells and the phenomenon of 'crossing over' that occurs when two chromosomes of a homologous pair exchange equal segments with each other at the first part of meiosis (sex cell division). He mapped the genes on the chromosomes of the fruit fly. In short, he **discovered that genes are responsible for heritable traits and that genes are linked in a series on chromosomes.**

In the 1930's a Russian geneticist, Theodosius Dobzhansky, worked in the laboratory of Thomas Hunt Morgan. He studied the genetics of populations by examining the differences that emerged between different isolated populations of the same species. He discovered that all members of a specific species do not have identical genes. He believed that what kept a species distinct was simply sex. Members of the same species could mate and have offspring. Organisms outside the species could not. In 1937 he published *Genetics and the Origin of Species* in which he explained that mutations occur in the alleles of genes all the time. And, most importantly, he observed that:

"the process of mutation is the only source of raw materials of genetic variability and hence of evolution."

By the middle of the 20th century much had been observed and learned about the nature of proteins at work in living cells. By then it was known that cell chromosomes consisted of both proteins and DNA. Most scientists believed that proteins held the key to understanding the chemical basis of genetics, and were responsible for passing on genetic information through cell reproduction. Oswald Avery then **discovered that DNA, not proteins, was the carrier of genetic information.** His discovery prodded Francis Crick and James Watson to further investigate the mysteries of DNA.

In 1953 they used X-ray diffraction to discover the structure of the DNA molecule. Their three dimensional model of DNA shows the coiling of two sides of a genetic ladder into a double helix. The sides of the ladder (backbone) repeat a phosphate-sugar bond over and over. The rungs of the ladder comprise the variable part of the molecule. The sequencing of the base pairs of nucleic acids along the rungs of the ladder provides the genetic information. They further discovered that the exact sequencing of the four bases in DNA and in RNA served to provide construction instructions for building functional proteins out of 20 discrete amino acids. In short, they broke the genetic code. Crick and Watson were awarded the Nobel Prize in Medicine in 1962.

The Modern Synthesis of Evolutionary Theory

When the brilliant life scientists today talk about evolution it seems almost like they are trying to make the matter confusing. The terms are often used in different contexts and are used to describe different aspects of natural phenomena in seemingly contradictory ways. The problem lies not with these brilliant scientists. The problem results from the immense amount of knowledge that they have gained about how living things operate and change. It is really hard to explain all this stuff in a simple manner.

Darwin's theory of evolution by natural selection has been expanded to today include not only random mutations, but such things as 'genetic variation', 'random genetic drift', 'gene flow', 'speciation by reproductive isolation', and 'punctuated equilibrium'.

Darwin's idea was that all organisms that have ever lived have evolved from a common ancestor. Virtually all biologists agree with that today. Darwin believed that the mechanism for evolutionary change was natural selection through random mutations. Today, biologists have expanded the mechanism. In 1986 D.J. Futuyama, a renowned biologist, published *Evolutionary Biology* which includes this synopsis of the modern synthesis of evolutionary theory:

"The major tenets of the evolutionary synthesis, then, were that populations contain genetic variation that arises by random (i.e., not adaptively directed) mutation and recombination; that populations evolve by changes in gene frequency brought about by random genetic drift, gene flow, and especially natural selection; that most adaptive genetic variants have individually slight phenotypic effects so that phenotypic changes are gradual...that diversification comes about by speciation, which normally entails the gradual evolution of reproductive isolation among populations; and that these processes, continued for sufficiently long, give rise to changes of such great magnitude as to warrant the designation of higher taxonomic levels (genera, families, and so forth)."

It is most obvious that modern evolutionary biology has discovered a great deal about the evolution of living things since the days of Charles Darwin. But, most importantly, the modern mainstream evolutionary belief maintains that the evolution of all adaptive change is the result of random mutations of genetic DNA. These scientists believe that random mutations of genetic DNA produce all adaptive traits which result in either:

- increased survivability within a species population, or
- on a large-enough scale, in the development of a new species.

All adaptive change is, therefore, the result of random mutation that occurs in one of two ways:

- **a mistake during DNA replication in a sex cell, or**
- **an alteration of genetic DNA in response to some radioactivity or cosmic rays or some poison in the environment.**

In short, the very foundation of the modern mainstream evolutionary belief is **random mutation caused by an accident**.

An Attempt to Gain Understanding Through Simplification and Clarity

To me, much of the confusion regarding evolution results from commingling the examination of two entirely different aspects of living things:

- the evolution of the completed living organism; and
- the evolution of the component parts of the completed living organism (the complex organs and metabolic pathway systems).

It is not too difficult to follow the theory of evolution by natural selection presented by Charles Darwin as he discusses the change over time of completed living organisms. It makes incredibly good sense that those organisms who possess the heritable traits (traits passed-on to offspring through reproduction) that allow them to better-survive and produce more offspring will end up blessing the entire species population with those traits. Explanations regarding the swiftness of wolves and cheetahs, the size of birds' beaks, and the coloration of peas or moths' wings are pretty easy to follow. Darwin does not know the actual mechanism that would result in such traits being produced, but he is quite sure that the mechanism is random. Mutations somehow develop to meet the challenges of the environment that the organism lives in. The mutations are random. Those mutations that result in a heritable trait that proves valuable for species survival are retained in the species population. Now the hard part.

Darwin then extended his theory of natural selection by random mutations to include the evolution of each and every part of each and every living organism on Earth. The theory proposes that the evolution of every component part within each and every living organism came about through the same natural selection of random mutations. The same reasoning regarding 'adaptive change' and retention of favorable 'adaptive traits' that is applied to the completed organism is now applied to all of the component parts of the organism. So natural selection of random mutations is pronounced to be the mechanism for the evolutionary development over time of complex organs like:

- hearts
- lungs
- gills
- blood
- roots
- stems
- shelled egg
- mammal placenta
- kidneys
- liver
- eyes
- ears
- skin
- pancreas
- spleen
- stomach
- intestines, to name but a few;
  and also complex internal systems like:
- blood circulatory system in mammals
- xylem and phloem circulatory systems in plants

- hormonal regulatory systems in plants and animals
- blood clotting system
- immune systems in plants and animals;
  and finally the organs that make us uniquely human:
- a larynx and epiglottis that give us the power of speech
- a human brain that gives us the power of reflective thought.

Evolution by natural selection is solidly founded on the challenges presented by the environment.

Applying the same evolutionary process that affects the completed living organism to each and every component part of the completed living organism presents this first conundrum: the **internal organs and metabolic pathways operate in an internal environment that is entirely different from the external environment that presents challenges to be met by the completed living organism.** The cellular structures that make up these organs and pathways operate within a fluid milieu that is far removed from the external environment and is regulated to maintain a proper balance between chemical bases and acids. Elite and mainstream scientists seem to simply ignore this fundamental difference.

A second conundrum results from the modern scientific discoveries of protein structure and RNA and DNA genetic development. Scientists now acknowledge that all of the internal physiological development of living organisms is directed by the information reposing in the DNA of each and every living cell of the organism. Simply stated: How does the cellular information reposing in the nucleus of each living cell sample the external environment and then make the necessary modifications to deal successfully with the environmental challenges presented to the completed living organism?

Again and again, the elite and mainstream scientific answer to these conundrums is the same. These scientists maintain that all is explained simply as the result of undirected random mutations of genetic DNA.

Read again the partial listing of the complex internal organs and metabolic pathways that these scientists maintain developed over time by the process of undirected random mutations. **These scientists actually believe that each of these complex organs and metabolic pathways evolved over time through the same step-by-step process of natural selection of random**

mutations. You may be surprised to discover that **there is absolutely no scientific evidence in support of that belief.** That discovery surprised me as well.

Ponder that thought for a while as we proceed to further examine the science of evolution and the hypothesis of natural selection by random mutations.

The Science of Evolution

Darwin's earth-shaking book was published twenty years after his five-year ocean voyage. His journey took him to the far reaches of the planet and provided him with a naturalist's dream laboratory to study exotic creatures in their native habitats. During his voyage aboard the *HMS Beagle* he observed and cataloged wondrous features about the nature of life.

When he visited the Galapagos Islands off the Ecuadorian coast in 1835 he collected 31 different species of finches from three separate islands. Each type of these birds was strikingly similar to each other except for the size of their beaks. Darwin hypothesized that each beak had been modified through natural selection to achieve different ends necessary for survival. The species of finches with larger beaks have adapted to a primary food source of hard seeds that need to be cracked open. Other species with narrower beaks are suited for eating insects as their primary food source. Still other species have evolved beaks that are matched to the primary food source of soft fruit for one and cactus for another. **Each species had evolved and adopted the appropriate beak to comport with the evolutionary challenge of food supply.** The finches on the island with the hard-shelled seeds that had inherited the trait of a larger and stronger beak had a greater chance to survive and pass on that trait to their offspring who would retain and pass-on that favorable trait, and so on, and so on. He concluded that such heritable traits that prove valuable in the struggle for the survival of the species will be retained. In Darwin's own words:

"Seeing this gradation and diversity of structure in one small, intimately related group of birds, one might really fancy that from an original paucity of birds in this archipelago, one species has been taken and modified for different ends."

Darwin's observations concerning fairly trivial changes, such as beak size or a bird's coloration or plumage, is the stuff of what is termed **micro-evolution.**

But Darwin did not stop there. Not by a long shot. No, he then proceeded to expand such observable and plausible explanations of micro-evolution by natural selection and proposed a grand extrapolation. Charles Darwin was not shy about extending his theory of evolution by common descent based on natural selection. Indeed, in *The Origin of Species* he included this far-reaching extrapolation:

> "Throughout whole classes various structures are formed on the same pattern, and at an embryonic age the species closely resemble each other. Therefore **I cannot doubt** that the theory of descent with modification embraces all the members of the same class. I believe that animals have descended from at most only four or five progenitors, and plants from an equal or lessor number. Analogy would lead me one step further, namely, to **the belief that all animals and plants have descended from some one prototype....Therefore I should infer from analogy that probably all the organic beings which have ever lived on this earth have descended from some one primordial form, into which life was first breathed."** (emphasis added)

This Darwinian **macro-evolution hypothesis** extolled that **all living things had evolved from a common ancestor.** Such a hypothesis he contended was supported by the fossil record and by the homology of living organisms.

Is Darwin's Theory of Evolution by Natural Selection scientific? The answer is certainly yes, based on scientific evidence. Let's take a look at some of the evidence.

Scientific Evidence of Evolution by Natural Selection

The Darwinian theory of micro-evolution and certain aspects of his theory of macro-evolution are solidly founded on good science.

Darwin's theory of evolution by natural selection was and remains brilliant. Aspects of his theory have been observed for heritable traits like the color of grouse, the swiftness of wolves and the stripes of zebras.

Experimental Tests of Evolution by Natural Selection

Aspects of Darwinian evolution have been tested by experiment. Let's review two of those.

(1) The English peppered moth is a species of moth that is found in and around the area of Manchester, England. Because butterfly and moth collections have been kept intact by enthusiasts for the last 200 years or so there is a **continuum of peppered moths in museum collections**. The continuum collection shows that before the 1850's most of these moths were light-colored with dark specs. A few were dark-colored. By 1900 the tables had turned. Most of these moths were dark-colored and but a few were light colored. In the 1950's H.B.D. Kettlewell, a physician and moth collector, proposed an answer to how the tables turned in terms of evolution by natural selection. In the 1850's the environment around Manchester changed quite dramatically. Soot from the smokestacks of new factories collected on the tree trunks where the peppered moths settle. Before that time the tree trunks were light-colored and provided the moths with camouflage as protection from bird predators. After the soot turned the tree trunks dark the light-colored moths became easy prey. The dark-colored moths were now camouflaged and thereby survived in greater numbers to produce more offspring. Natural selection in action. The result that the Darwinian hypothesis predicted.

Kettlewell **conducted his own experiment** by pinning an equal number of light-colored pepper moths to the trunks of trees in a forest of light-colored trees and also in a forest of dark-colored trees. The better-camouflaged moths evidenced a greater survivability rate. **The results of his experiment conformed with the hypothesis. The heritable trait of wing color in insects conformed to the theory of natural selection with variation.** The gene pool of peppered moths retained the favorable trait of dark coloring that aided survivability.

(2) Princeton University researchers Peter and Rosemary Grant have observed and analyzed populations of Darwin's finches in the Galapagos Islands for over thirty years. They concentrated on one small island with a species of medium ground finch having a population of less than 1,000. They observed that the size of the finch beaks corresponded to the size and hardness of seeds that they eat. When a drought killed 80% of the population they observed that those finches who survived had a larger beak size than those that died. The drought-

resistant plants produced large, hard seeds that required large beaks to open. The Grants have since observed an increase or decrease in beak size corresponding with drought years and wet years. Darwinian evolution in action documented by scientific observation.

Evolution by Natural Selection in the Fossil Record

Darwinian evolution in nature has been observed and documented in the fossil record.

The **fossil record** is largely incomplete. But it does provide compelling **evidence through radioisotope dating that there has most certainly been a progression of life through evolutionary time. The evidence shows that early eukaryotic creatures preceded fish; fish preceded amphibians; amphibians preceded reptiles; reptiles preceded mammals; pongid ape preceded hominid man.** This is strong circumstantial evidence in support of common descent.

The fossil record contains abundant data supporting the Darwinian evolution of anatomical structure. The term used by scientists for conformity of type is **homology**. And, homology is a foundational cornerstone of the macro-evolutionary theory. Homology shows the similarity of features found in different types of organisms and species. It evidences a fundamental plan of structure that has been inherited. For example, all vertebrate creatures are animals having a backbone. All are characterized by a vertical column. As these animals evolved over time, from fish to reptile to bird to human, they continued to contain a similar fundamental anatomical structure. Although these structures look very different in form and may perform very different functions, **the same basic bones are the same in porpoises and frogs and horses and bats and us. Evolution by natural selection is a highly convincing hypothesis to explain why these very different structures are composed of the same bones. This is compelling evidence of Darwinian evolution found in the fossil record.**

Evolutionary Evidence in Embryonic Development

Homology of embryonic development strongly supports Darwinian evolution as well. For example, in the very early stages of embryonic development there is a striking similarity in the appearance of the embryos of developing fish, reptiles, birds, and humans. And, **human embryos seem to have gill slits like a fish, then a long bony tail and then a fine fur**. We, of course, lose these features later in the embryonic process, but their existence in the early stages is **compelling circumstantial evidence of descent with modification - Darwinian evolution.**

In short, the scientific evidence supports Darwin's hypothesis that all living organisms likely came from a common ancestor. **Scientific circumstantial evidence clearly supports evolution through natural selection.** Evolution through natural selection is a scientific fact.

However, the explanation that evolution occurred through natural selection is a classic example of circular reasoning. **Evolution through natural selection simply tells us that the organisms that survived possessed the traits that made them best able to survive. The fact that those organisms that survived were best suited to survive gives us exactly zero information about how they became best suited to survive.**

The real issue is the mechanism that causes adaptive change to occur in natural selection. Darwin and the vast majority of scientists who support the modern evolutionary synthesis contend that the only mechanism that causes adaptive change to occur in natural selection is an accident. The accident is a genetic mutation.

Let's first look at what problems are encountered by that mechanism. Later, in Chapter 21, we will explore another mechanism that might be a more realistic foundation for natural selection in evolutionary adaptive change. But for now let's more closely examine the random mutation mechanism, for that is, after all, the centered heart of modern evolutionary theory.

Problems Encountered by the Random Mutation Hypothesis for Evolving Adaptive Traits

Mutations are random. Mutations are directionless. Mutations are mistakes. Mutations are accidents. **Most mutations are far from useful.**

Charles Darwin knew nothing about DNA and RNA or genetic sequencing. To Darwin the inside of a living cell was filled with a mushy protoplasm. Only through the painstaking efforts of brilliant scientists throughout the

20th century was Darwin's evolution by natural selection refined to include further scientific discoveries about the wonders of cellular structure and operation. Yet, today the mainstream scientific view is that adaptive evolutionary change is, at core, the result of pure accident. **Mistakes in DNA sequences are made through transcription errors in our germ (sex) cells, or those DNA sequences become altered due to an environmental pollutant or poison or cosmic radiation. The view of mainstream science is that these are the sole sources of genetic mutations that lead to adaptive evolutionary change**. Most importantly, the contention of mainstream science is that mutation changes are not directed. They are random.

Let's take a look at the magnitude of the issue involved with the **hypothesis that only random mutations are responsible for adaptive change in living organisms.** Let's first look at our human genome.

Our Human Genome

Our human genome is that of our species, Homo sapiens. For each of us the genome consists of 46 chromosomes (23 pairs) containing proteins and approximately 30,000 genes. In turn those genes are made-up of about 3 billion base pairs of DNA. Where did our genome come from?

We each inherited our genome of 46 chromosomes from our parents. The germ (sex) cell of our mother is the egg and contains a single chromosome strand containing 23 chromosomes. The chromosome that determines sex on mother's single strand is always designated as an X chromosome. The germ (sex) cell of our father is the sperm and contains a single chromosome strand containing 23 chromosomes. The chromosome that determines sex on father's single strand is either an X (female) or a Y (male). The father's sperm cell always determines the sex of the child. So, that is where we came from and how our sex was determined.

When the sperm fertilized the egg it transformed it into a new embryonic zygote cell containing 46 chromosomes (23 chromosome pairs – ½ from mother and ½ from father). Then over a period of 9 months that one cell divided and divided and divided and divided until a human embryo was formed and finally a baby was born. The baby grew, became a parent and, in turn, passes on the human genome to a child of the next generation.

Same Genome, Different Variations

Each of the 6+ billion people now living on planet Earth has a human genome of 23 pairs of chromosomes with 3 billion base pairs of DNA. Obviously, we are not all identical. **The great variation that exists in members of the species Homo sapiens today is the result of genetic recombinations of mothers' and fathers' DNA through sexual reproduction.** These recombinations by pairing one chromosome strand of the father with one chromosome strand of the mother provides the great variety that we see in people. **Yet, such sexual recombinations do not alter the sequence of the genes, and do not result in adaptive change.**

In order for an adaptive trait to evolve, the gene sequences in DNA must be altered. The alteration of gene sequences in DNA is the result of evolution by natural selection. The prevailing scientific belief is that the natural selection mechanism for the alteration of gene sequences in DNA is random mutations. The mutations that provide for the evolution of adaptive traits occur in the reproductive germ (sex) cells of the mother or father. Such mutations in the reproductive sex cells are the only mechanism of evolution that produces adaptive traits.

Evolution of Our Genome

Our parents inherited their genome from their parents and their parents from their parents, and so forth and so forth backwards in time generation before generation until the first members of the species of modern Homo sapiens evolved roughly 100,000 years ago.

Evidence from the fossil record and DNA analysis has determined that the Primate family split into two orders, Pongidae and Hominidae, about 5 million years ago. Gorillas and chimpanzees are pongids while we and our extinct ancestor relatives, like Homo erectus and Archaic homo sapiens, are hominids. The Archaic homo sapiens were our direct-line ancestors. Archaic homo sapiens apparently evolved into two separate species of Homo sapiens in the Near East about 100,000 years ago. One species was Homo sapiens Neanderthalensis. The other species was us, modern humans, known as Homo sapiens sapiens. The Neanderthals were much stronger than us and probably just as brainy. The two species lived in the same neighborhood for about 70,000 years, and then the Neanderthals became extinct about 30,000 years ago.

Shortly thereafter fully modern humans rapidly progressed toward the beginnings of civilization.

Hold that reference point in mind as we look a lot further back to trace our roots.

Prokaryote and Eucaryote

Evidence from the fossil record indicates that the first life on Earth occurred about 3½ billion years ago and took the form of a type of bacteria, a single-celled prokaryote. A single-celled organism without a cell nucleus is called a prokaryote. All bacteria are prokaryotes. An organism with a cell nucleus is called a eucaryote.

During the first two billion years of the history of life on Earth all living organisms remained single-celled and did not contain a nucleus. In that period single-celled bacteria grouped together to form large masses of blue-green algae and developed the process of photosynthesis that made the emergence of higher life forms possible. But, these bacteria never developed a nucleus. Yeast, the first single-celled living organism to evolve a nucleus was found on Earth about 1½ billion years ago. Yeast was the first eucaryote.

Multi-celled Organisms

In the next billion years following the evolution of single-celled organisms with a nucleus, simple multi-celled organisms began to evolve. The evolution of adaptive traits in living organisms was not rapid during this time. The first multi-celled organisms were invertebrate (had no backbone) and seem to have evolved through an aggregation (called a **flagellate**) of single-celled organisms. Scientists believe that the origin of virtually all the members of both the plant kingdom and the animal kingdom may be traced to such flagellates assembled into large groups. By this manner, cells comprising both plants and animals developed an interdependence and an elemental form of **symbiosis** (mutual dependence).

Up to about 540 million years ago the most advanced creatures to have evolved on Earth still lacked distinctive features, had no mouth or anus, no head or tail, and were not capable of self-locomotion. The most advanced form of life, the jellyfish, simply drifted along at the whim of ocean currents.

From Cambrian to Current Day

What scientists term the Cambrian Period explosion of life began about 540 million years ago and lasted for about 30 to 40 million years. During that time the basic morphological design for all living organisms to follow emerged. In the time from the Cambrian to the present very few new body plans have been added to the ones that apparently originated in the Cambrian Period.

The following is a very rough timeline for the evolution of species since the end of the Cambrian Period.

- 500 million years ago (mya)   first fish
- 450 mya   first land plants and fungi
- 390 mya   first amphibians, insects and reptiles
- 250 mya   1st great mass extinction (90% of all species)
- 220 mya   first dinosaurs
- 200 mya   first mammals and birds
- 130 mya   first flowering plants
- 65 mya   2nd great mass extinction (50% of all genera, including the dinosaurs)
- 50 mya   first primates
- 5 mya   first hominids
- 100,000 years ago   Neanderthals and Homo sapiens evolve separately
- 30,000 years ago   Neanderthals become extinct, only Homo sapiens remain
- 15,000 years ago   Homo sapiens (modern humans) begin agriculture

If indeed mutation is the mechanism for all adaptive evolutionary change, the timeline of history of life on Earth reveals that either:

- random mutation upon random mutation necessarily results in increased functional complexity occurring at an ever-accelerating rate (that seems to be a scientific hypothesis that is at odds with scientific evidence and common observation); or,
- the evolution of life as we know it today should not have occurred until billions and billions of years from now.

Two billion years was the time required for mistakes and accidents in DNA sequences to make the **one adaptive change** of providing a living cell with a nucleus. An **additional one billion years** of mistakes and accidents was then necessary in order **to accomplish the second adaptive change of making two-celled organisms.** Following that major accomplishment, over the next 600 million years the adaptive traits of all of the approximately 1½ million distinct species of plants and animals that have been discovered on Earth today were produced by the same process of mistakes and accidents in DNA sequences. Then between 100,000 and 50,000 years ago an enormous number of accidents and mistakes must have caused the DNA of our ancestors to quickly beneficially mutate in an exponential manner to provide us with the adaptive traits of both speech and reason. And, those adaptive traits set us apart from all other species who have ever lived on the planet.

Science defines evolution simply as the change in the gene pool of a species' population over time. We Homo sapiens, as a species, therefore, have to evolve over time since our gene pool constantly changes. Scientists almost universally agree that racial groups within our human species are biologically equivalent, that they possess the same adaptive traits. So, the **human evolution of '***adaptive biological traits***' must have stopped about 50,000 years ago**, before the races diverged. That is not to say that evolution in human beings does not occur. **We simply do not seem to any longer evolve adaptively.**

The modern mainstream evolutionary belief says that adaptation is a positive character trait that results in a living organism through a random change in its genetic makeup (DNA). In effect, adaptation is the result of an accident or mistake in the DNA sequences among our some 3 billion base pairs of DNA that just happens to be beneficial to us as a living organism. How can we explain the vast changes that occurred by random mutation of our DNA around 50,000 years ago that resulted in the unique adaptive traits of speech and reason? How can we explain the fact that no random mutations since then have resulted in any further adaptive traits? These are unanswered questions to ponder as we peruse the next five chapters and attempt to gain a fuller perspective of the wonders of the living world that have evolved through natural selection. Let's take a look.

CHAPTER 14
The Plant Kingdom

Modern organizational terms such as 'genus' and 'species' originated with Aristotle's biological classification scheme. Living organisms were arranged in accord with the 'scale of nature' from the simplest to the most complex. Each rung on the ladder of life was full and represented a perfectly formed 'type' that did not change. In the 18th century a German biologist, Carolus Linnaeus, expanded Aristotle's classifications into the seven-part structure that is widely used today in science. **'Kings play chess on fairly green spaces'** is the mnemonic for descending biological classifications from highest to lowest (kingdom, phylum, class, order, family, genus, and species). That may prove helpful as we proceed with our foray into the wide varieties of living things.

The first life on Earth was neither plant nor animal. It was pretty inglorious. The first life on Earth took the form of single-celled organisms without a nucleus. Our name for those organisms is **bacteria**. After those first bacteria (technically known as archbacteria) appeared about 3½ **billion** years ago, another ½ **billion** years elapsed before the next significant step of bacterial life - **photosynthesis** - occurred. It then took about another 1½ **billion** years for those single-celled creatures to develop a nucleus. And then another 1 **billion** years for those simple nucleated yeast cells to develop a more complex structure and to develop multiple-celled offshoot organisms whose cells could communicate with each other. We call those first complex organisms **plants.** Plants were the first living creatures to evolve as complex organisms.

Plants are wondrous creatures. Plants purify our air, provide our food, and protect us from the ravages of the atmosphere and the inner-Earth. Plants can exist in many forms without animals. But, no animal can exist without plants. More than 400,000 species of plants exist in our world today.

The first plants were creatures of the ocean sea. These first blue-green algae of the sea played a most significant role in evolving the Earth's atmosphere, through the process of photosynthesis. Although some 7,000 species of algae still thrive today, they never developed more than a few differentiated cells, and their fertilized eggs never advanced to develop into embryos. They grow from spores, not seeds. There are about 170,000 species of plants in the world today (like algae and mushrooms) that grow from spores instead of seeds.

About 470 million years ago plant life began to appear on the newly-dry, moist land of the Earth. Today we call these plants mosses. Mosses do produce a number of differentiated cells and their fertilized eggs do develop into embryos. Some 16,000 species of moss exist today, and they still thrive on dampness. But, the mosses never developed a vascular circulatory system and they never developed woody tissues.

The first vascular seed plants with woody tissue first appeared on Earth shortly after the appearance of moss and grew to a height of no more than six inches. But, the appearance of the land quickly changed. By the time that the first amphibian animals appeared on Earth, about 370 million years ago, the first vascular seed plants with woody tissue had grown to evolve into great forests with 100-foot-tall trees crowning the tropical swamps.

Vascular seed plants with woody tissue are by far the most abundant in our world of the 21st century. And, they are basically of two types:

- **gymnosperms** (like pines, spruces, and fir trees), and
- **angiosperms** (all other trees, shrubs, and flowering plants).

Gymnosperms form seeds in an exposed manner, usually in cones. Angiosperms form seeds in a protected manner, safe within their flowering fruits.

Angiosperms comprise about 240,000 species of plants today. They comprise most all of our food, most all of our hardwood forests, and all that grows in our lawns and flowering gardens. When we marvel at the beauty of flowers, and when we talk in general about plants and trees, we are usually referring to angiosperms. Such being the case, the remaining discussion in this primer will deal only with angiosperms.

Angiosperms are **flowering plants**. We are drawn to flowers principally by their beauty and fragrance, not dissimilar to the birds and the bees. However, we must keep in mind the true purpose of flowers in the plants we admire. The sole function of flowers is sexual reproduction. And, since all plants come into being through reproduction, let's take a brief look at how they perform this reproductive function that is critical to species' survival.

Sexual Reproduction in Flowering Plants

Just as in human sexual reproduction, flowering plants reproduce when a sperm fertilizes an egg to create a single-cell offspring called a **zygote**. However, plants do not directly engage in sex. A plant forms microspores bearing male characteristics that then produce sperm. The method of fertilization is quite a wonder.

Flowering water plants are fertilized when the sperm swims through the water to reach the egg. In flowering land plants, the sperm swims through the liquid-filled **pollen tubes** of the plant.

Plant Reproductive Organs and Reproductive Cycle

The male reproductive structures of a plant are called **stamens**. Each stamen consists of a stalk with an **anther** located at the tip of the stalk. **Pollen** develops within the anther as a fine granular substance. Each pollen grain contains two cells. One of the two cells divides in due course to form **sperm cells**.

The female reproductive structures of a plant are called pistils. Each pistil consists of three sections:

- a sticky surface area to which pollen adheres, called the **stigma**;
- a long **style** that holds up the stigma into a proper position to collect pollen; and
- an **ovary** at the base of the pistil (the ovary contains the eggs called **ovules**).

When pollen adheres to the flower's stigma, the reproductive cycle begins. First, one of the two cells within the pollen grain proceeds to grow into a long pollen tube that extends through the pistil and finds a tiny opening in an ovule contained within the ovary at its base. The second cell within the pollen grain then divides to become two sperm that move down the pollen tube and enter the microscopic opening in one of the undeveloped seeds, called ovules, located in the ovary. The egg within the ovary now becomes fertilized by one of the sperm. In multiple-seeded plants the process happens not once, but many times as other pollen grains perform the same feat of growing other pollen tubes that precisely locate other ovules to fertilize within the ovary. The second sperm that traverses the pollen tube combines with another cell in the ovule to create a food-storage tissue, called the **endosperm**. The endosperm provides nourishment to the new zygote as it grows into an embryo encased in a seed. Fertilized ovules mature inside seeds which are, in turn, protected within the ovary. The ovary then grows to become a fruit.

Embryo and seed growth seem to stimulate the production of growth hormones in the plant that cause the fruit to grow. While we find some mature plant fruits delicious to eat, the actual function of the fruit is to protect the seeds until they are ready to be dispersed.

But, we are getting ahead of our reproduction story. Before we take a look at plant seed dispersal, let's return to take a closer look at just how that pollen got stuck on the flower's stamen to start the reproductive process. For flowering plants, this act of **pollination** occurs only through the wonder of symbiotic relations between plants and animals. Each needs the other.

Pollination

Remember, the sole function of flowers is sexual reproduction.

As a flower blossoms, the beauty that unfolds serves as an attractant to the literal 'birds and bees'. Invitation is extended to visitors from the animal kingdom. Different colors attract different visitors. Bees are attracted to blues and hummingbirds are attracted to red, for example. And, other birds, and insects, and butterflies, and bats, and wasps, and flies, are attracted by other colors and scents and flower shapes. Because of their unique shape, some flowers are only able to be pollinated by a specific species of insect or bird.

The animal visitor is an unwitting participant in the plant reproductive process. What the animal visitor is in fact attracted to is food. The food is in the form of a very nutritious liquid called **nectar**. The nectar is produced and reposes in special glands, called **nectaries,** which are located at the base of the pistils and stamens. As the animal visitor attempts to reach the nectar reposing in the nectaries, the pollen is transferred to him. The hungry visitor from the animal kingdom can't help but get pollen on him. He literally cannot miss. A flower's contrasting colors seem to serve as a **nectar guide** actually directing the visitor to the pollen source.

As the avian or insect visitor reaches the nectar source, as it hovers and feeds, the pollen is transferred to its body. As it then visits a second plant in further search of nectar, it thereby inadvertently transfers the pollen onto the sticky stigma of the second plant. Pollination starts at that point and ends with the fertilization of the ovule reposing in the plant's ovary.

Sexual reproduction can only occur between plants of the same species. And, in many plants, chemical barriers in the stigma prevent self-pollination. They do this by treating the plant's own pollen as if it were from a different species. An exquisite system of cross-pollination results, with the 'birds and bees' providing the methodology to ensure a healthy gene pool. But, just in case insects or birds fail to do their job, for whatever reason, some plant species possess the ability to self-pollinate. They do this by then **not** treating the plant's own pollen as if it were from a different species. Amazing.

Seed Dispersal

As we have examined, following successful pollination, a tiny plant in the form of an embryo is now encased in a seed reposing in the ovary. And the ovary grows into a ripened fruit.

To complete the process of plant reproduction, the seed (or seeds) within the ripened fruit need to be transported away from the parent plant to a place that has the proper conditions present to support a growing life. To prosper, the seed must be deposited in a location with adequate light, water and soil nutrients. That step is provided through **seed dispersal**.

In some species, like apple trees or pear trees, seed dispersal occurs when the fruit simply drops away from the parent plant and then exposes the seeds within as it rots or opens-up from dryness.

In other species, like the dandelion, where each fruit contains a small seed with a parachute attached, seed dispersal occurs as the seed is taken for a long ride by a dispersal agent called the wind.

In yet other species, like the coconut palm, the hard but buoyant shell provides a well-adjusted vessel for water travel in search of a proper location.

Finally, for many plants, animal symbiosis again plays a key role. Birds and animals act as seed dispersal agents as they eat the fruit of the plant and later deposit the seeds (with a little welcome fertilizer) at a location only determined by the end of their digestive processes.

Once dispersed, seeds need to germinate. Let's see how that happens.

Seed Germination

Inside a seed is a plant embryo, a tiny plant complete with a miniature stem, root, and tiny leaves. Upon **germination** it will proceed to develop and grow to become a mature plant.

In the life cycle of a green plant, a seed is the stage in a plant's life that can withstand the harshest environmental conditions. In order for a seed to prosper it needs very little. A seed is indeed a wonder to behold. Seed cells show no signs of metabolic activity. They are in a complete state of suspended animation. That state is called **dormancy**.

In order for a green plant to 'sprout' from a seed and then grow and prosper it needs a lot. It needs an ample supply of water, the proper temperature, and oxygen-providing, loose-textured soil, and, lastly, sunlight. A seed apparently 'knows' when the lay-of-the-land is just right. When such is the case, seed germination occurs.

A seed patiently waits in the dormant state until the conditions are right to germinate. It awaits adequate light, water, temperature, and soil nutrients. If these conditions are not present a seed may lie dormant for a long, long time awaiting the right conditions. Mimosa seeds have been documented as germinating after a dormancy period of more than **two centuries**.

The final critical requirement for seed germination is **scarification**. The outer layer of a plant seed is called a **seed coat**. The thickness of the seed coat varies greatly among plant species. In order for a seed to germinate, the seed must be **scarified**. Scarification usually occurs when the seed coat is scraped by shifting rock or soil particles. However, some scarification occurs again through symbiosis when a seed is pecked-at or ingested by members of the animal kingdom. By whatever method, scarification is essential in order to create minute openings in the seed coat so that water may enter.

Upon scarification, in the presence of the right conditions, the dormant seed starts to soak-up water like a dry sponge. The cell structure of the seed softens, swells-up to about twice its initial size, and then splits open, thereby allowing water from the soil to enter.

As water enters the seed, germination actually starts when the tiny root of the embryo rapidly grows and bursts through the seed coat to anchor the new living organism into the soil.

During germination, the embryo inside the seed develops into a young plant called a **seedling**.

In order for the embryo root to burst through the seed coat and anchor the plant into the soil it needs energy. And, the young plant then needs more and more energy to grow and prosper.

Just like animals, plants derive energy through the wondrous process of cellular respiration that we reviewed in Chapter 12. Just like animals, plants need to 'burn' food (high-energy organic compounds like sugars, proteins, fats, and starches) in the presence of oxygen. Unlike animals, plants do not breathe-in oxygen. Rather, that essential gas is diffused through their pores from the soil.

During the early stages of the growth process, the plant is entirely dependent on the food supply contained in the endosperm within the seed. But, as the seedling breaks through the soil it is no longer dependent on the endosperm supplied by its parent. As soon as the green plant seedling reaches the sunlight it becomes a fully independent organism. It thereafter produces its own food supply through the wondrous process of photosynthesis. And, in turn, it becomes part of the food supply for the creatures of the animal kingdom.

Upon reaching the light, germination officially ends. We will now take a look at how the young seedling plant further develops by examining the structures that combine to produce a mature green plant: **roots, stems, and leaves**.

Roots

As we have seen, the first important task of a root is to anchor the plant into the soil. The anchoring system employed by a root is actually of two types, and, both types may be present at the same time in many plants. The two types of roots are (1) **tap roots**, and (2) **fibrous roots**.

In some green plants, only one or two thick tap roots drive deep into the soil to secure water and minerals from the water tables well below the Earth's surface.

In other green plants, an ultra-thin network of fibrous roots comprise a vast system to capture surface water and mineral supplies before the water and minerals leach into lower ground. One steadfast botanist actually counted the root segments of an adult rye plant and reported **14 million root fibers** adding up to over **300 miles in length**.

Roots absorb water through root hairs projecting from the outer or epidermal root layer. Beneath the epidermis of the root is the cortex. Cortex cells are loosely-spaced, thereby allowing oxygen and water to move about freely.

Beneath the cortex lies the **vascular tissues** in the center of the root: the **xylem** and the **phloem**.

Phloem tissues are comprised of the plant's food-conducting cells. These cells form long, narrow cylinders called **sieve tubes**. Amazingly, the nuclei of the sieve tube cells are located in adjoining **companion cells**. This tidy arrangement leaves a clear opening through the cell's cytoplasm for the unobstructed transport of food throughout the plant.

Xylem tissues are comprised of the plant's water-conducting cells. The cells of these tissues are dead at maturity and form long cylindrical tissues called **vessels**. The absence of protoplasm in these cells provides a clear passage for the transport of water throughout the plant.

Water is transported through xylem from the root, through the stems, to the leaves. Food is transported through phloem in the opposite direction from the leaves, through the stems, to the roots.

Leaves

You will recall that the leaves of green plants are, in essence, an energy factory. It is in plant leaves where the wonder of photosynthesis occurs to provide the source of all usable food energy that is essential for all life on Earth that is larger than a single cell.

Photosynthesis actually takes place in the middle layers of a plant leaf. In the upper-middle layer (directly below the epidermis) **palisade cells** catch the sunlight as it first enters the leaf. Then, in the lower-middle layer called **spongy cells**, the chloroplasts reside and perform their energy transformation work.

In order for photosynthesis to work, the chloroplasts must be supplied with sunlight, water, and carbon dioxide. They are provided with sunlight as it is passed to them through the palisade cells. They receive water through their **veins**. Each leaf vein is comprised of a few xylem cells to provide the necessary water, which has been transported originally from the plant's roots. Carbon dioxide is supplied from the atmosphere and enters the spongy cells through microscopic openings called **stomata**. With these necessary ingredients in place, the chloroplasts then produce the high-energy organic molecules that we call sugars. Each leaf vein contains a few phloem cells that then transport the newly-created sugar foods back down the plant stems and throughout the organism and to the roots.

Stems

By far the most important areas in growing green plants are called the **meristems.** At the very tip of each plant stem (and each root) is an **apical meristem**. The apical meristem adds cells to **elongate** the plant's structure. This is called primary growth.

After stems have experienced a significant growth spurt in this manner they begin to **thicken** in order to provide a **stable structure for the leaves**. This is called secondary growth, and occurs through **lateral meristems**. Both the apical and lateral meristems coordinate to determine the size and shape of the plant that they are building.

The **shoot system** of a plant begins at the apical meristem. The **apical bud** at the growing tip controls the orderly arrangement of leaves on the stem and provides then for the development of branches. The primary growth of each branch, in turn, is prompted by auxiliary buds with an apical bud, leaves, and additional auxiliary buds to provide for yet further branching and growth of the shoot system. All is orchestrated in a manner that provides for properly-spaced branches with each leaf thereon given the proper exposure to atmospheric circulation and sunlight.

A green plant's shoot system can produce, in this fashion, an unlimited number of branches. Most of the auxiliary buds, however, lie dormant for years and years. In this manner they provide for a ready source of additional growth should the plant's apical buds be destroyed by a harsh environmental turn.

A green plant that contains a large number of branches is classified as either a shrub or a tree, mostly depending on its size. The thickened branches of shrubs and trees are both comprised of **wood.**

Wood first occurs at the point on a stem where it turns from green to brown. This happens as a covering **bark** begins to form. As the bark ages it thickens. The outer layer of bark is called **cork**.

Keep in mind that the support structure for stability provided by a plant's stems (comprising the entirety of its shoot system, including its branches) is not the only important function served. An equally important function of the shoot system is to provide for the transport of the necessary food, minerals, and water throughout the organism. And, this is done through the **xylem tissues for water and minerals**, and though the **phloem tissues for food**.

Xylem cells and phloem cells comprise the vascular tissues that serve as the literal pipelines for the plant. The pipeline system, with water flowing up from the roots, and food flowing down from the leaves, is contained within the plant's stems and circulates throughout the plant. Water and food are then provided to each part of the plant as the main bundles of xylem and phloem tissue get smaller and smaller as secondary systems to pass nutrients and moisture to each plant cell. The system is akin to the system of capillaries employed in our bodies.

A tree becomes a tree through the processes governed by two lateral meristems called **cork cambium** and **vascular cambium**. The outward divisions of the cork cambium develop the cork which makes up the bark of the outer layer. **The porous cork tissues of a tree's bark allow for the easy exchange of oxygen and carbon dioxide between the inner tissues of the tree and the atmosphere.**

The vascular cambium cells form the dividing line between bark and wood. The **inner portion of bark** is **secondary phloem** while the **outer portion is cork**. The vascular cambium is comprised of **wood cells.** The wood contains the water-conducting cells of **secondary xylem**. Wood cells continue to be added during the life of a tree.

The Water and Mineral Distribution System

A plant derives the water and minerals it needs for life directly from its native soil. The cells of a plant's root system obtain water when water moves from the soil to the interior of the roots' cells by the process called **osmosis**. Through osmosis the water in the soil diffuses across the cell membrane. The membrane is selective in its permeability, thereby trapping the water within the cell structure. As more and more water is taken-up by the root cell through osmosis, the internal cell layers work together to move the water through the cortex of the cell to the **xylem** at the center of the root.

As more and more water is taken-up the cell becomes turgid (inflated like an automobile tire). In this manner water that is initially taken-up by the root cell is under great pressure as it reaches the xylem tissue in the center.

You will recall that xylem tissues extend in continuous chains from the roots through the stems to the leaves of the plant, thus allowing water and minerals, which have been diluted into the water from the soil, to be distributed to each living cell in each and every part of the plant.

The resultant root pressure pushes water up the xylem to the stems and leaves. But, in order to reach the upper branches in towering trees more than just root pressure is needed. A product of photosynthesis itself occurring in the leaves supplies that need.

You will recall that simple water (H_2O) is a basic necessity for photosynthesis. Electrons from water molecules fill the electron 'holes' created by redox reactions and thereby donate electrons to build the high-energy glucose molecules, and produce oxygen as a waste product.

Photosynthesis takes place in the middle layer of the leaf, a layer called the **mesophyll**. The mesophyll uses osmosis to draw water from the xylem. The mesophyll thereby becomes turgid with water pressure. The heat of the Sun then turns the pressurized liquid water into water vapor, which is then pulled into the atmosphere through open stomata. This is a process called **transpirational pull**. And, the **transpirational pull thereby extends as a drawing force throughout the entire plant.**

In order for the suction-like system of transpirational pull to continue, water must be continuously vaporized through the stomata into the atmosphere. **For most green plants, all but about 2% of the liquid water entering the roots through osmosis is transpired as water vapor through the open stomata of the leaves.**

If water is not abundant in the soil, the green plant will simply close the stomata to avoid this enormous water loss. However, closing the stomata to reduce water loss can only be done at the cost of reduced photosynthesis. The stomata must be open in order for the plant to gain the carbon dioxide from the atmosphere necessary for photosynthesis to occur.

The Food Distribution System

The wondrous process of photosynthesis produces the food that is necessary for all living creatures, not the least of which is the living plant itself that just photosynthesized the food.

Through photosynthesis hydrogen and carbon dioxide combine to form glucose sugar. Then thousands of glucose sugar molecules unite into long chains to form the even-larger organic molecules of starch and cellulose. Starch is the principal food that is stored in plant cells, and can later be broken-down as needed as an energy source for cellular respiration, or for conversion into other products through employing various enzymes. Cellulose, on the other hand, is seldom broken down but, rather, is used to form the stable structure of plant cell walls.

Once the process of photosynthesis has produced the high-energy food molecules like glucose, that food energy must be distributed to every living cell of the green plant in order for the process of cellular respiration to occur. The food distribution system in green plants is done through the vascular tissue system that is companion to xylem. That companion vascular tissue system is called **phloem**. Water is conducted upward from the roots toward the leaves through the tissue vessels called xylem. Food is transported in the opposite direction from the leaves to the roots through the companion vascular tissues called phloem.

The amazing bio-chemical pathways of plants branch out in numerous directions and manners. Enriched by nitrogen, sulfur, phosphorous, and other minerals from the soil, the seemingly miraculous products of bio-chemistry result from the cellular metabolism of life. And, plant life leads to animal life, including our life. Green plants supply all that we need for food, shelter, clothing, and the very air that we breathe.

Cellular Development and the Plant Hormonal System

All of the cells in a plant start out life as **parenchyma cells**. As **the plant equivalent of human stem cells**, parenchyma cells have the capacity to differentiate into other cell types that proceed to form the various tissue systems of green plants.

As plant embryonic development proceeds, the cells become more and more specialized in forming the various parts of the plant. The cells of the roots, the stems, the leaves, the flowers, and the xylem and phloem vascular tissue systems that connect to everything, emerge as highly differentiated cells. Tissue development then results from cell mitosis, the elegant cell-splitting process that produces identical daughter cells from a single parent cell. Plant growth proceeds.

The growth and further development of a mature plant is controlled by a complex system of **hormones**. These hormones, in essence, trigger the biological processes that result in physiological responses, the same as with animals. But, where animals have highly specialized organs, called **glands**, that regulate hormonal production and release, plants do not. Plant hormones are produced in the cells of the general plant organs - the roots, the stems, the leaves, and the flowers. And, the orchestration of the unique physiological responses of green plants is the result of many hormones working together.

Gibberellin is a hormone that, among other things, controls seed germination. It also stimulates flowers to form and fruits to grow.

Cytokinin is a hormone that promotes cell division.

Ethylene is a hormone that initiates flower development in some plants and promotes thickening in plant stems.

Yet another hormone, called **abscisic acid**, causes some plants to enter into a state of dormancy.

But the most diverse plant hormone is **auxin**. Auxin controls the plant phenomenon called **phototropism**. It orchestrates cell growth in such a manner that a green plant always grows toward the sunlight. By deftly regulating hormone concentrations, auxin allows for plant stems to defy gravity and grow upward and outward by stimulating the cells on the lower side of stems to grow more rapidly than the cells on the upper side. And then auxin, conversely, allows roots to grow downward in conformance with gravity by rearranging the concentration of the hormone in the plant's roots.

All of these plant hormones not only produce a great variety of physiological responses, they apparently work together in a finely-tuned dance that provides for the familiar phenomena that we observe with regularity as each year passes.

Auxin and gibberellin work hard to maintain the function and structural integrity of cells while ethylene and abscisic acid work hard to break down cell membranes and soften cell walls. And wondrously, the orchestration of which does what and when is provided by a conductor that we call the Earth's environment. **External cues from the environment, like the length of days and seasonal temperature variations, somehow shift the**

balance between the two sets of hormones. At just the right time, hard and green unripe fruits are hormonally altered. Green chlorophyll is broken down and mature brightly-colored soft fruits of red, orange, or yellow hues, become available for our table.

Auxin works hard to keep leaves attached to plant stems during the spring and summer months. Then the shorter autumn days and the cooler autumn nights signal the leaves to produce less auxin and to increase the release of ethylene to break-down the cell walls and the 'glue' called **pectin** which holds the cells together. And, ethylene is apparently directed to be quite selective. It works to break-down cells only at the precise place where the leaf is attached to the stem. And then the autumn leaves start to fall.

Photoperiodism is the term used for the response of plants to the changing length of days and nights as the Earth proceeds in its orbit around the Sun each year. And, not surprisingly, photoperiodism is influenced largely by hormones. As a result, at just the right time each year, the green plant begins to flower.

When a plant has reached a size sufficient to support the heaviness of blossoms and fruit, it is said to have obtained **ripeness-to-flower**. Once it has obtained ripeness-to-flower, it simply awaits the correct length of night and day. Once correct length of night and day then occurs, the green plant starts to flower.

This process of photoperiodism wondrously changes the activity of a green plant's apical meristems. They switch from leaf production to flower production. Apparently, a hormone called **florigen** moves from the leaves to the apical meristems to cause this physiological change. The reproductive process is thereby set in motion. **The plant has somehow figured-out, by measuring the length of the days and nights, how much time it needs to complete its reproductive process**. And, again, the timing has to be awfully good. Many green plants cannot survive in harsh environmental conditions. Only their seeds can survive a summer drought or the onset of a cold winter. The wonder of it all.

CHAPTER 15
The Bug Kingdom

Let's begin this chapter with a disclaimer. What we commonly refer to as 'bugs' do not fit neatly into any kingdom recognized by the scientists. In fact, 'bugs' in the vernacular comprise two entirely different varieties of creatures: (1) microbes, and (2) arthropods. But, in common parlance, we call them all 'bugs'.

The **first life on Earth** was neither plant nor animal. It was pretty inglorious. The first life on Earth took the form of single-celled organisms without a nucleus. They were a form of **microbes** that we now call bacteria or 'bugs'. And, since 'bugs' are our ancestors, we all may be understandably a little 'buggy'. Yeah, that is bad.

Microbes

All microbes are **single-celled** organisms. They were the first form of life on Earth, and they continue to thrive on our planet today.

The first life on Earth was **bacteria** (technically a bacteria look-alike called archaea). Bacteria have thrived on Earth since day one of life. **By whatever measure, numbers or weight, bacteria are the most abundant creatures on our planet**. They maintain our atmosphere, help us to digest food, and clean-up our garbage. And, they also cause us to get sick and sometimes to die.

The next living things on Earth were **fungi**, in the form of yeast. Yeast helps us today to recycle and decompose our waste products, and to bake the bread that we eat.

Following the fungi in the chain of life was the Protista family, in the form of photosynthesizing plant-like **algae**, producing much of the Earth's oxygen. This family also contained animal-like life in the form of protozoa, including the familiar **amoeba**.

The most harmful form of microbes that challenge human existence today are the **viruses**. Influenza, or the 'flu bug', is a virus that still kills over 20,000 people a year in the United States alone. The 'Spanish Flu' of 1918 killed world-wide far more people than all of the valiant soldiers that died during the five years of fighting called the first World War, that ended in November of that same year.

Viruses cannot live on their own. They must invade the body's host cells. They live off the host and reconfigure the host's cellular DNA. Such re-configurations often produce deadly results. However, through the painstaking research of dedicated scientists, the ability of viruses to move genetic information from one cell to another is now greatly assisting humans through cutting-edge medical gene therapy called **genetic engineering**.

These single-celled 'bugs', collectively called microbes, remain invisible to our naked eye. However the other type of 'bugs' that we generally can see are called **arthropods**. They are the creepy-crawly, flying, stinging, alien-looking types of 'bugs' that are technically **animals**. The arthropods, by far, account for more animal species than any other animal - about 900,000 species. Of these, about 800,000 species of arthropods are insects and about 100,000 are other than insects.

Arthropods Other Than Insects

There are roughly 100,000 species of arthropods that are not insects. They include crustaceans, spiders, centipedes, and millipedes. All have skeletons on the outside, called **exoskeletons**. They generally have jointed appendages, segmented bodies, a highly-developed nervous system, and sensory organs.

Lobsters, crabs, fairy shrimp, and crayfish are arthropods called **crustaceans**. Crustaceans have a combined head and thorax and no set number of paired legs. They have a hard external shell. Many are very tasty to other animals.

Spiders (and ticks and scorpions) are of the **Arachnida class** of arthropods. They have **four pairs of jointed legs**, but they do not possess antennae. Spiders have mouthparts that include a tiny fang to inject poison into their prey. Humans can become deathly ill from spider bites.

Centipedes are of the **Chilopoda class** of arthropods. They are from one to six inches in length and are known as 'hundred leggers', but may have more or fewer legs. Most of a centipede's body segments have a single pair of legs. Like spiders, they have venom glands to immobilize their prey. The female will guard the eggs of the newborn. The newborn-to-adult stage of life lasts for about ½ of a centipede's average life span of six years.

Millipedes are 'thousand leggers'. They are of the arthropod class called **Diplopods**. Each body segment of a millipede contains not one, but two pairs of legs. Once millipede eggs hatch the newborns go through about seven different stages as they mature into adults.

The remainder of this chapter on 'bugs' is devoted to the wonders of the by-far-most-numerous-and-diverse class of arthropods - the **insects**. Let's take a look.

Insect Anatomy and Characteristics

Both humans and insects are **animals**. Humans comprise but one species of animal. We generally consider ourselves to be the dominant animals on the planet. But, in **fact, the invertebrate animals classed as insects are, by far, the dominant form of animal life on Earth**. And, the **various species of beetle insects alone outnumber all of the plant species in the world put together.**

There are about 800,000 species of the **Insecta class of arthropods**. And all members of the class - all true insects - **have six legs**.

Identifying characteristics for insects, in addition to having six legs, include: a hardened body with no backbone; compound eyes of multiple lenses; antennae; and mouthparts.

Insect bodies are divided into head, thorax, and abdomen. Three legs are attached to each side of the thorax. The abdomen houses the breathing system of the insect, called the **tracheal system**. Air is inhaled and exhaled through the openings on each side of the abdomen. Oxygen flows through the tracheal tubes to all tissues of the body and waste gasses are expelled through these openings.

While the human circulatory system (and that of all other vertebrate animals) is interwoven with the respiratory system, such is not the case with insects. Insect blood is diffused throughout the creature's tissues and carries only food and waste products (no air). **Insects have no enclosed blood vessels like vertebrate animals have.**

Insects are incredibly strong animals. Insect muscles are attached to the inside of their exoskeletons, which gives insects great leverage and super-human strength. They frequently carry, with ease, many times their own body weight.

Flying Insects

Insects were the first creatures to fly. Insect wings seem to have been specifically designed for flight, while the wings of birds seem to have evolved as modifications of limbs of the body.

Of all the flying insects, only flies have but one pair of wings. All other flying insects have two pairs of wings. Some of the more advanced flying insects, like butterflies, wasps, and beetles, link their forewings and hindwings together in order to accomplish precision flight maneuvers.

Most ants are wingless. However, the reproductive members of an ant colony do have wings that they use during a brief connubial mating flight, after which the pregnant females lose their wings and the male fathers die.

The same murderous flight behavior holds true for the bees. On her maiden flight a queen bee acquires male sperm from her suitor while on the wing. She then stores the collected sperm on her body and uses that supply for the rest of her life to produce, further unaided, countless offspring. The 'lucky' drone father dies immediately following the wedding embrace on the wing. The queen returns to the hive to lay both unfertilized eggs, which produce male drones, and fertilized eggs, that produce female queens or workers. Unless the hive relocates, the queen bee never leaves the confines of the hive again.

Metamorphosis

All insects - all 800,000 species that we know of - go through amazing transformations from birth to adulthood. They undergo a metamorphosis.

Aphids, cicadas, cockroaches, and grasshoppers go through successive **molts**, whereby the look-alike, wingless infant, called a **nymph,** sprouts wings and grows larger and larger into an adult insect. Dragonflies and mayflies begin as aquatic **naiads**. All of these species are known as **hemimetabolous insects**, and their transformation is called an **incomplete metamorphosis.**

Other insects, like beetles, butterflies, flies, and wasps undergo **complete transformations of body and form.** These are known as **holometabolous insects**, and generally they develop in three stages.

The first, or **larval stage**, shows the hatched infant as a **wormlike** creature. Beetle larvae are called mealworms while fly larvae are maggots and butterfly and moth larvae are caterpillars.

Next, when the larvae are fully developed, the insect enters a **resting stage** called a **pupa**. At that stage the insect is usually enclosed in a protective covering, like a **cocoon**. It is during this resting stage that yet another and more wondrous change occurs. Much of **the body of the insect is actually reabsorbed and replaced with the adult body**. This transformation represents a **complete metamorphosis**.

Both incomplete and complete metamorphosis processes lead to amazing results. Let's take a closer look at both.

Incomplete Metamorphosis

Insects that emerge through incomplete metamorphosis perform amazing feats for which science provides little in the way of understanding. A few examples may prove helpful.

Cicadas

Cicadas are fairly large insects, measuring up to 60mm long. Their front pair of wings is about twice as large as their back pair and, when sitting, the wings resemble a roof over the body.

Female cicadas lay eggs in secure positions on the twigs of shrubs and trees. When the **nymphs** hatch from the eggs they fall to the ground and burrow into the soil. There they feed on the sap of plant roots as they grow, in that larval stage, sometimes for a long, long time. While some species of cicadas emerge from the ground in adult form on an annual basis, other species dwell in the soil and undergo numerous molts as nymphs for a 13-year or 17-year period. Then they burst forth at the same time as adult cicadas in veritable droves. And there is no scientific explanation for the fact that the cyclical hatches occur in prime numbers.

Dragonflies and Mayflies

Dragonflies and mayflies first appeared on the planet over 300 million years ago. They **were the first flying creatures to inhabit the Earth**. And, they remain a thriving part of the Earth's ecosystem today.

Dragonflies are incredible pilots. They can fly at speeds approaching 35 mph, can turn on a dime in mid-air, reverse directions, hover like a helicopter, and then fly off in a flash in yet another direction. Their two pairs of wings are independent of each other and enable their powerful and elegant flight maneuvers.

They are spooky but beautiful creatures, with gigantic eyes that contain as many as 30,000 individual facets. At the height of each summer in the South, dragonflies mate on-the-wing over fresh-water lakes and rivers. The female then deposits her fertilized eggs into the water. When the eggs hatch in the water the **larvae, called naiads,** quickly become voracious and skilled predators of other insects. As naiads they actually breathe-in water through tracheal gills located along their backsides. For their final molt before adulthood, the naiads crawl out of the water onto a surface structure, shed their skin, begin to breathe air through their abdomens, dry their newly-developed wings, and fly away to begin anew the reproductive cycle.

Mayflies mate in much the same manner as dragonflies. However, they do so by the droves and in swarms, through the phenomenon of the **mayfly hatch**, wherein huge numbers of naiads emerge as adults all at the same time. A male grabs a female in flight, has his way with her and then immediately dies. The female then immediately flies into the water and attaches her eggs by filament 'anchors' to underwater objects. Then she dies.

The naiads bust out of the fertilized eggs and live underwater in a larval stage, preying on insects and undergoing several molts over a period as long as three years, before emerging as an adult in a new mayfly hatch. The adult life is then devoted exclusively to reproduction. And that adult life lasts but for a single 24-hour day.

Complete Metamorphosis

The story of the ugly caterpillar turning into a beautiful butterfly is the storybook example of complete metamorphosis. And, it is a true story.

The primary function and task of adult moths and butterflies, as is the case with many insects, is reproduction. The adult life span in some species of moths and butterflies is about a week. For the old-timers the life span can be as long as eight months. For most, about two to three weeks.

Butterfly and moth courtship behavior is exotic with actual mid-air-copulation lasting literally for hours. After successful copulation, the fertilized eggs of the female are painstakingly laid on a plant that is the favorite food of the soon-to-be-hatched larvae.

The wormlike **larvae**, called **caterpillars**, then embark upon their primary task of eating and growing. At the end of the larval stage, **pupa** emerge from caterpillars. Pupa do not eat. The moth pupa is encased in a cocoon of silk woven by the caterpillar. The butterfly pupa is housed in a hard-shelled cocoon. The wondrous transformation of body and form occurs in the pupa stage as **tissues are dissolved and reassembled into new body parts.** The eyeless and antenna-less caterpillar develops new digestive organs, compound eyes, antennae, and wings.

As the adult moth or butterfly emerges from its cocoon it bears virtually no resemblance to the caterpillar that it once was. These amazing insects seem to have developed two entirely different sets of body cells - one set for embryonic development with the other set held in reserve to build the adult organs needed to continue the reproductive cycle yet again.

Insect Reproduction

Most insect species reproduce sexually. The male copulates with the female and thereby transfers his sperm to a sac usually located in the female's abdomen. The male most often dies shortly after mating.

How soon the female lays her eggs after sexual mating is highly species-dependent. Some insects, like the mayflies, do so immediately, and that act is shortly followed by death. Other insects, like the queen bees, make egg laying their life-long occupation. Once the eggs are laid most species of insects leave the eggs unattended. The social insects are a notable exception.

Some insects, like the aphids, **reproduce without sexual fertilization of the eggs**. This reproduction process is **called parthenogenesis**, whereby exact genetic replicas of the mother are born. These aphids in fact do not lay eggs. Rather, the aphid embryos are nourished by the mother's own tissues preceding their live birth.

The queen bee lays both fertilized and unfertilized eggs. The eggs that the queen leaves unfertilized develop into male bees, called drones, whose only purpose is to mate with future queen bees. The eggs which the queen chooses to fertilize develop into females. A select few females are fed 'royal jelly' throughout their larval stage and thereby become queen bees, whose only duty is to produce more offspring. The members of the remaining vast horde of females become the non-reproductive (but very, very productive) worker bees.

Social Insects

Most species of insects are quite anti-social, ignoring others of their species except for mating. Social insects, however, comprise a fascinating minority.

Social insects live cooperatively with others of their same species in large colonies. In such colonial life they care for their young, divide their labors into specific castes, create nests of exquisite complexity and tend to the common needs of food and shelter.

While social insects engage in a high degree of specialization of work, the work is actually done by members of the non-reproductive caste. The only duty of the reproductive caste, consisting of the males and the queen, is to produce more offspring.

The non-reproductive caste consists of all the other members of the colony. Some of this caste care for the offspring eggs, larvae and pupae. Others build and repair the nest. Others guard the nest. Others forage and provide food and supplies to the nest. And **all of these workers are females** who develop from fertilized eggs.

The most widely studied social insects are the **social bees** and the **ants**. Let's take a look at some amazing stuff about these social insects.

Honeybees

There are about 400 species of social bees. The species longest known to man is the honeybee, with renderings of these bees and their hives discovered on cave walls.

The nest of a honeybee colony is called a **hive**. At the height of summertime activity a hive may contain up to 20,000 individual bees.

Again, the **reproductive caste** of the hive are the **queen and the male drones**. The queen bee may lay over 1,500 eggs each day and may live up to eight years. The male drones, in contrast, live about eight weeks. Their only purpose is to mate with a new queen if one is necessitated by the departure of the old queen from the hive with a swarm. If a swarm does not occur during the drones' lifetime they are killed by the workers of the hive.

The **non-reproductive caste** are the **female worker bees**. Worker bees who do not over-winter live about six to eight weeks, while those that are born in the fall of the year live until the next spring. Various groups of younger worker bees, called **house bees**, tend to the needs of the queen and the drones; tend and feed the larval brood; clean the hive; ventilate the hive; defend the hive; and construct the hive cells. The older worker bees, called **field bees**, fly out of the hive on daily forages to gather the nectar and pollen from flowering plants as well as water and resins to supply the hive. They carry the supplies to the hive in their abdominal pollen baskets. The **nectar** becomes the basic source for their over-wintering food supply, called **honey**.

When a honeybee colony increases in population to the point of over-crowding, the hive begins to buzz with activity that anticipates finding a location for and constructing a new hive. **Swarming** occurs, and in due course the swarm of worker bees leave with the old queen bee to establish a new colony. At the old hive the house bees become busy continuously feeding some of the larvae the 'royal jelly' so that they may mature into new queens. The first new queen to hatch as an adult from the pupa stage proceeds to kill the other developing queens. Then she ventures forth from the hive on her once-in-a-lifetime nuptial flight.

The swarm of worker bees leaving the old hive is led to the new hive location in the following way. The field bees who have hitherto been wholly engaged in foraging for food and water supplies now, for the first time in their lives, begin to perform an entirely novel behavior. They begin to search widely about for a new dark hollow cavity of a tree most suitable for constructing a new hive. They become **new hive scouts**.

Upon finding what they feel to be a good candidate location, the hive scouts report their findings back to the swarm on the old hive. They want to select the best dry, dark, cavity of proper size that can be found within a reasonable distance from the old hive.

Each scout presents her competing location to the members of the hive through **waggle dances.** Each waggle dance describes the direction, distance and desirability of the candidate hollow cavity that has been found by the scout. As each scout describes the desirability of the candidate cavities, an elimination 'waggle dance-off' occurs. One by one the candidates drop out until a consensus of the hive selects the one location judged to be the best on the basis of dancing vigor. Once the swarm has made its communal decision it heads straight-away to that location to begin a new hive. The winning waggle dance itself apparently describes with great particularity the direction and distance necessary to travel in order to reach the location selected for constructing a new hive.

When the swarm of honeybees arrives with the old queen at the location of the new hive, construction begins. The first step is hive orientation. Quite wondrously the orientation of the bees in the construction of the new hive is influenced by the Earth's magnetic field. No one knows how, but thousands of worker bees cooperate as construction begins and, without the aid of foreman or architect, they orient their new combs in space exactly as in the old hive.

The house bees who construct the hive perform a wonder to behold as they construct geometrically-perfect hexagonal cells into combs for both honey and broods. The cells within the combs indeed serve a dual purpose: brood combs serve as a home for feeding and raising the larvae and pupae from the eggs that the queen deposits into the **brood cells**; and **honey cells** are used to store honey for over-wintering as a food source.

The building material from which the combs are constructed is called **beeswax**. Beeswax is secreted from glands located on the underside of just four specific segments of the abdomens of the house worker bees. These bees use their mouths to form the beeswax into an intricate comb structure.

The walls of the main body of the interconnected cells of the combs form perfect hexagonal prisms. The construction design is the most economical possible for a storage house. The least beeswax possible is employed to construct the sturdiest possible comb. A 9" by 15" comb can hold about four pounds of honey, yet the bees use only about 40 grams of beeswax to construct it.

Honeycomb construction begins at the ceiling. The worker bees begin at many places on the ceiling all at once and commence constructing cells that taper downward. The bees attach themselves, one to another, in chains, to form a building **cluster** of bees. The cluster produces a temperature of 95° F, the exact temperature necessary for the beeswax to be secreted from the abdomens of the worker bees.

As the workers secrete beeswax from their abdomens they work the wax with their mouths to form the cells of the combs. As triangular sections are enlarged laterally they intersect from the top down. The individual worker bees cannot continue their frenetic construction activity for more than about 30 seconds. So, about every 30 seconds a new worker bee relieves his predecessor. Most wondrously, the relief bee picks up the construction project at the exact spot where the first bee finished, and continues the task forward from that point without error. And so on. And so on.

The cell walls are built with a gradient of 13 degrees, just sufficient to prevent the honey from running out. The thickness of the cell walls is maintained at 0.073 mm for the honey cells and 0.094 mm for the brood cells. And, construction tolerance is limited to no more than 0.002 millimeter.

A developing bee goes through the pupa stage in a brood cell before emerging as an adult honeybee. After the queen deposits eggs into the brood cells the larvae is fed continuously by the worker bees until the pupa stage begins. Brood cells are then covered with a wax lid. Under the lid of wax the larvae spin a cap of silk threads from which an adult bee emerges in due course.

Honey cells are also closed by the worker bees with a lid of wax when they are full. The honey serves as the over-wintering food supply for the honeybees until the green plants flower the next spring and thereby provide a nectar and pollen supply to begin the wondrous process yet again.

Ants

All species of ants - **all ants - are social insects**. While they have been known to ruin many a picnic, they also help us by being natural re-cyclers who aerate and fertilize the soil.

Like honeybees, all ants live in colonies. And, all ant colonies have well-defined castes. The reproductive cast consists of the winged males and females (queens). The non-reproductive cast consists solely of wingless and sterile females (workers). Workers will live for about seven years and the queen will live as long as fifteen years.

In the spring and fall the queens mate with males on the wing. Shortly after mating the male drones die and the queens lose their wings. Like the honeybee hive, the ant colony has but a single queen who collects the sperm she needs to last a lifetime for egg fertilization just once on a nuptial flight.

Following her nuptial flight the queen locates a nesting place and, according to its species, either excavates a chamber or locates a natural protected chamber to serve as a nest. The queen then starts to lay eggs and grow the ant colony. The queen decides which eggs are fertilized and which are not. The fertilized eggs produce sterile female workers and the unfertilized eggs produce the drones and future queens.

At first the queen must perform all of the necessary duties by herself. She feeds herself and the larvae, and maintains and defends the nest. However, after the first female workers are born the queen quickly restricts her activity to her egg-laying task. The workers now feed and care for both the queen and the larvae.

As the ant colony grows in both size and area the workers devote more and more of their activities to searching for food and other provisions. The extent of their needs and the manner of satisfying them differs from species to species. For example, leaf-cutter ants are strictly vegetarians, while army ants are strictly carnivores. Over a period of years the colony grows in size, sometimes increasing to millions of individuals sharing a communal life.

When the queen of the colony dies the colony generally can survive for only a few months. The queens are seldom replaced in an old colony. A new colony is usually formed and this must then begin with a new queen's nuptial flight.

The colonial worker ants become quite specialized in performing their tasks of scouting, feeding the queen and the brood, protecting the nest, cleaning the nest and securing food. **Odor plays a large role in ant communications**. When food is located the scouts lay down a chemical trail, called a **pheromone**, for other ants to follow. And, each worker caste seemingly has a different size head and jaws, with the jaws being powerful and ferocious.

Ant ferocity is exemplified by **army ants**. An army ant colony is seemingly always on the move and army ants, **like all ants, never seem to sleep**. They must continually change their hunting grounds when the meat supply becomes exhausted.

Foraging scouts for the army ants seek out fresh prey. Small columns push out on the flanks of the main body of ants and flush out the spiders and cockroaches and grasshoppers into the main body. The prey is then caught, bitten into pieces with the ants' strong jaws, and carried off. The thus-dismembered bodies are carried in pieces to the rear where they are further processed for food by rear guard specialists.

Countless battalions of army ants attack every living creature that is in their path. These ants perform incredible feats of strength and cooperation for the good of the colonial hunt. The **soldiers use their very bodies to form living bridges of ants stretched to the breaking point in order to allow the army to advance and conquer.** This selfless conduct, in part, enables the colony of army ants to feed itself and live for yet another day.

As you examine the marvelous activities of the social insects, the honeybees and the ants, you may be filled with a sense of awe and something very large but unknown at work here. You are not the only one. Nobel prize-winner Maurice Maeterlinck, over a hundred years ago in his book *The Life of the Bee* made these observations about what he aptly termed '**the spirit of the hive**'.

"What is the spirit of the hive - where does it reside? It disposes pitilessly of the wealth and the happiness, the liberty and life, of all this winged people; and yet with discretion, as though governed itself by some great duty. It regulates day by day the number of births, and contrives that these shall strictly accord with the number of flowers that brighten the country-side....At other times when the season wanes, and flowery hours grow shorter, it will command the workers themselves to slaughter the whole imperial brood, that the era of revolutions may close and work become the sole object of all....It regulates the workers' labours, with due regard to their age; it allots their task to the nurses who tend the nymphs and the larvae, the ladies of honour who wait on the queen and never allow her out of their sight;...the architects, masons, wax workers, and sculptors who form the chain and construct the combs; the foragers who sally forth to the flowers in search of the nectar that turns into honey....It comes to pass with the bees as with most of the things in this world; we remark some few of their habits; we say they do this, they work in such and such fashion, their queens are born thus, their workers are virgin, they swarm at a certain time. And then we imagine we know them, and ask nothing more. We watch them hasten from flower to flower, we see the constant agitation within the hive; their life seems very simple to us, and bounded, like every life, by the instinctive cares of reproduction and nourishment. But let the eye draw near, and endeavour to see; and at once the least phenomenon of all becomes overpoweringly complex; we are confronted by the enigma of intellect, of destiny, will, aim, means, causes; the incomprehensible organization of the most insignificant act of life."

Symbiotic Relations

Symbiosis is a mutually beneficial relationship between dissimilar organisms. The bug kingdom is rife with examples of such symbiotic relations.

Pollination, of course, is the head-liner for symbiotic relations. As insects harvest the sweet nectar from the flower of a plant, the plant's pollen adheres to the insect's body. When the wasp or fly or butterfly or bee then visits another flowering plant of the same species to consume more of the nectar, it inadvertently transfers the pollen onto the sticky stigma of the second plant. That pollen transfer begins the elaborate process of cross-pollination that ends with the fertilization of the ovule reposing in the second plant's ovary. Sexual reproduction in most flowering plants is thereby accomplished.

Bees are the foremost pollinators of flowering plants. Worldwide about 150 agricultural crops are pollinated almost entirely by bees. But, insect pollinators are of all shapes and sizes. They range in size from the world's largest bee (about 2 inches long) to tiny, tiny thrips (about 0.04 inch long). Many different species of insects engage in pollination: 30 families of beetles; 45 families of flies; wasps; butterflies and moths; and, of course, bees. Scent is an important player in the pollination business. But, so is color. Butterflies are attracted to yellow, red, pink and purple while bees are attracted to blue, purple or yellow. And, moths do their best work at night but require that the flowers on plants be in bloom.

The symbiosis is achieved by what the flowering plants give in return for the gift of pollination. Very simply, flowering plants provide these insects with the complete food supply that they need for life and living.

However, mutuality is not always easy to see. In two of the most interesting examples of what has to be mutual benefit, the mutuality is sometimes quite difficult to discover. Let's take a look.

Slave Ants

Large red ants, called Amazon ants, found in the western United States, steal the larvae of other ant species and raise them as slaves. This behavior is not isolated and seems to be geographically wide-spread.

Charles Darwin in *Origin of Species* dwelt at some length on the behavior of ants who keep slaves, comparing the behavior of two species he observed in England and Switzerland. One of the species, called *Formica rufescens*, is absolutely helpless and dependent on its slaves. These ants cannot build their own nests, collect food for themselves or their offspring, or even feed themselves. The slave ants that they capture provide them with all of these things. In fact, when the old nest must be abandoned, the slaves not only determine the migration route to a new nest, they actually carry the masters to the new nest in their jaws. Darwin conjectured that such slave behavior had evolved in this way:

> "It is possible that pupae originally stored as food might become developed; and the ants thus unintentionally reared would then follow their proper instincts, and do what work they could. If their presence proved useful to the species which had seized them - if it were more advantageous to this species to capture workers than to procreate them - the habit of collecting pupae originally for food might by natural selection be strengthened and rendered permanent for the very different purpose of raising slaves....I can see no difficulty in natural selection increasing and modifying the instinct - always supposing each modification to be of use to the species - until an ant was formed as abjectly dependent on its slaves as is the *Formica rufescens*."

Charles Darwin's explanation does little to explain any mutual benefit in this master/slave relationship.

Aphid Ranching

All ants are social creatures and live together in colonies. And, some of those colonies may contain a million or more individuals. The colonies of many an ant species engage in an apparent symbiotic relationship with other insects called **aphids**. This behavior is aptly called aphid ranching.

There are about 4,000 species of aphids in the world. They infest about 25% of all plant species.

Aphids are small, placid insects that **feed directly on the phloem of plants.** They may attack the phloem in the roots, stems, or leaves. The proboscis (like a nose) of the aphid contains long stylets that it pushes through the epidermal layer of the plant directly into the phloem tubes. It is a painstaking process that takes up to 24 hours to accomplish before the aphid actually gets something to eat.

The saps from the phloem that the aphids tap are rich in sugars but poor in proteins. This results in the aphids excreting large amounts of a sugary liquid called **honeydew**.

It just so happens that the aphid honeydew is a nearly perfect food for many ants (adult ants do not need proteins, just sugars). And, ants seemingly cannot get enough of the stuff.

Several ant species have developed quite a close relationship with aphids. These ants collect aphid eggs in the late autumn and store them in their ant nests over the winter. In the spring the ants move the aphid eggs to the appropriate plant food source so that the plant phloem will be immediately available to the aphids as soon as they hatch. During the aphids' developing stages and throughout their adulthood the ants will protect the aphids from other insect predators, like hoverflies.

The ants provide these aphid-tending services because the ants are addicted to the aphids' honeydew. The ants obtain the honeydew by '**milking**' the aphids. They milk the aphid by stroking its abdomen with their antennae. Some aphids will not excrete their honeydew unless they are so stimulated by the ants to do so. The stimulation also causes the aphids to feed at 2-3 times their normal rate, thus providing an ample food supply for themselves and the ants. And, if an old plant begins to wither from over-usage, the ants will carry the aphids to a new plant in order to maintain their supply of honeydew.

The mutuality of benefit, however, begins to become quite blurred when there arrives an abundance of aphids and consequently an abundance of aphid honeydew. At that point the ants forthrightly proceed to thin the aphid herd by slaughtering many of them and feeding their carcasses to the ant larvae.

Insect Camouflage and Architecture

Camouflage

Many species of insects have gone to great lengths to protect themselves by blending into the natural world around them. Stick insects, like walkingsticks and timemas, are the most notable. Walkingsticks look simply like leafless twigs on a branch. Timemas look just like the green leaves on the branch.

Other species of insects use camouflage of a different sort. They mimic the look and behavior of other insects in order to protect themselves or otherwise gain a life advantage. Birds, like the flycatcher, consider flies to be a delicacy but are loathe to be stung by a toxic bee or wasp. Robberflies mimic the bees and wasps in an effort to remain uneaten by the birds. Viceroy butterflies, in a similar vein, have learned to mimic the look and manner of monarch butterflies. Because the larvae of monarchs feed on milkweed plants the adult insect body retains a milkweed taste that is toxic to potential bird predators. Those birds learn to leave them alone and dine elsewhere.

Still other insects use camouflage not as a defense but, rather, as a part of their offensive arsenal that assists in their predatory behavior. The 'praying' mantis perfectly mimics green plant stems and leaves and uses this mimicry to stalk and attack its insect victims.

Architecture

We have already taken a brief look at the architectural designs and building skills of the honeybee combs and nests. Ants and termites also employ advanced construction skills in developing their nests.

The African continent is home to many exotic species of insects. One species of termite, commonly called **white ants**, have been observed to be, indeed, master builders. What are first seen to be 'ant hills' prove to be, upon further examination, the exterior of extremely detailed building structures. The exterior structure stands up to twelve feet above the common surface of the ground and is made of well-tempered clay to provide a warm and moist interior climate necessary for hatching eggs and raising brood. This exterior is actually one large domed-arch shell. The interior structure is divided into a vast number of differently-sized chambers. The 'royal chamber' houses the king and queen of the nest and is located in the exact center of the structure. This central chamber is in a state of constant modification and is surrounded in closest proximity by many other smaller chambers that serve to house the royal attendants. Chambers for the brood and supplies are next adjacent and are connected by arched hallways. Because the queen continues to grow and grow in size during her lifetime, the workers continually must modify the chambers within the domed structure. The adult workers seem to innately 'know' when and where to demolish, repair, or rebuild according to the changing needs of the royals and the colony.

But, insects don't have to be adults to perform amazing feats of architecture and construction. The larvae of caddisflies are water-dwellers and build protective cases around themselves. These house-like structures are spun by the glandular secretions from their mouthparts. This sticky substance is used to construct these protective cases from twigs and pine needles, or sand and small pebbles, depending on what the stream bed has to offer. Throughout their larval stage the caddisflies carry these cases with them and they never outgrow them. They continually add new materials. Finally, they modify their protective cases to also serve as a trap for food, by spinning ultra-fine threads used to capture microscopic bacteria.

The 'other-than-insects arthropods' that we call spiders are, indeed, the master web-spinners, constructing delicate but immensely strong webs of silk to ensnare their insect prey. However, they are not the only master web-spinners. The larvae of certain species of flies that live in dark caves have also mastered the web-spinning art. But, their webs are different than those of any spider. Their webs are made not of silk but, rather, of slime. The webs of these **cave flies** are simple, and are strung down from the ceilings of caves as structures from one to two feet long. The webs are supported from the roof of the cave by perpendicular strands about 2" long and 2" apart and are attached to a horizontal cable along the roof line. In order to construct the uniformity of structure observed, it would seem the supporting framework would have to be built **before** the webs are used for trapping insects for larval food. Such ordered construction in any society seemingly requires planning and planning, of course, infers thought.

Insect 'Thought' and Adaptability

Is it really possible that the larvae of the web-spinning cave flies can think? One cannot imagine how cognitive reasoning could possibly be employed by such diminutive creatures. Their behavior is surely limited strictly to that programmed by instinct. But, their actions as well as many other instances in the insect world give pause to wonder. Let's consider but one example - the **burying beetle**.

Burying beetles use the dead carcasses of small animals as a food supply for themselves and their larval young. They work in pairs.

Upon finding the carcass of a dead bird or a mouse on the ground, either the male or the female of the burying beetle pair burrows under the carcass and begins the tedious process of moving the remains of the dead animal under the ground. He or she bulldozes headfirst into the soil and pushes it away. Soon after, the other member of the pair arrives to assist in the arduous task. They work both together and in shifts to move the carcass an inch or so under the ground.

Once buried, the male and female secrete enzymes onto the carcass and proceed to work the body mass into a compact ball of food. The male and female then mate and the female lays her eggs in a chamber excavated above the food. When the eggs hatch both the male and female then transfer some of the liquefied food to the larvae. After the adults have fully provided for the larvae to the point that they can survive on their own, the adults burrow upward through the ground and fly away.

What makes the behavior of the burying beetles so remarkable is that their many tasks require them to be **adaptable**. And, **adaptability is not provided by instinct.**

In order for the burying beetles to be successful they must be able to modify their tactics. If instinct simply impelled 'dig' they would fail miserably in their task. They must adjust to the vicissitudes of their environment in order to succeed. And, various experiments with burying beetles disclose that they have the ability to figure out solutions to taxing problems that they face in completing their burying tasks.

Lorus and Margery Milne recorded this account from their experimental observations of the work of burying beetles that fills one with wonder.

> "Once we drove a good-sized stake into the ground at a 45-degree angle and tied a strong cotton string around its upper end. We tied the dangling end of the string around a hind leg of a dead mouse lying on soft ground. A pair of *Nicrophorus* beetles pushed away the soil below the body until the mouse hung from the tethered leg over a cup-shaped depression. The insects cleared a space the thickness of their bodies between the mouse and the soil and then kept swiveling the carcass in wide arcs. The tail of the mouse dragged on the rim of the depression until one of the beetles chewed it off. That did not solve the problem, and so both beetles explored the surface of the carcass. Only about six hours after they had begun to work did one of them discover the tether. In less than a minute the insect settled down to gnaw through the cotton fibers. By dawn the carcass had been liberated and buried."

To state the obvious, insects are highly adaptable, even if they don't think.

Insects generally have very short lifespans and they produce enormous numbers of offspring. This combination, in Darwinian evolutionary terms, provides insects with the greatest advantage of adaptability to their environment. Large numbers of offspring produce more frequent mutations and some of the random mutations will prove to help the species adjust better to its changing environment. Favorable mutations thereby become firmly entrenched in the species' **gene pool**. This is the gist of Darwinian evolution and the modern synthesis of evolutionary theory.

One of the classic tributes to insect adaptability was penned over a century ago. William Jacob Holland in *The Moth Book* made this fanciful prediction as a testament to the fortitude of insect life:

> "When the moon shall have faded out from the sky, and the sun shall shine at noonday a dull cherry-red, and the seas shall be frozen over, and the ice cap shall have crept downward to the equator from either pole, and no keels shall cut the waters, nor wheels turn in mills, when all cities shall have long been dead and crumbled into dust, and all life shall be on the very last verge of extinction on this globe; then, on a bit of lichen, growing on the bald rocks beside the eternal snows of Panama, shall be seated a tiny insect, preening its antennae in the glow of the worn-out sun, representing the sole survival of animal life on this our earth,- a melancholy 'bug'."

Indeed, in five billion years from now, with the death of the Sun, the 'bugs' that we call bacteria may be the last living organisms on planet Earth.

CHAPTER 16
Worms and Fish

One of the really enjoyable pastimes for both kids and adults is going fishing. On a lazy summer day in the country, baiting a hook with a red worm or a nightcrawler and floating it in the water underneath a cork bobber in hope of catching a sun-fish is an age-old delight. As you see the bobber disappear under the blue-green surface and feel the line tighten, you are treated to an 'ounce for ounce' contest that can't be beat. And, those little blue-gills or red-ears aren't just fun to catch. They are great to eat. We are just like all of the other living creatures in the animal world. We need to eat living things or things that used to be living in order to survive.

The sun-fish ate the worm and we ate the sun-fish. In the evolution of living things we know that we came last in the food chain. But, which came first in the chain, the worm or the fish?

Worms (Phylum Annelida)

Worms are animals that are invertebrate. They have no backbone.

The fossil record evidences worm 'tunneling tracks' that date the origin of **segmented worms** to about 650 million years ago. And, at the time that they first emerged there were no fish yet in the sea to eat them.

At that time all life on Earth was sea life. Living creatures were found either on the surface, within the liquid water, or upon or within the ocean floor. Segmented worms first lived a burrowing life in the soft mud of the ocean bottom.

The linkage of the body segments of these worms allows them to work together and thereby provide the function of locomotion. The intestines and blood vessels pass through the segments to provide for nutritional input and waste disposal.

Many of the species of segmented worms live in the ocean while others are fresh-water dwellers and still others are landlubbers.

The water inhabitants have bristles that stick out from side flaps called **parapods.** These bristles serve as **rudimentary gills** to provide for the gas exchange of oxygen and carbon dioxide necessary for cellular respiration.

The circulatory system contains hemoglobin that is dissolved directly into the blood.

These fascinating creatures normally obtain nourishment from either ingesting detritus that settles on the surface of the bottom or filtering plankton and detritus from the water. (Note: **detritus** is the term for **organic waste particles**; **plankton** is the term for **minute marine life** that is passively floating and drifting in a body of water.)

Some species, like bloodworms, are predators and may eat their neighbors. If you have ever tried using bloodworms as bait in an attempt to catch 'croaker' fish in salt water bays or along sea shores you probably know by experience that these worms have 'teeth'. Their bite is induced by piercing hooks that they inflate with their mouths.

In contrast to species of worms that dwell in the water, earthworms are the landlubbers. They breathe through their skin and thereby ingest oxygen through their capillary blood beds. They obtain nourishment by ingesting soil (and this greatly assists our agriculture production by aerating and improving the soil).

Sexually, earthworms are hermaphrodites (having both male and female reproductive organs). But, amazingly, two hermaphroditic worms mate by copulation, with their heads pointing in opposite directions. Eggs are then shed into a cocoon, from which the next generation arises.

The varieties of worms of the water and earthworms comprise a great many species on Earth that today are descendants of those early sea-bottom-dwellers of 650 million years ago. The number of living species exceeds 13,000.

Fish (Phylum Chordata)

Fish are vertebrate animals. They have a backbone. In fact fish were the first vertebrate animals on Earth. The fossil record indicates that fish appeared on the planet toward the end of the Cambrian Era explosion, about 510 million years ago. The first fish were jawless and some of their descendants are still around today. If you have ever encountered a lamprey eel you have no doubts that even a jawless fish can look pretty fearsome.

About 100 million years later, about 410 million years ago, the first jawed fish evolved. That fish had a skeleton of bone that somehow evolved into just cartilage. The cartilaginous sharks and rays that roam the world's oceans today came on the scene about 370 million years ago.

Modern bony fish appeared about 390 million years ago.

Fish are cold-blooded animals. Their ecomorph is generally cigar-shaped which allows them to move gracefully and easily through their environment of water.

Fish move through the water by swimming. This is accomplished by the contraction and relaxation of groups of muscles, called **myomeres,** located alternatively on each side of the long body. Starting at the head and working down to the tail the muscle groups alternate to produce a series of waves traveling backwards. As the waves reach the tail they cause a whipping action back and forth that thus propels the fish forward. Fish are great swimmers. Most cruise at a speed of 4 to 5 times their body length per second. The really fast tunas can swim at speeds exceeding 50 mph while some sailfish can accelerate to almost 70 mph.

If we were successful in using that worm as bait and caught some fish for dinner, before we could tastefully eat them, we would have to clean them. In doing so we would learn a lot about basic fish anatomy. As you slice open the body cavity to begin the process, you observe an intricate array of internal organs. And, they are all connected in a manner that allows fish to adapt perfectly to a generally cold-water environment.

Fish Anatomy and Physiology

As with other vertebrates, including us, fish have blood that serves to transport oxygen and nutrients throughout the organism's body and to carry waste products away. The blood circulatory system consists of a closed-circuit of vessels transporting blood from the heart-to-gills-to-body-to-heart. The heart has two chambers, an upper and a lower atrium. And the circulatory vessels progressively step down in size from the heart to the cells where cellular exchange is made through minuscule capillaries.

As with other vertebrates, including us, oxygen has to be extracted from the environment to be placed in the bloodstream before it can be circulated throughout the body. Without a continuous supply of oxygen cellular respiration ceases and death of the organism soon occurs.

Fish have a lot tougher time than we do in extracting the oxygen that they need for life from their external environment. A water environment contains only about 3% as much oxygen as an equal volume of air that we use. To meet this need, fish have developed gills which are highly efficient mechanisms that are able to extract as much as 80% of the oxygen that is contained in the water passing through them. By contrast, we extract only about 25% of the oxygen in the air that we breathe into our lungs. Here is how they do that.

Fish gills arch open to present a double-row of folded filaments as the water moves over them. The folds of the filaments are so intricate that their water surface area is comparatively huge, more than ten-fold the rest of the whole body surface of the fish. Blood that circulates into the folds of the filaments in the gills becomes barely separated from the water flowing through. The blood flows forward and the water flows backward. The imbalance between the lower amount of oxygen in the blood and the higher amount of oxygen in the water allows an exchange. Oxygen is thereby diffused into the blood.

As cold-blooded animals, fish generally maintain a body temperature that is within one degree of the temperature of the surrounding water. As with all animals, heat is produced by 'burning calories' through the process of cellular respiration. However, in fish, that heat is almost entirely lost from the body as blood moving through the gills rapidly loses heat to the passing water.

As fish swim through the water they have a type of built-in-radar. That sensory organ is called the **lateral line system**. It detects pressure differences in the water. As a fish swims it sets up pressure waves in the water that are detected by other fish nearby who also have a lateral line system. At the same time, the pressure wave difference is detected by the swimming fish itself which allows the swimming fish to avoid collisions with nearby objects, such as other school fish or predators. Perhaps you have observed a school of literally hundreds or thousands of fish who instantly react to avoid a perceived threat. They never run into each other. They scurry-off in a new direction all in unison without ever touching each other. This wondrous 'sense' is provided through bundles of sensory cells located along the fish's sides and heads that send out the necessary nerve impulses to accomplish this feat.

Fish also are assisted in maneuvering through the water by two types of fins - median fins and paired fins. Median fins are single fins and are located along the centerline of the body. They are three in number. The one on top is the dorsal fin, underneath is the anal fin and behind is the tail fin. Sets of paired fins are the forerunners of the arms and legs of land vertebrates. The pectoral fins are located behind the gills, and the pelvic fins are located along the bottom side.

If you have ever tried scuba diving you have gained a lot of first-hand knowledge about **buoyancy** in water. A buoyant body is a weightless body that can, without effort, hang suspended and weightless in the water. It need expend no energy in doing so. As your dive proceeds deeper through various depths of water, the atmospheric pressure changes greatly and you must add air to your **buoyancy compensator (**BC). For each dive you wear a BC and you add or bleed-off air within it as necessary to retain a buoyant body. You would not want to go diving without a BC. If you do so you will soon become worn out by all the effort you find you must exert simply to avoid sinking deeper and deeper into the abyss. Fish must do the same thing. And, they have a built-in BC to assist them. It is called a **gas bladder**.

A gas bladder allows a fish to vary its body density relative to the density of the water. As a fish swims deeper the water gets denser and the water pressure increases. A fish's gas bladder allows more air to be added as depth increases. This is done through the transfer of gasses from the bladder through the connecting blood vessels. When too much pressure builds-up in the gas bladder the gas is forced into the adjoining blood capillaries that then carry the gas away.

Fish Reproduction

Fish reproduction is a fickle process. Most fish reproduction is heterosexual. Yet some is hermaphroditic. And, some young are born live while most fish hatch from eggs.

In some heterosexual reproduction, with separate male and female partners, the female is able to store the male's sperm for months at a time. In others, the male fertilizes the eggs with his sperm after they are laid.

In hermaphroditic reproduction both the male and female gametes reside in a single fish. The fish produces both eggs and sperm, yet it mates with another hermaphroditic fish companion.

Parental responsibilities also vary greatly among fish species. Some fish build nests and care for both the eggs and the offspring as they are newly hatched. Others simply abandon their eggs.

From Fish to Amphibians

By quantity of numbers, 80% of the world's fish live in the ocean while 20% are freshwater dwellers. By species the corresponding proportions are 60% and

40%. The number of species of fish living on Earth today is estimated to be over 20,000.

During the Devonian Period a certain fish that had developed lobed fins for locomotion crawled out of the water onto dry land. The next chapter proceeds with the story of those **amphibians**.

CHAPTER 17
Amphibians, Reptiles, and Birds (oh, my)

Fish were the first **vertebrate animals** to evolve in the struggle for life on Earth. Vertebrates are animals that **have a backbone**. The backbone is the foundational structural element of the **internal skeleton** that is found in all vertebrates. The skeleton is made of bone and cartilage. It maintains the body shape and protects the vital internal organs of the vertebrate organism.

All amphibians, reptiles, and birds are vertebrate animals of the **Phylum Chordata**. All mammals, including us, are also of this same phylum.

Amphibians - Class Amphibia

About 100 million years after the first plant life appeared on dry land, the first amphibians arrived. **Amphibians** are found in the fossil record at a time about 365 million years ago. They are thought to be **descendants of** a lobe-finned crossopterygian **fish.** These fish had developed air-breathing lungs. The four-lobbed fins evolved into four legs. That **four-legged** body structure became a part of the **ecomorph** for all vertebrate animals that followed them, including us.

Amphibians live in the water, like fish, and they live on the land, like the reptiles to follow them. Because they must keep their skin moist they must periodically return to water. There are about 6,600 species of amphibians on the planet.

Because their **eggs lack a hard shell**, they must lay them in water in order to protect the embryo from drying out. Eggs are laid in gelatinous clusters and are attached to sticks, vegetation, or other submerged matter. Because they are so vulnerable to predators, the eggs are laid in clusters of hundreds or even thousands. Eggs hatch into **aquatic larvae that have gills**, which allows them to breathe in water like fish. **As the larva develops into a mature amphibian, air-breathing lungs replace the gills.**

Amphibians are **cold-blooded creatures** whose skin has no scales or fur. And, these dwellers of both land and water are distinguished primarily by the presence or absence of a tail upon reaching maturity.

Salamanders and Newts

Salamanders and newts are the amphibians with a tail. They have large mouths and eyes with a sense of sight. They have short bodies and smooth brightly-colored skin that may be stripped or spotted. These creatures have teeth on the upper and lower jaws and their carnivorous diet consists mainly of worms and insects.

These amphibians smell and taste with their tongue and they excrete toxic solutions to provide protection from predators. They themselves are the prey of fish and birds.

Most salamanders and newts have gills only in the larval stage and replace the gills with lungs upon reaching adulthood. However, some salamanders retain their gills throughout both their larval and adult lives. On land they take-up oxygen through their skin.

The skin of a newt is rougher and not as slimy as a salamander. A newt is born in water. Before reaching maturity, some newts go through a larval stage as a red eft. They become a land dweller for a year or two and then return to the water to become a mature and aquatic adult.

Frogs and Toads

Frogs and toads are the amphibians that lose their tail upon reaching maturity. Some are herbivores and others carnivores. Carnivorous frogs can quickly nab an insect prey with a long, sticky tongue.

From gelatinous egg clusters laid in the water, frogs and toads enter a larval stage as tadpoles. As **tadpoles** they breathe in the water through gills and propel themselves by their tails as they swim through the water. By maturity the gills have been replaced by fully-functioning, air-breathing lungs, and the tails have fallen off.

As cold-blooded creatures whose internal temperatures nearly match the outside environment, both frogs and toads have developed control mechanisms for surviving in harsh environments. In extremely cold weather toads and frogs burrow into the ground and hibernate. In some extremely cold climates some frogs can survive even after more than one-half of the water within their bodies has frozen into ice.

Frogs are short and squat and have very long hind legs that provide great propulsion for jumping. They are smooth-skinned and bland in color.

Toads differ from frogs by having a thick, 'warty' skin and short legs. With those short legs they are not the celebrated leapers lauded by Mark Twain.

Both frogs and toads are heterosexual and lay their shell-less eggs in water. In the spring after a male frog has attracted a female he climbs on top of her and fertilizes the eggs in the water as she lays them. Tadpoles that emerge from eggs first eat algae and then eat plants and dead insects in the water. **Metamorphosis produces an adult frog normally in about four months,** however, the process in some bullfrogs will take two years or so. An adult life span will average 5 to 10 years, with some species living up to 40 years.

For toads, mating time is normally the only time that they will venture back into the water.

Reptiles - Class Reptilia

There are about 8,000 different species of reptiles. These cold-blooded creatures come in all shapes and sizes and types. They range in size from the two-inch long gecko to the 30+ foot-long anacondas and crocodiles. The now-extinct dinosaurs were gigantic. Reptiles are thought to have evolved from amphibians. But, while most amphibians have smooth, moist skin, most reptiles have dry skin covered with scales. They no longer need to keep their skin wet like the amphibians do. Some reptiles have strong limbs with claws while others, like the seemingly limbless snakes, have fangs.

Reptiles are heterosexual and the females' **eggs are fertilized internally.**

Reptiles are the first living creatures to have developed and **amniote egg.** This is a **hard-shelled egg** that provides a stable, fluid internal environment to protect the embryo. The egg has a yolk sac that provides food and gets smaller as the embryo matures. The shell prevents the moist interior from drying out while at the same time it allows air to come in. When the hatchling emerges from the egg it **does not proceed though a larval metamorphosis**. Let's take a look at some of the different reptiles.

Turtles

Turtles are slow and plodding creatures on land and quite fast swimmers in water. They are distinguished primarily by their hard protective shell. The upper shell is called the **carapace** and is constructed with over fifty bones. The lower shell, called the **plastron**, has only about one-fifth that number. Most turtles can pull their head and legs into their shell for protection. A layer of skin covers the shell and allows the turtle to breathe through the tissues of the skin if deprived of surface air. Some sea turtles can hold their breath for hours and can hibernate under water for literally weeks at a time without breathing surface air directly.

Some species of turtles are sea dwellers (fresh water turtles are called **terrapins**) and some are landlubbers (called **tortoises**). Some types can hibernate in winter and estivate (summer hibernation) during dry, hot spells. And, all lay their amniotic eggs on land. When the young emerge from hatched eggs they grow to adulthood and live a long time, having an average life-span of about fifty years. Some individuals have been documented to have lived over 120 years.

Turtles are omnivores, eating both plant and animal matter, living or dead. They range in size from about 4 inches to the giant leatherback sea turtles up to eight feet in length.

While most turtles live their entire life very near to where they were hatched, some sea turtles migrate for thousands of miles, presumably by the use of some kind of built-in magnetic detector.

Alligators and Crocodiles

Alligators and crocodiles are the largest of the reptiles and exhibit the most elaborate courtship and nest-building rituals. They care for their eggs, hatchlings and juveniles. These reptiles are very similar in appearance, with a long body, immensely strong jaws and a powerful tail. There are 23 different species, but in general they grow to about 7 to 15 feet long and weigh over 1,000 pounds. Their normal life-span is in the range of 50 years.

Alligators have a broad snout and a rounded head and the upper jaw overlaps the lower jaw, thereby hiding the teeth. Crocodiles differ by having a more slender snout and displaying a showy 4th tooth on their lower jaw when their mouth is closed. Both are carnivores who normally eat fish, small rodents and birds. They are opportunistic feeders and will eat carrion if no live prey is available.

During mating season both male alligators and crocodiles engage in elaborate rituals. They slap the water with their heads and emit olfactory cues from musk glands. They emit low frequency 'bellows' that travel a great distance through water. Once a partner has been successfully lured, actual courtship can last for several hours.

Female crocodiles and alligators build nests for their young. Some species dig 'alligator holes' on dry land while others build large mounds of dirt and vegetation about 3 feet high and twice as wide. Eggs are thereby protected from predators and water, for the eggs will die in about 12 hours if the nest is flooded.

Generally between 20 to 30 eggs are deposited in the nest. Alligator eggs incubate for about two months and crocodile eggs about three. The sex of offspring is determined entirely by incubator temperature (less than 86° F produces females and more than 93° F produces males). After incubation is complete the hatchlings call upon mother for help. Upon hearing their calls the nearby mother breaks-open the nest of hardened mud and vegetation. She thereafter tends to their care as newborns and juveniles for a period of one to two years. This ritual is thought to evidence a close alliance to birds and their dinosaur forebears.

Lizards and Snakes

Lizards and snakes are both of the same biological order, Ophidia. Lizards usually have four legs, external ear openings and movable eyelids. Snakes have none of those features. Although a snake is deaf, it can 'hear' through ground vibrations.

Most lizards and snakes are carnivores, eating insects and rodents. Some lizards are omnivorous or herbivorous, like the iguana. Both are prey for large birds, raccoons and foxes.

Both lizards and snakes hibernate in winter and estivate in summer and live off of stored fat. Snakes often hibernate in a large group called a **hibernaculum.**

While all lizards know how to swim if need be, they generally avoid water. Some species of snakes are water dwellers and others landlubbers.

Most lizards and snakes are non-poisonous. However two species of lizards, the Gila monster and Mexican bearded lizard, and several species of snakes, most notably the pit vipers, inject poison into their victims through fangs. **Snake venom is actually a protein enzyme that is used to break-down parts of their prey**, thereby killing them.

For protection **many lizards can change colors**, a characteristic called **metachrosis.** The skin color responds to light, temperature change, and mood. The infamous **chameleon** is noted for quickly changing color under stressful conditions to exactly match its environment. A snake's primary defensive weapon is its tough scaly skin which it periodically must replace. This 'molting' is done by the snake loosening the skin around the lips and literally crawling out of its skin. A growing snake sheds its skin about once a month. An adult molts at least once a year.

When approached by a predator a lizard will hiss, lash out with its tail, puff up his body and bite. If a predator nabs his tail the lizard will break it off in order to escape. That results in no harm to the lizard and he proceeds to regenerate a new tail to replace the one he lost.

Both lizards and snakes recognize prey primarily by scent and taste. A snake's long forked tongue samples particles from the air and carries them back to an olfactory gland on the roof of the mouth.

Lizards and snakes do not chew their food. They swallow their prey whole. A snake can intake a prey up to three times its size. Long tendons allow their jaws to literally stretch apart and allow a victim to be ingested within.

While lizards travel along on four legs, snakes crawl on their bellies. Snakes move by special muscles attached to their ribs. The scales on their bellies allow them to gain traction.

These amazing creatures have a heart, lungs, kidneys, blood, and essentially all the major organs that humans have. However their brains lack the large cerebral hemispheres that birds and mammals have. They are incapable of learning.

Birds - Class Aves

The more than 9,000 species of birds living on Earth today far outnumber the species of amphibians and reptiles. Modern birds are descendants of the reptiles. The first bird-like creature was a **theropod dinosaur** that is found in the fossil record of about 150 million years ago. The first 'maybe' bird, Archaeopteryx, dates to 140 million years ago and has long been extinct. It differed from modern birds principally by having teeth and claws on its wings. By 35 million years ago most bird orders that we know today had appeared.

Birds are the only creatures on the planet that have feathers. They are thought to have evolved from reptile scales, and all birds have them. Birds and the mammals called bats are the only truly flying creatures around today (excepting insects). Birds are warm-blooded creatures, like us, who must maintain an internal body temperature within a fairly narrow range. While most birds are excellent fliers some species, like the emus and ostriches, are flightless. Yet other species, like the sooty tern, can fly continuously literally for years without ever touching down. Birds range in size from the 2½ inch hummingbird to an albatross with a wingspan of over 10 feet.

Birds have a poor sense of smell but generally excellent senses of sight and sound. A flying eagle can spot a rodent on the ground from a mile in the air and an owl can locate a mouse moving on the ground over 100 feet away through his sense of hearing alone. The metabolism of birds progresses at a fast rate and they, therefore, require a great deal of food proportionate to their body weight. A hummingbird for example: must feed all day about every 10 minutes or so; consumes more than 60% of his body weight in food each day; and, efficiently uses about 90% of the energy found in carbohydrates, fats, and proteins. And, one-fifth of all bird species, like the hummingbirds, pollinate flowering plants.

Bird brains are proportionally larger than reptiles but smaller than mammals, and provide little reasoning ability. And, scarce as hens teeth is correct. Birds have no teeth.

Some Bird Anatomy

The internal layout of birds is in many respects very similar to our own, with a heart, lungs, kidneys, bones, and skeletal muscles.

Their heart is four-chambered like ours, but it proportionally weighs about six times as much. It must beat faster to send oxygen and nutrients via the bloodstream quickly throughout the body in order to power flight. The high oxygen-carrying capacity of a bird's blood is enhanced by a great concentration of red blood cells. And, some migratory birds actually have two different forms of hemoglobin in their blood that differ in oxygen-carrying-and-releasing capacities. That allows these birds to adapt to different levels of oxygen in the atmosphere at different altitudes that they must encounter during migration. Through its high metabolism a bird maintains an internal body temperature ranging from 107 to 112 degrees F, depending on species.

The muscles make up over ½ of a bird's weight, with the largest being located in the center of the body, thereby providing a good center of gravity essential for flight efficiency.

A bird's lungs are the most efficient of any of the vertebrates. They are proportionally smaller than a mammal's, but they have nine unique air sacs that act as bellows to provide the lungs with a constant flow of fresh air needed for rapid breathing during flight. The lungs and air sacs together take-up about 20% of a bird's body volume compared to about 5% for us.

Birds have fewer bones than reptiles or mammals, but those that they have are lightweight and very strong. Some bones are hollow to aid in flight and are supported by internal struts, similar to the struts that support the wings of a small airplane.

We have all heard lots of birds sing. Below their larynx is a 'voice box' called a syrinx. It vibrates to create the wide array of bird songs that we enjoy. As muscles relax and contract, different tones are created as air flows through and produces sound. The sounds may be quite varied from species to species. Some of us have spent many a sleepless night due to a mockingbird perched outside our bedroom window and 'crooning' his inharmonic romantic and varied repertoire. And, we have all marveled at the ability of a parrot to 'talk'. Mostly, we just enjoy bird songs.

Birds store a great amount of energy in their bodies as fat. They eat a lot of high-energy foods, for they need a great deal of energy to fly. Flying takes about fifteen times as much energy as perching. And, stored fat can be used when food is scarce or during long migratory flights.

Bird Eyes

Bird eyesight is much better developed than humans. Most birds have eyes located on the side of their head, a feature that allows them to have excellent peripheral vision to the side as well as keen vision to the front. The drawback to this **monocular vision** is a lack of depth perception and difficulty in judging distances.

Raptors, like hawks, eagles, and owls, have eyes in front giving them **binocular vision** like we have, whereby the field of vision from each eye overlaps the other providing for depth perception. To compensate for the reduced peripheral vision that results, some raptors, like the owls, can actually turn their head to face backwards without otherwise moving.

Birds that hunt at night, nocturnal birds like owls, have many more **rod cells** in the retina of their eyes than we do. This feature, along with having larger pupils to admit more light, allows them to see much better at night for hunting. Some non-nocturnal birds have more than ten times the number of **cone cells** in the retina than we do. That allows daytime feeders to easily spot colors that identify feeding opportunities from far away. A hummingbird can spot its favorite red flowers from thousands of feet away.

Bird Beaks and Feet

A bird's beak (or bill) has a core of bone comprising the upper and lower jaws, called **mandibles.** While we move but one jaw, birds move both.

As Darwin's finches disclosed during his journey to the Galapagos Islands, a bird's beak (or bill) is often tailored to its principal source of food. Short, strong beaks crack open hard-shelled seeds. If the diet consists primarily of flying insects, the beaks are flat with a wide base. Long bills are adapted to nabbing insects on plant leaves or, as in the case of woodpeckers, within tree bark.

Some water birds, like herons and egrets, are excellent fishers, using their sharp, pointed beak to spear fish. They will even use bait by dropping small seeds or insects into the water as they fly over and then spearing the fish that rise to the bait. Eagles and hawks have sharp, hooked beaks that they use to rip their prey to shreds.

Pollinating birds, like hummingbirds, have long beaks that enable them to reach deep within the flower petals to reach the rich nectar that lies below. And, while a duck bill is used as a seine to filter food directly from the water, a pelican uses the pouches on his lower jaw as a fishing net.

Bird feet are covered with skin that is, in turn, covered with scales. They help birds to run, wade, paddle, grasp, and perch. They come in a wide variety. Each is adapted to fit a specific need of the particular species. Most of them have toes with a claw at the tip.

Eagles and red hawks will seize prey in their talons. Ducks have webbing between their toes, and they use their web feet to swim through the water with alternating left and right strokes. Other web-footed birds, like diving loons, paddle with both feet at once.

Songbirds, that make-up over one-half of all bird species, have four toes, usually with one facing back and three facing front. Other birds have three toes, with two facing front.

The muscles on the legs and feet of many birds allow them to automatically lock around a branch. This allows them to sleep while on a perch.

Bird Feathers

Feathers are unique to birds. They are the only animals that have them, and all birds have feathers. Feathers are very light, normally about 6 to 7% of a bird's total weight. They provide: insulation for warmth; color for camouflage protection and mate attraction; and, strength and stability for flight. If feathers were too heavy a bird could never become airborne.

Down feathers provide insulation. They are the fluffy feathers without shafts. Contour feathers cover the bird's body and provide shape and color. Flight feathers provide essential aerodynamics for flying.

Flight feathers are the strong wing and tail feathers. They are straight and smooth, with a slight arc. These feathers overlap each other, providing for a curved and streamlined shape that is excellent for flight. The feathers are attached to the skin or bone. The central shaft of a feather is hard, and parallel rows of barbs are attached to it to comprise the feather vane. Within the vane of a feather are interlocking lateral sets of barbs that prevent the air from rushing right through the feather. The overlapping provides a surface to push against the air on the down-stroke and then separates to reduce drag on the up-stroke.

Mating and Reproduction

The most powerful instinct in birds, as in all animals, is to reproduce and leave off-spring. In birds the reproductive cycle is a seasonal one. Birds mate in the spring and summer, not during the cold of winter nor the fall migration time. Mating is preceded by the ritual of courtship.

Males court females. Songbirds, like robins and mockingbirds, sing love songs. Hummingbirds perform aerial dances saying 'notice me' and swallows accompany their aerial dances with love calls. Other species, like prairie chickens and peacocks, puff themselves up and strut their stuff to attract the elusive female. The most romantic seems to be the bowerbirds. They build a honeymoon suite as large as nine feet high, complete with flower petals and decorative shells, for their lady love. Not surprisingly, the dullest-colored males build the fanciest bowers and even paint them with colored fruit pulp by employing a bark brush.

The mating season is announced when the pituitary glands signal the male's testes to enlarge and fill with sperm and the female's ovaries to enlarge and produce ovum. After a successful courtship the male initiates an act of internal fertilization.

Even though most male birds do not have a penis, the couple does have intercourse. He has a flat area, called a **cloaca**, into which the sperm collects. She also has a similar cloaca connecting to her ovaries. She moves her tail to one side, and he climbs up on her back and presses their cloaca together. Intercourse occurs. This is affectionately termed a 'cloacal kiss'.

The female stores the sperm and, at the appropriate time, releases it to fertilize each ovum as it moves from the ovary to the oviduct through which an egg is laid. A **clutch** of fertilized eggs is thereby accomplished. The clutch may contain from one to twenty eggs, depending on the species.

Some birds are monogamous. Some are not. Some birds, like geese, eagles, hawks, and owls, mate for life and only take a second mate if they become widowers. Some males, like the red-winged blackbird have a harem of 10 to 15 females in his territory. And, many birds are very territorial and will chase off all others of their species that try to invade their de-marked area. And, one particular species of hawk in Darwin's Galapagos archipelago mates in threesomes, with both males helping the female tend to the eggs and hatchlings.

Most birds sit on their eggs in nests to incubate them. Proper temperature for embryonic development is maintained in this way. The incubation period prior to hatching is species-dependant, varying from 10 to 70 days.

Most hatchling chicks are entirely dependent on their parents for food and survival. Sometimes both mom and dad share the duties, as with robins and jays. Sometimes just mom, as with hummingbirds. Sometimes foster parents are employed.

A very strange practice of child rearing is employed by a few species of parasite-like birds. These species, like the cowbird, lay their eggs in the nest of another bird and let the 'foster parents' raise the hatchling chicks. Even when the chicks destroy the natural offspring of their new parents and grow larger than their new parents, the foster parents continue to nurture them. While natural selection would strongly disfavor this practice from the point of view of the foster parents, it would certainly favor it from the point of view of the natural parents. Like all creatures, birds are in a race to reproduce successfully as many times as possible. If the natural parents can save the energy normally expended on feeding and nurturing offspring, they will have more energy to produce more eggs. But, why any of the foster parents that fall for this ruse would not become extinct is a mystery.

How Birds Fly

When I struggled to learn how to fly an airplane I quickly found out how difficult it is for us in our machines to crudely do what birds efficiently and effortlessly do when they fly. The physics of flight are the same for a bird as they are for an airplane. But our machine imitations are not nearly as elegant as feathered flight. Four forces work together to make airplanes and birds fly: thrust, drag, lift, and weight. If thrust balances drag and lift balances weight, then the bird will fly straight and level, just as predicted by Newton's law of inertia.

Thrust is the forward and upward force that is accomplished by a bird flapping its wings and creating a forward momentum. The primary lift feathers act much like the propeller does on an airplane. These feathers are shaped so that the low-pressure area is toward the front of the bird pulling him forward.

Lift occurs when the forward motion through the air overcomes the bird's weight. This forward momentum creates lift. The thick, rounded leading edge of the bird's wing tapers to a thin trailing edge. Because the wing is curved in this concave fashion, air traveling over the wing creates a low-pressure zone along the upper wing surface that sucks-up the wing. The higher pressure created below the wing causes the wing to be lifted up. The combination of these two creates total lift.

Drag is the force that is created in the opposite direction of the airflow. A bird increases or reduces drag by the shape of its wings and the angle that he presents to the wind.

The main lift and thrust forward occurs on the down-stroke of the bird's wings. On the up-stroke the secondary flight feathers help to create lift. On the up-stroke the feathers at the end of the wings twist to let air slip through with little drag resistance.

The shape of bird wings varies greatly among species.

Soaring wings allow eagles, pelicans, geese, and hawks to maneuver with precise control at slow speeds. The spacing of the feathers at the end of the wing helps to reduce drag. Some soaring eagles and vultures use natural lift and wind currents to soar on rising columns of warm air called thermals. Soaring geese will often fly in energy-saving 'V' formations. The updraft produced by these birds flying in formation helps to lift each neighbor.

Large seabirds, like the albatross, have long, narrow **gliding wings** used to travel long distances effortlessly. The length of the wing generates tremendous lift, while the narrowness reduces drag. They glide along with strong winds over the ocean. They coast downward and downward until they are just above the tops of the waves and then use their forward momentum to soar up and up again.

The fastest flyers, like falcons and swifts, have a **high-speed wing** shape that allows them to be flapped very rapidly. This provides great thrust with little drag, resulting in increased speed.

And, of course, the hovering wing of the little hummingbird is the most wondrous of all. It allows the hummingbird to fly like a miniature helicopter. Their unique shoulder physiology allows the wings to turn upside down on the backstroke, enabling the path of the wing to follow the pattern of a figure-8. The speed of flapping is an amazing 50 to 100 times per second. Thereby these creatures can hover stark-still and then instantly dart down, sideways and then up and away.

Syncopated Fish and Flight School

Huge schools of thousands of fish dart through the water in waves **without ever touching each other. Flocks of starlings numbering in the hundreds of thousands** undulate and wheel in fits and starts without colliding. How do fish and birds do that?

The birds seemingly use just their sense of sight to accomplish these precision maneuvers, while the fish use their eyes and a sensory organ called the lateral line system that detects and interprets pressure waves from the water. What the birds and the fish do, however, with the sensory input is quite amazing.

The movement of a whole flock and a whole school results from the behavior of the whole emerging from the individual actions of each of the members.

When birds flock and fish school they do so without a leader. They apparently gain an advantage from this group behavior. They are better able to secure food and better able to protect themselves from predators. To accomplish this syncopated group behavior, each individual has the ability to:

- maintain a minimum distance from all other members of the group,
- match the speed of adjacent members by steering toward the average heading of nearby members of the group, and
- attempt to move to the center of the group.

Random movements by individuals can quickly generate changes in direction in waves, and once a wave begins it spreads through a school or a flock more rapidly than can be explained based on the reaction times of individuals. The result is amazing syncopated swimming and flight.

Migration

All sorts of living creatures migrate.

Some insects migrate. Termites and earthworms migrate vertically, moving from near the surface of the soil to deep underground. The monarch butterfly navigates her way over 2,000 miles from Canada to central Mexico each fall.

Some reptiles migrate. In fact, the largest reptile, the leatherback sea turtle, migrates thousands of miles each year. She finds her way through an ocean area larger than the total land mass of Earth to return to the exact sandy beach where she was born.

Some fish migrate. Salmon eggs hatch in fresh-water rivers and the fledgling 'fry' develops in freshwater and then transitions to adulthood in brackish estuaries before proceeding out to sea. They will travel for thousands of miles in the open ocean over a period of several years before returning to their birthplace to spawn and leave new offspring.

Some mammals migrate. The caribou migrates further than any of the other land mammals, and the longest-migrating mammal in the world, traveling over 12,000 miles each year is a marine mammal, the California Gray Whale.

But, of all the creatures on Earth that engage in true migration, the most widespread and varied are the birds. Let's take a look at why and how they migrate.

Why do Birds Migrate, Where do they Go, and How do They do It?

Migration is generally between breeding and non-breeding areas. Migration destination is normally determined by the availability of food, water, and shelter. And, true migration is a two-way trip taken each year from one place to another and back again.

Birds fly south in the fall to over-winter and then fly north in the spring to over-summer. And, many birds fly a long, long way to do this.

The tiny arctic tern migrates the furthest of any bird. This tern migrates about 22,000 miles each year, roughly equivalent to flying around the world at the equator. They do not do so non-stop. Terns are seabirds who naturally feed on aquatic life at sea and can rest and refuel along their trek.

Some land-birds and shorebirds must migrate non-stop. The Pacific Golden-Plover flies from Alaska to Hawaii without ever touching down. The Blackpoll Warbler flies over-water in the fall months from New England to South America, staying aloft continuously for as long as 90 hours, the longest of any bird observed to date. And, even the tiny Ruby-throated Hummingbird makes a spring flight from Mexico's Yucatan Peninsula to Texas non-stop in 24 hours.

In the spring birds migrate from south to north, with the opposite in the fall, along migratory routes. However, a migratory route defines only general north-south pathways, not proverbial bee-lines. In the western hemisphere, migratory birds winter in the Caribbean, Mexico, or Latin America and return to the United States or Canada to reproduce in the spring.

We all understand why birds would migrate south in the fall in order to have an adequate food supply and tropical sun during the winter. Why they fly north is a little more complicated. Birds fly north in the spring in order to mate and raise their young. The real advantage is because the summer days up north are about 16+ hours long while the daylight hours at the equator remain an even 12. The additional daylight allows more time for the birds to gather food for their hungry offspring. Hatchling weight will increase over 50 times in the two weeks following their birth. The longer days and abundant food supply in spring and summer is good reason to migrate north when spring arrives. If all the birds decided to stay south all year long the food supply would soon dry up and famine and extinction would result for many species.

Small songbirds migrate by night in order to avoid predators. They stop in the early-morning hours to forage for food and then rest at usually pre-determined stops along the way. They generally stay at altitudes below 2,000 feet and travel at speeds of 20-30 miles per hour.

Swifts and swallows migrate by day so that they can feed off of insects that are plentiful along their route. Soaring birds, like hawks and other raptors, also migrate by day to take advantage of favorable thermal updrafts that allow them to expend less energy on route.

Geese and ducks are the highest flyers. At a speed of 30-50 miles per hour they fly as high as 29,000 feet as they cross over the Himalayas, remaining higher than Mount Everest.

The winds are very important for most migratory birds. Birds will usually wait until the weather and winds are favorable before beginning their journey. Some actual migratory routes look very haphazard until the wind is factored into the equation. Then it is observed that the more indirect route is in fact the faster one due to favorable tail winds.

Birds seem to have an internal clock that initiates both a physical transformation and a transformation of activities. The internal clock seems to set its alarm based on the length of days and the weather. The bird begins to exhibit restlessness, evidenced by increased fluttering of the wings. The bird begins to 'bulk-up'. A transition called **hyperphagia** results in increased appetite, food consumption and change in diet. This begins about two to three weeks prior to migration take-off.

Bulking-up occurs by increasing fat deposits. The bird will often switch its diet from insects to fruits and berries. Fruits and berries contain a large amount of carbohydrates and lipids that are easily converted to fat. Fat is not only lighter than proteins, it also supplies about twice as much energy. Normal fat deposits will increase from two to five times their normal amount. And, the very long-distance flyers, like the Blackpoll Warbler, can actually double their weight, which enables them to stay continuously aloft for as long as 90 hours.

How Do Migratory Creatures Navigate their Route?

The length of days, called photo-periodism, triggers a migratory bird's internal clock that calls the bird to migratory activity. After it then 'bulks-up' in preparation for its long journey, it takes to the skies with its companions. Other cues and senses are then used to help migratory birds find their way back to their seasonal homes.

Daytime migrants use visual clues. They seem to possess a precise sense of time tied to the location of the Sun. Indeed, some migratory birds can detect polarized light as it occurs perpendicular to the setting Sun. And, they use land-forms such as mountain ranges, rivers, shorelines, and valleys to guide them along their north-south transit. These land maps are apparently learned on a bird's first migration and serves him throughout his life. Geese use these land maps to find their usual resting places and feeding grounds on route as well as their ultimate destination. While other birds can rely on a 'sixth sense' for direction, geese cannot. That's why they frequently get lost in the fog.

Nighttime migrants use celestial navigation. They guide themselves to their Earthly destination by following star patterns in the heavens. They somehow recognize the position of star constellations relative to the North Star. As the Earth moves through space in its orbit around the Sun, the constellations appear to move to different locations in the sky during different seasons of the year. Birds use this 'star map' to navigate home. In one experiment, caged indigo buntings were placed in a room and then different night skies were projected up on the ceiling. When a spring night sky was projected the buntings hopped into the northern part of their cages. When the fall sky was projected they did the opposite. In another experiment migratory birds were placed within a darkened planetarium. They were observed to follow the northerly route guided by the spring constellations projected on the planetarium ceiling, and not the actual fall night sky that spread above the Earth outside the planetarium.

Some birds, like homing pigeons, seem to have a 'homing instinct' whereby they can find their way home after being transported in crates for great distances. They have tiny pieces of **magnetite** located above their nostrils that they somehow use to navigate. In a famous experiment, magnetic caps were placed on the heads of pigeons that actually reversed magnetic polarity. Again and again, when released those pigeons flew in the opposite direction that they should have.

Many living organisms use the Earth's magnetic field, from bacteria to salamanders to fish to birds to mammals. Molecular biologists have discovered a magnetic particle in the brain cells of migratory birds that enables them to employ this 'sixth sense.' We lack this sense to locate things by using the Earth's magnetic field even though our brains apparently do contain some magnetite.

Other animals take this sense of magnetism to a whole new level. They use the magnetic field like we use a Global Positioning System. During their 8,000-mile migration around the Atlantic Ocean, loggerhead sea turtles can detect the intensity of the Earth's magnetic field. And, they also can detect the angle at which the magnetic field lines intersect the Earth. They use these two points of information, which vary at every point on the Earth's surface, as navigational markers to guide them home.

The California Gray Whale migrates some 12,000+ miles across the oceans primarily using the Earth's magnetic fields for guidance. It's a wives tale that birds migrate on the backs of these whales. They do not. But, the whales do use an advanced sense of magnetic detection for their trips. They can determine where magnetic anomalies written into the seafloor differ from those produced by the main magnetic field. They apparently have a degree in advanced magnetism.

As we conclude this chapter, we will use the whales as an appropriate segue to the next. For whales are the largest mammals on Earth. Let's turn to find out more about those amazing mammals.

CHAPTER 18
The Mammal Class

Mammals comprise a large and diverse category of creatures living on the planet. Remember the mnemonic for descending classification: Kings play chess on fairly green spaces:

- Kingdom   Animalia
- Phylum   Chordata
- **Class**   **Mammalia**
- Order   Primates
- Family   Hominidae
- Genus   Homo
- Species   Homo sapiens

As a class mammals comprise 29 orders, 153 families, 1,200 genera and 5,400 species. We are but one single species in the classification scheme of things that has been used by science to classify over one and one-quarter million animal species discovered to date on the planet.

All mammals have backbones, are warm blooded, have three middle ear bones, and a neocortex region of the brain. The offspring of most mammals develop internally in the mother's placenta, have a four-chambered heart, and have hair.

Since we are one of these animal creatures and since we know the most about ourselves, let's review the subject of basic mammal anatomy by using human anatomy as the example. While the human anatomy is in many ways unique to us it is, on the other hand, quite similar to the anatomy of many of the animals, called mammals, that we are most familiar with. Keeping firmly in mind that this is a primer, let's take a look at human anatomy.

Basic Human Anatomy

Molecules combine to form cells. Many cells combine to form body tissues. The human body is comprised of trillions of cells. About 10% of the fluids within our bodies surround the cells in body tissues as interstitial fluid.

Each human cell is enclosed within a plasma membrane and is filled with a fluid called cytoplasm. Each cell contains, within its nucleus, a DNA molecule composed of two strands forming a double helix. Each strand of the helix contains 23 chromosomes, providing for 46 in total (23 pairs). Each chromosome is comprised of a string of genes. Genes contain the information necessary to construct proteins and protein enzymes that serve as biological catalysts for chemical reactions. The historical central dogma of genetics informs us that each gene controls the manufacture of a particular protein or protein enzyme.

All genetic information is carried in the DNA molecule in the cell. DNA provides the coding mechanisms for certain proteins called enzymes that enhance cellular reactions that control the cells. Amino acids are the building blocks of proteins and their order is dictated by codons, which are the basic coding units of the DNA molecule.

The chemical changes in living cells release energy to provide for vital processes and activities and, further provide for the assimilation of new materials into living cells. This is called **metabolism**.

The energy stored in chemical bonds and released through such metabolic reactions is called adenosine triphosphate (ATP) and comprises the energy system for all living organisms. That's where human power comes from.

And, the human body all starts off from a single cell.

Trillions from One

Each of us began in exactly the same way. The egg cell of our biological mother (containing 23 single chromosomes) was fertilized by the sperm cell of our biological father (also containing 23 single chromosomes), thereby producing a single diploid cell (containing 23 pairs of chromosomes). That single diploid cell is called a zygote. We are that resulting zygote. That's us.

Each of us began as one single cell. A zygote. That zygote is a stem cell. All stem cells have three distinct properties:
- they are unspecialized,
- they can divide and renew themselves, and
- they can develop into the specialized cells of the body.

Within three to five days from conception the zygote has grown and includes a group of about 30 stem cells. Those stem cells give rise to the thousands of specialized cells needed to complete our journey to adulthood.

The process whereby unspecialized stem cells give rise to specialized cells is called differentiation. A cell's genes control the internal signals for differentiation. The genes are interspersed across long strands of DNA. The genes carry the coded instructions for all of the structure and functions of a cell. Heart cells thereby know where and how to become heart cells and what to do when they so become. The same for bone cells, and skin cells, and liver cells and eye cells and toenail cells, and all other differentiated cells that make up the wonder that is the human body.

Guided by the instruction map encoded in each cell's DNA, the cells that perform specific body functions assemble into groups of cells that are called tissues. The human body has over 25 different types of tissues that group to develop the body organs and connect everything together in a coherent manner.

We grow and develop within our mother's body as dependent beings for about nine months. Then we are born as a unique individual into the Earthly world. We then further grow and develop as independent beings for many years through childhood and adolescence until we became fully adult human beings. Of course, we have to experience the physical and psychological pain of adolescence to finally reach adulthood.

It is far beyond the scope of this primer to review the stages of human development. A brief sketch of some of the adult finished product is all that will be attempted.

Each of the trillions of living cells in the adult human body now performs a specific function that is determined by the genes on the chromosomes of the DNA molecule that sprang to life when our mother's egg cell was fertilized by our father's sperm cell.

Each of the trillions of living cells in the adult human body must receive chemical nourishment in order for it to perform its proper function. For each cell's life, the source of nourishment is ultimately the food that we eat, the fluids that we drink, and the air that we breathe. That's it. But, of course, that nutrient process is quite complicated. We will briefly examine that process as part of our cursory examination of the various organic systems of the human body. Let's take a look.

Muscle and Blood and Skin and Bone

Skeletal muscles account for about 40% of our body weight. Smooth muscles, like the heart and kidneys, account for another 3%. The skin, the hair, the nails and the glands (known as the endocrine system) account for yet another 7%. That's ½ of us.

There are about 200 bones and 700 skeletal muscles in our body. The muscles attached to the bones provide the mechanism for our body movements. Muscles are made of muscle tissue which, in turn, is composed of muscle cells.

The bones are organs of the skeletal system. Bones are made of bone tissue which, in turn, is composed of bone cells. In addition to supporting our body and protecting our internal organs, the bones provide a storage center for calcium, phosphorus, magnesium, sodium, and lipids.

The bones also serve as the production center for our blood cells. The red marrow in the bones produces red blood cells, white blood cells, and platelets.

Blood transports oxygen and nutrients through our arteries, veins and capillaries throughout our body. It also transports hormones from our endocrine glands to the tissues of the body that need them. Blood serves to regulate our body temperature, transports healing white blood cells to the sites of injury or bacterial invasion, and performs a marvelous 'clotting cascade', involving discretely-coordinated steps of twelve distinct clotting factors, to mend us when the body is cut or bruised.

Our body contains in total about 1½ gallons of blood that constantly circulates to all parts of the body to both nourish and cleanse our cells. It carries oxygen and nutrients to the body tissues and carries carbon dioxide and nutrient waste products away for elimination from the body.

And, none of these complex processes requires a single conscious thought.

The Blood of Life

Our 1½ gallons of blood flows throughout our body constantly during our life. It is an essential of life. Blood is pumped continuously, at the rate of about five gallons a minute, by a smooth muscle located roughly in the center of our chest. The heart.

The heart is composed of four chambers that regulate the flow of blood in two completely separated systems. The chambers on the left side of the heart fill with blood that has received oxygen from the lungs. The chambers on the right side fill with blood that has been used by the body's tissues and is carrying carbon dioxide waste to be expelled from the body as the right chambers pump the 'used' blood back to the lungs. And, this process occurs continuously at the rate of normally 60 to 100 times each minute as our heart 'beats' throughout our life. And the beat goes on, effortlessly, without a single conscious thought.

As the heart pumps, the chambers on the left side of the heart regulate the flow of oxygenated blood through a highway network that connects from the heart to each and every tissue and cell of the body. The system is the arterial system. It starts off with the large artery called the aorta, decreases in size to smaller and smaller arteries called arterioles which further divide into microscopic **capillaries**. The exchange of air and nutrients takes place at the capillaries, which **are only one cell thick.** The highway of blood vessels, including capillaries, is rather extensive. It is about 50,000 miles in length, or about twice the circumference of the Earth.

The tissue cells expel the nutrient and gas waste products of cell metabolism through the capillaries. Proceeding back toward the heart, the capillaries increase in size to venules, which join to form larger and larger veins which carry the waste air back to the right chambers of the heart. This system of veins drains the blood back to the heart. The pulmonary artery, attached to the lower chamber on the right side of the heart, carries the waste air in the 'used' blood through a series of smaller and smaller vessels to the lungs where the carbon dioxide passes through the blood capillaries into the microscopic one-celled air pockets in the lungs called the pulmonary alveoli. The alveoli then join to form larger and larger brachial tubes until the carbon dioxide is finally expelled from the body through the trachea, mouth, and nose. Again, not a single conscious thought is necessary.

The Breath of Life

Breathing is the natural process whereby we inhale air containing the oxygen gas necessary for cellular life, and exhale the carbon dioxide waste gas from the body's metabolic processes that is necessary for plant life. Breathing is an essential of life.

Air from the atmosphere is transported through the nose and mouth through the big trachea tube to the lungs. The tubular system in the lungs gets smaller and smaller until, at last, the air is contained in one-cell-thick gas chambers called alveoli. The air transport part of breathing is only the beginning of the body's respiratory system, which is actually quite a wonder. It consists of three distinct gas exchanges.

The first exchange is called **pulmonary respiration**. This **is the exchange of gas in the lungs with the body's blood.** The site of the exchange is in the one-celled alveoli in the lungs and the one-celled capillaries of the blood located at the wall of the lungs. Oxygen from the lungs passes at that point into the blood. The pulmonary vein drains the oxygenated blood into the left chambers of the heart for pumping throughout the body. Carbon dioxide in the blood passes at that point into the lungs for exhalation, having been carried to that site by the pulmonary artery attached to the lower right chamber of the heart.

The second exchange is called **internal respiration**. This **is the exchange of gas in the bloodstream with the cell tissues throughout the body.** The arteries carry the oxygenated blood to each living cell of the body. As the arteries become smaller, arterioles reduce to one-celled capillaries. The oxygen is then passed through the capillaries for use by the cell.

The third exchange is called **cellular respiration** that we reviewed in Chapter 12. This simply **describes the cell's use of the oxygen in the metabolic processes**. The carbon dioxide waste gas resulting from the cell's metabolic processes is passed from the cell through the capillaries into the veins that drain the 'used' blood back to the right chambers of the heart. The heart pumps, and the process continues again, and again, and again....

And, not a single conscious thought is necessary for the entire process. It's as natural as breathing.

In One End and Out the Other

Our digestive system includes the processes of the human body whereby we obtain the nutrition necessary for the business of living. The system is about forty feet long. It starts at the mouth and ends at the anus. In between are many wonders.

The entire system is comprised of a **tubular gastrointestinal tract containing an upper and a lower portion**. The upper gastrointestinal tract contains the esophagus, the stomach, the liver, the gallbladder and the pancreas. The lower gastrointestinal tract contains the 'guts' of the small and large intestines. Here's how the whole thing works.

The mouth takes in food and fluids and passes them along to the esophagus. The esophagus transports the food and fluids into the stomach. In the stomach the food is processed into 'chyme' by the mechanical work of the stomach's smooth muscles and by the chemical secretion of gastric juices. The partially processed 'chyme' is passed from the stomach into the **first section of the small intestine**, which **is called the duodenum.**

The liver, gallbladder, and pancreas work in harmony to secrete hormones, digestive juices and enzymes that aid in digestion and absorption. These organs process worn red and white blood cells and bacteria and remove toxic compounds in an effort to protect the body. Here proteins are converted back into amino acids; vitamins are synthesized, stored, and released; toxic compounds and bacteria are removed; and bile is produced.

Bile is secreted through common ducts from the liver, gallbladder, and pancreas into the duodenum of the small intestine. The contribution of bile to digestion is enormous. Through bile fatty acids, cholesterol, and various vitamins are absorbed into the blood and the lymphatic system. And, again, all exchanges between the digestive organs and the blood takes place at the one-cell-thick capillary level. **The capillaries are the only physical exchange agents of the human body.**

The small intestine is about ten feet in length. The primary function of the small intestine is to further process the 'chyme'. **The nutrients from the food are passed into the blood through the capillaries in the small intestine.** The undigested waste material is passed on to the large intestine for elimination from the body.

The large intestine (or colon) is about five feet in length. The undigested 'chyme' is passed through the colon, without significant further processing, to the anus where it is eliminated from the body.

And, the only conscious thought we give to the entire process is choosing what we eat and drink. The body takes it from there. Automatically.

Hold Your Water

Under normal conditions, the amount of fluids that we take into our body each day is about equal to the amount that we eliminate. And the fluids we take in are mostly water.

The human body is made up mostly of water. About one-half of our body weight is water. About 70% of our body's water is contained *inside* the cells of our body. About 30% of our body's water is *outside* the cells in the form of interstitial fluid (fluid surrounding the cells) and blood plasma (the fluid part of the blood).

Water helps to maintain the chemical balance in the body. Elements that form molecules by ionic bonds (share electrons) dissociate when they dissolve in the body's fluids. Electrically charged ions, called electrolytes, are thereby created which form acids, bases, and salts. Electrolytes help to maintain the acid-base balance in body fluids, control osmosis in fluid compartments, and take part in all cellular activities. And, electrolytes metabolize as essential minerals necessary for our health.

Water serves to lubricate body organs, regulate body temperature, and is a universal solvent and suspending medium. Water is an essential of life.

Our Recycling Center

We have discussed the respiratory - cardiovascular system whereby the body delivers oxygen to the bloodstream for transport to the cell tissues and transports carbon dioxide, as a gas waste product of cell metabolism, back to the lungs for elimination. And, we have discussed the digestive tract whereby the body processes the food and fluid that we ingest, transfers the chemicals that we need for energy from the food and fluid into the bloodstream for delivery to our cell tissues for metabolism, and eliminates the undigested solid and semi-solid waste through the colon to the anus. Quite a process. But only half of the story.

The other half takes place in **the body's recycling center - the kidneys**. That's where the body recycles all of the non-gaseous waste products of cell metabolism that it possibly can and sends on that which it cannot to the bladder and in turn to the urethra for liquid elimination.

The kidneys are so important to our bodily processes that we have two, just in case something goes wrong. Each kidney can perform the required functions on its own, so we can donate one to a friend or relative in case both of theirs quit working.

Each kidney is composed of over one million nephrons. Each nephron is a highly complex looped structure of tubules and collecting ducts **that serve to filter the bloodstream.** Most of the solutes are reabsorbed completely or nearly completely, depending on the body's need for a particular substance. The kidneys regulate acid-base balance by secretion of hydrogen ions into the tubules and by reabsorption of bicarbonate. The phosphate and ammonia buffer systems carry excess hydrogen ions into the urine for elimination and thereby maintain the acid-base balance. **About 99% of the filtrate is reabsorbed** from the renal tubules and returned to the bloodstream with the remaining 1% being excreted as urine.

The volume of blood plasma passing through the kidneys is incredible. At the rate of 180 liters per day, that amounts to **the entire supply of blood in our body being passed through the recycling center of the kidneys every thirty minutes, 24 hours a day, 7 days a week. And, the recycling efficiency rate is 99%.**

Body Armor and Protection from Invaders

Our bodies are constantly subjected to hostile invasion from **pathogens** (bacteria and viruses) in our environment that will infect us with disease and may kill us. Fortunately, we are well protected from such invasions by an amazingly-complex defense system.

First Line of Defense

Our first level of protection against unwanted invaders of our body is our skin, which is actually comprised of three layers:

- the epidermis,
- the dermis, and
- subcutaneous tissue.

The **epidermis** is the outer layer of skin, which itself consists of five distinct sub-layers. The innermost layer has cells that look like columns. As cells divide in the innermost layer they push neighbor cells into higher and higher layers and they become flatter and flatter. When they reach the outermost of the five layers, called the stratum corneum, they are quite flat and quite dead. We shed those flat, dead skin cells every two weeks or so. But within the epidermis certain specialized cells, called Langerhans' cells, maintain to provide a front line defense for our immune system.

Below the epidermis are yet another two sub-layers of the skin known as the **dermis.** The dermis contains many specialized body cells and glands, such as hair follicles and sweat glands. Coursing through the dermis are the blood vessels and nerves that provide transit for both nourishment and sensation.

The innermost layer of the skin is comprised of **subcutaneous tissue**. It is made-up of connective tissues and fat that provide housing for nerve cells and the larger blood vessels. The subcutaneous tissue also serves to regulate body temperature.

Assisting the skin in repelling unwanted invaders are various chemical barriers such as antibacterial peptides and enzymes. Secretions from saliva, tears, and the respiratory tract, as well as from the skin itself, all work to provide a first line of defense system against pathogen invaders. And, if we eat or drink nutrients containing pathogens, gastric acids in the stomach automatically intervene to provide another protection in our first line of defense against pathogen invasion.

A **pathogen** is a biological agent that **causes disease or illness to the host organism** (like us). The **most common pathogens are bacteria and viruses**, but they also include fungi and parasites, like tapeworms, that may invade our body.

The Innate Immune System

If a pathogen gets through the first line of defense, it immediately encounters our innate immune system. This immune system is a non-specific defense against invaders that we share not only with other mammals but with plants and invertebrate animals as well. While non-specific to invading pathogens, the innate immune system does depend on the ability of the system to distinguish between self and non-self molecules. It must kill the bad but spare the good.

Immune cells are **white blood cells** known as **leukocytes**.

At the site of infection or inflammation from a wound, specialized chemical signaling agents, called cytokines, are called into action. Cytokines signal for help to certain leukocytes called **phagocytes.** Phagocytes are **white blood cells that constantly patrol the body looking for pathogens.** When a phagocyte then encounters a pathogen it kills it by eating it. The pathogen is engulfed within the body of the phagocyte cell and then killed by its cellular digestive enzymes.

Other leukocytes, **known as mast cells**, reside in mucous membranes and attack pathogen invaders associated with allergic reactions. When activated, these mast cells **release histamines, which dilate blood vessels and signal phagocyte specialists to come to the site and kill the invaders.**

Yet other leukocytes, known as **natural killer cells**, attack and **kill host cells that have been infected** by invading pathogens. Natural killer cells accurately distinguish between healthy and infected host cells, then **automatically kill the bad while sparing the healthy.**

In general, the white blood cells of the innate immune system prevent the growth of many harmful bacteria within the body. However, many pathogens have developed the ability to evade the innate immune system. So, wondrously, we have another major line of defense against pathogen invasion that is present only in mammals, the acquired immune system. But, before we take a look at that system let's take a look at the **bridge between the innate immune system and the acquired immune system** that is known as **the complement system.**

The Complement System

The complement system is made-up of over 35 different 'protector proteins', either bound within cells or traveling throughout the bloodstream and other body fluids (the humoral system). **The proteins circulate in an inactive form until signaled by a messenger agent.** Different biochemical pathways are used by the innate immune system and the acquired immune system to activate the complement system. Therein the bridge.

The pathway used by the innate immune system is activated on the surface of the cell of an invading pathogen. When a pathogen is encountered by a specific 'protector protein', initially produced in the liver, the protein is transformed through a series of successive chemical alterations (called a cleavage cascade) to attach itself to the plasma membrane of the invader. Once attached, a further cleavage cascade produces what is known as a membrane attack complex (MAC). The MAC pokes a hole through the pathogen's cell membrane and then, through yet a further cascade of chemical transformations, serves to kill the pathogen cell.

The pathway used by **the acquired immune system** also results in the death of the pathogen cell through a cascade of cleavage and activation steps. But, importantly, this **pathway requires antibodies for initial activation**. And, **antibodies are part** of the story **of the acquired immune system.**

The Acquired Immune System

A clarity of terms is helpful to understanding the fundamental nature of the acquired immune system.

An **antigen** (<u>anti</u>body <u>gen</u>eration) is a molecule that calls forth an immune response from an antibody.

An **antibody** is a large Y-shaped protein that identifies and kills foreign pathogen invaders, like bacteria and viruses. **Each antibody recognizes a specific antigen unique to its target.**

T cells (unique white blood cell lymphocytes) serve to make antibodies against cell-bound antigens.

B cells (other unique white blood cell lymphocytes) serve to make antibodies against soluble antigens.

Both B cells and T cells are produced by the stem cells in the marrow of our bones. B cells circulate through the blood and tissue fluids, while T cells migrate to and develop in the thymus. **Both T cells and B cells express unique receptors that will cause antibodies to bind to a unique antigen and, thereby, then proceed to neutralize a specific pathogen.**

When T cells and B cells are activated, most are used to kill the specific pathogen but some of them are destined to become **memory cells** used to ward off potential future invasions from that specific pathogen. **Thereafter, for the rest of our lives, the memory cells remember the specific pathogen that had caused the previous infection.** So if that unique pathogen is again encountered at some future time in our lives the memory cells will become active and secrete literally millions of copies of the antibody required to kill the specific pathogen by devouring its cells. Such represents the **acquired** immunity of this defensive system.

The **acquired immune system** is able to distinguish between many different antigens. The receptors that allow for such differentiation are produced in very large numbers. **The human body is capable of producing in excess of one trillion different antibody molecules.** So an amazing degree of differentiation is required. And, the most important differentiation feature of the acquired immune system is the same as for the innate immune system: **each must be able to accurately distinguish between self and non-self molecules in the body.** Otherwise, the antibody response that kills the invading pathogen would kill us as well.

In summing-up it is important to recognize that the innate immune system and the acquired immune system are part and parcel of a total package. They work together. The innate system would be swamped by rampant infectious agents without the acquired system. On the other side of the coin, the B cells and the T cells of the acquired system could not become functionally active in the first place without the help of the innate system. Together they protect us from a multitude of invading pathogens trying to kill us. And, they do so without as much as one conscious thought. Our immune system is indeed a wondrous protective system.

But nothing is as wondrous as the human electrical information processor comprised of our brain and our nervous system. Let's take a look.

The Central Nervous System

The **central nervous system** in mammals serves to coordinate instinctual behavior, body orientation, muscle coordination, and learning. The system resides in the brain and spinal column and is composed of **a network of nerve cells called neurons.**

The spinal cord is the portion of the central nervous system that is involved with reflexes and conducting nerve impulses from the peripheral nervous system (PNS) to and from the brain through 31 root pairs of spinal nerves. Each spinal nerve is attached to the spinal cord by a root of sensory fibers and a root of motor fibers.

The adult human brain is our on-board computer that weighs about three pounds. To run the computer requires about 20% of our body's total energy production. Our adult brain and nervous system contain about 100 billion specialized nerve cells, called neurons. Most of the brain is contained within our skull, but it is integrated into our spinal cord to provide us with a complete central nervous system (CNS). The CNS receives messages from the nerves that are in the tissues of the peripheral nervous system (PNS) and sends messages to the motor receptors of the PNS.

Neurons are cells that transmit electrical nerve impulses and respond to stimuli. Neurons look like and **act like specialized cellular wires, much like electrical wires.** Neurons run from the body of the cell located in the central nervous system to each and every part of the body through long processes (called axons) extending away from the body of the cell. So, the neurons that run from my central nervous system to my toes are several feet in length. At the end of each axon is a synaptic terminal that allows for the transmission of a signal to another cell, and so on and so on.

The messages to and from neurons are transmitted through an elaborate electrical circuit.

Motor neurons conduct electrical nerve impulses away from the brain or spinal cord. Sensory neurons conduct electrical nerve impulses from sensory receptor cells to the brain or spinal cord. And these **electrical nerve impulses transmit information**.

Neuron cells come in a wide array of shapes and sizes. Most are quite small, but some (like motor neurons that stretch from the base of our spine to the tip of our toes, and sensory neurons that stretch from the tip of our toes to the base of our spine) can be over three feet long. And, neurons run to and from every part of the body, from just under the skin to the inner organs of the body, like the heart and lungs.

In many respects neurons are very much like other body cells. They contain cellular membranes, cytoplasm, organelles, and a nucleus housing the DNA library. However, neurons have certain characteristics that make them special. And, these special characteristics largely involve electricity.

The neuron nerve cells consist of four major parts:

- **soma** - the central part of the neuron which contains the nucleus of the cell;
- **dendrites** - the multiple input branches of a neuron that attach to the soma;
- **axon** - the long wire-like extension of the cell that carries electrical signals away from the soma; and
- **axon terminal** - the cellular output structure at the end of the axon that is used to release neurotransmitter chemicals and thereby communicate with other neurons.

Neural Electrical Circuitry

At core, the story of neurons is an 'electrifying' tale. For neurons are electrically excitable cells in the nervous system that serve to process and transmit information.

Like all living cells in the human body, nerve cells (neurons) have **ions** (molecules that have either gained or lost an electron and thereby become **electrically charged**) in the fluids both within the cell and without the cell. And, like all living cells, there are **ion channels** transversing the cell's plasma membrane through which ion passage is accomplished. Ion channels are pore-forming proteins that control the small voltage gradient that exists across the plasma membrane of all living cells.

A cell is said to be 'at rest' when it is not stimulated. When a neuron cell is at rest it has a high concentration of potassium ions on the inside and a low concentration on the outside; and a low concentration of sodium ions on the inside and a high concentration on the outside. Ions passively flow through selective ion channels for potassium and sodium, and are actively transported by sodium-potassium pumps through ion channels until a proper balance is reached. The balance is called the resting potential. **In neurons, the proper balance results in a resting electrical charge across the cell membrane of negative 70 millivolts (−70mV).**

Resting potential is really misleading, for the neuron cell, like all other cells, must constantly expend energy through work in order to maintain the resting potential of −70mV, by actively transporting ions through sodium/potassium pumps imbedded across the plasma membrane. Thereby 3 ions of sodium are pumped out of the cell for every 2 ions of potassium pumped in.

When a neuron reaches it's resting potential state of −70mV, it is **ready to 'fire'** (scientists refer to this as reaching its **action potential**). It is ready to send an electrical signal.

An action potential occurs when a neuron sends information down the axon, away from the cell body. The action potential is actually an explosion of electrical activity that is created by a depolarizing current. Here's what happens.

Depolarization occurs when some 'excitatory stimulus' causes the voltage-gated sodium channels in the cell's membrane to open wide and allow sodium ions to rush headlong through the channel down their electrochemical gradient. If the stimulus is strong enough to cause the membrane voltage to drop to −55mV, more and more voltage-gated sodium channels spring fully open and allow sodium ions to gush into the cell.

When depolarization reaches about −55mV the neuron 'fires' an action potential. This is an 'all or nothing event'. If the stimulus results in a drop from −70mV to −55mV the neuron 'fires'. If depolarization does not reach −55mV, nothing happens. So, the 'threshold' is always −55mV. And, the size of the action potential is always the same. Either the neuron reaches the threshold and a full action potential is fired, or it doesn't and no signal is sent. If the threshold is reached, the membrane voltage **instantly raises to a peak of about +45mV**. At that peak point, the voltage-sensitive sodium gates slam closed and the voltage-gated potassium channels open up. As potassium ions then begin to rush outside, the reverse occurs as the voltage within the cell falls back into negative values. This process is completed as the sodium/potassium pumps are actively employed to regain the resting potential of −70mV inside the cell. The neuron is ready to fire again if stimulated. The time necessary for this wondrous process of neuron 'firing' to be accomplished is measured in milliseconds.

The depolarization process from −70mV to +45mV to −70mV again represents the generation of an **action potential**. The action potential or 'firing a neuron' thereby sends an electrical impulse charge (current) down the entire length of the axon transmission 'wire' until it reaches the end of the cell's axon terminal.

The electrical impulse signal that is sent forth as an electrical wave down the axon 'wire' is called an 'action potential'. It has been generated when the polarized cell's 'resting potential' becomes depolarized as the result of being 'excited' by some stimuli, like touching, or stretching, or a chemical transmitter. More often than not the exciting agent is a chemical transmitter. And, that chemical transmitter usually originates when the action potential electrical impulse signal reaches the end of an axon terminal, which is adjacent to a dendrite on the next neuron.

When an electrical impulse signal reaches the end of an axon terminal it encounters **a gap between the terminal end of the cell and the dendrites of an adjacent neuron**. This gap is called a **synapse.** The electrical impulse signal has run out of 'wire' and cannot be electrically transmitted any further. However, the electrical transmission story does not end there. The electrical signal is converted to a chemical signal, a most remarkable event.

After traveling the entire length of the axon and reaching a synapse at the end of an axon terminal, the action potential (electrical signal) causes the release of chemical transmitters at the end of the axon terminal. These **chemical neurotransmitters** are released at an axon terminal when the action potential (electrical signal) opens voltage-gated calcium ion channels, thereby allowing calcium ions to enter the axon terminal. The calcium causes neurotransmitter molecules to fuse with the cell's membrane and then be released outside of the membrane to activate receptors in the dendrites of the neuron next door, adjacent to the synapse. Chemical neurotransmitters released by other neurons at the end of their adjoining axon terminals may join forces to depolarize the neurons next door to the threshold level of -55mV. If the combined stimuli reach threshold for the next door neurons, they induce a further action potential 'firing' in those neighboring neurons. That may, in turn, serve to propagate the electrical signal on down the line in the next neuron cell. And so on. And so on. (Note: I am not making this stuff up.)

In this manner, a change in a neuron's state, from resting potential to action potential, usually occurs as the result of stimulation by a chemical neurotransmitter at the synapse of a nerve cell.

There are three major types of chemical neurotransmitters that serve to either stimulate or inhibit:

- **amino acids (**mostly the workhorses, glutamic acid and GABA);
- **peptides (**amino acids link together to form peptides and peptides further link together to make a polypeptide called a protein); and
- **monomines** (amino acids joined to a two-carbon sugar). Monomines include dopamine, epinephrine (also called adrenaline), serotonin and histamines.

The usual result of releasing chemical neurotransmitters is **excitatory**, causing depolarizing currents in the post-synaptic neuron. However, some neurotransmitters can produce an inhibitory response resulting in hyper-polarizing currents, or a modulating response (as with dopamine and serotonin).

It must be noted that the wiring system in our nervous system is both quite complex and well insulated. Neurons have a myelinated cell membrane. The cell membrane is encased in a special protein coating called **myelin.** This myelin sheath both protects the cell and inhibits the flow of ions between the fluid within the cell and the fluid outside the cell. It also **insulates the membrane, thus allowing the electrical signal to be transmitted nearly instantaneously and with virtually no loss of signal strength.**

Let's Keep Everything in Balance

The body's **endocrine system** really serves to keep things in balance.

The **endocrine system consists of glands that secrete hormones,** which are in essence **chemical messengers**. The hormone messengers are secreted directly into the spaces around the cells (interstitial) where they may be picked-up and circulated by the blood stream.

Hormones generally regulate growth, reproduction, and metabolism. **Most of the fifty plus hormones affect only a few of the body's cells although they reach all the cells of the body through the bloodstream.** And, the target cells are well armed to receive the hormones that they need. **Each target cell has as many as 100,000 receptors for a certain hormone.**

Among many other functions, the **endocrine system glands** serve to regulate our blood sugar balance, our body temperature, our thirst, our hunger, our survival reactions in general. They **all aim for proper body balance.** Balance is maintained by continual adjustment of hormonal output in response to changes in the body's environment.

If the body can achieve a **stable internal balance** then the **metabolic functions of the cells can proceed with the greatest efficiency possible.** That **is homeostasis.** And, homeostasis is what the body is continually attempting to maintain. The body wants to be in balance.

Let's Maintain a Syncopated Rhythm

Biological systems are non-linear (i.e., the whole is more than just the sum of the parts). And, biological systems are characterized by **synergy** (i.e., working together as a combined action or operation) among the parts of the systems. So, reductionist science simply does not work to fully explain living organisms. The whole has to be examined all at once as a coherent entity. **Life depends on such non-linearity**.

Synergy within and between the parts of a living organism occurs through rhythm and syncopation. **Rhythm** occurs when something repeats itself at regular intervals. **Syncopation** occurs when two things happen simultaneously. In living systems we always have syncopated rhythm - **sync**.

So, how do parts of living organisms get into sync? Let's turn again to the fundamentals of our electrical circuitry.

An oscillation is a flow periodically changing direction. In electricity, the flow changes from a maximum to a minimum. Cells in the human body act as oscillators.

Frequency measures the number of events occurring per unit of time. Frequency has an inverse relationship to a wavelength. The more oscillations, the shorter the wavelength, and the higher the frequency. Many cells of the human body produce electrical oscillations sending signals at different frequencies.

Within our central nervous system, signals are sent through the electrical circuitry of brain/nerve cells called neurons. Individual nerve cells transmit electrical signals without losing signal strength along their cell membranes through three separate neural states:

- quiet (until triggered into action as an action potential by a stimulus);
- excited (as the signal is transmitted along the cell membrane without attenuation); and
- refractory (incapable of being excited for a brief rest period).

This simple process within the human brain somehow produces all of the wonders that derive from human consciousness. The electrical interplay of the billions of neurons in the brain somehow produces consciousness and allows us to recognize the face of a friend that we have not seen for years and years, in spite of the fact that we all change as we age. As Steven Strogatz explains in his book entitled *SYNC, the Emergence of Spontaneous Order*:

> "Neurobiologists have discovered that such acts of cognition are linked to a brief surge of neural synchrony, in which millions of far-flung brain cells suddenly switch on and off in precise lockstep at about 40 times a second, and then just as rapidly unravel to allow the next thought or perception to occur. If this view is right, a flash of insight is literally a burst of electrical synchrony, an instant when separate parts of the brain begin to harmonize."

A similar electrical process occurs within the human heart. A cluster of about 10,000 cells comprise the heart's pacemaker. It generates a rhythmic beat that instructs the rest of the heart cells to beat. These pacemaker cells oscillate automatically and provide for about three billion beats during an average lifetime.

Each heart pacemaker cell is an oscillating electrical circuit. When voltage reaches a certain threshold the cell 'fires' the current (or action potential). Then the cell voltage immediately drops to zero. Then the voltage starts to rise again until it reaches the firing threshold. Whenever one oscillator fires it pulls all the others up to the threshold. As one oscillator kicks another one over threshold they thereafter remain synchronized, as if one had absorbed the other. Once the two oscillators fire together they continue as one. More and more absorptions lock more and more oscillators together as they grow into one giant grouping of 10,000. In this way the pacemaker of the heart sets off a wave of electrical signals that spread out along conduction fibers causing the heart's pumping chambers to contract and send forth the blood of life throughout the body.

The principle remains the same for all oscillating systems in the body. A firing of one oscillator brings the others toward threshold, advancing their phases.

Oscillators in living organisms synchronize by pulling on each other's frequencies, a process of 'mutual cuing'. The science of sync studies these coupled oscillators. All biological oscillators share this ability to send and receive signals.

We have briefly looked at the pacemaker cells of the heart and glimpsed at the neural structure of the brain. But oscillators are ubiquitous. Even the lowly intestine operates as a tube of oscillating muscle and nerve cells as the segments rhythmically squeeze together to achieve body waste expulsion.

The most fascinating and mysterious aspect of sync is that the individual cells of the human body have a sense of 24-hour rhythm. The cells within each organ are mutually synchronized. And, the organs in turn are internally synchronized to work in harmony with each other. How they obtain this ability is a mystery to science. There has to be a fundamental biochemical basis for all this synchrony, but science does not know what it is.

The master controller for our body's clockwork seems to reside in the hypothalamus of our brain. This master controller pacemaker regulates body temperature and hormonal secretions and the other wonders of syncopation among our organs and bodily processes. It is, indeed, a wonder. As Dr. Strogatz explains:

"The details of how the pacemaker works are still sketchy. It's known that many of the thousands of neurons in the suprachiasmatic nuclei are oscillators. They spontaneously cycle through a cadence of electrical firing each day, driven by the waxing and waning concentrations of molecules called clock proteins. These molecular circadian rhythms are themselves generated by an interlocking set of biochemical feedback loops, involving DNA transcription and translation of something like eight clock genes (at last count - this research is in constant flux). Then, somehow, thousands of these oscillating 'clock cells' manage to synchronize their electrical activity, coupled perhaps by chemical diffusion of a neurotransmitter called GABA. Finally, the collective electrical rhythm of the pacemaker is conveyed - again, through unknown means - to the peripheral oscillators in the liver, kidney, and other organs throughout the body, disciplining them to run at the same period as the master clock."

While science has discovered much about the ability of biological organisms who use electrical syncopation to achieve the wonder of self-organization, no one can explain how they acquired this ability. That remains - scientifically unknown.

A Perspective Thought

This completes our brief look at human anatomy. It is most fitting to conclude human anatomy with an examination of the master controller pacemaker of the human brain whereby coordinated electrical signals are continuously transmitted from and to our onboard computer. What is actually transmitted is information.

For perspective, we need to consider the magnitude of this information wonder. **Several thousand miles of interconnected nerve cells** interlink to control our every movement, sensation, thought and emotion. Our brain, spinal cord and nervous system contain about **100 billion neurons of 10,000 distinct varieties**. Each neuron has a multitude of neural synapses, on average 7,000 per neuron (and literally trillions in aggregate). And, our brain continues to generate new neurons with multiple synapses until the day we die.

Each and every part of this highly-orchestrated system has but **one primary purpose**: to **convey information**. Let's turn now to Part V to consider the significance of all this information.

PART V

THE SIGNIFICANCE OF INFORMATION

Renowned quantum physicist, Edwin Schrodinger, in his book, *What is Life*, marveled at the power of information contained in DNA:

"Since we know the power this tiny central office has in the isolated cell, do they not resemble stations of local government dispersed through the body, communicating with each other with great ease, thanks to the code that is common to all of them?

Well, this is a fantastic description, perhaps less becoming a scientist than a poet. However, it needs no poetical imagination but only clear and sober scientific reflection to recognize that we are here obviously faced with events whose regular and lawful unfolding is guided by a 'mechanism' entirely different from the 'probability mechanism' of physics....It results in producing events which are a paragon of orderliness....The situation is unprecedented, it is unknown anywhere except in living matter."

The information of life reposes in the DNA of all living organisms. Evolutionary success relies on altering the processing of DNA information in a manner that results in adaptive change. Mainstream science steadfastly maintains that the only mechanism that can result in adaptive change is stochastic (i.e., random) mutations of DNA. In order to assess whether that is a reasonable explanation it is necessary to examine the significance of information itself.

What exactly is information and what does information within living things signify? That is the query of Part V.

CHAPTER 19
Biological Physics and the Fate
of Humpty Dumpty

Humpty Dumpty sat on a wall. Humpty Dumpty had a great fall. All the king's horses and all the king's men couldn't put Humpty Dumpty together again.

Is this bit of classic wisdom from childhood fantasy really true? It depends.

Sometimes Humpty is depicted as a giant egg made of plaster sitting atop the wall. Sometimes he is depicted as a quite-rotund living being sitting atop the wall. Science discloses that his fate is quite dependent on whether the stuff that Humpty is made of is **living or non-living**.

In Part I we explored how everything in the universe is ultimately made of the same stuff - electrons and quarks. Those are the irreducible particles that combine together to make up the atoms that comprise all things - living and non-living. But, the invariant laws of nature seem to treat non-living things differently than living things. How can that be? The answer itself is mysterious. They don't. They just seem to.

1st and 2nd Laws of Thermodynamics

When a glass falls off the kitchen counter or when a plaster egg falls off a wall, it breaks into hundreds of pieces when it hits the ground. It is transformed from a state of high order to a state of lower order. The disorder of the material object has increased. Increasing disorder is called entropy.

Science has discovered the laws of thermodynamics that are inviolate and fundamental for all things.

The first law says that the amount of energy in the universe is a constant, always the same. Never changes. And, with $E = mc^2$ Albert Einstein discovered that, at core, energy and matter are equivalent and can morph into each other. Energy changes readily from one form to another and energy is equivalent to matter, but the total amount of energy is always perfectly conserved. And, heat can be viewed as a property of the energy of matter.

The second law says that in a closed system heat will always flow from a hot region to a cold region. The temperature difference between the hot region and the cold region is the measure of available energy. Dynamic changes of energy are caused through heat exchange and that heat energy can be used to do work. Heat will continue to flow from a hot area to a cold area until a final state of equilibrium is reached, which will be at the lowest temperature possible and which will be uniform throughout the entire system. As heat flows from a hot area to a cold area the total amount of **available energy decreases**. Since the 1st law requires the total amount of energy to remain constant, the **unavailable energy (entropy) must increase**. Non-equilibrium distributions of atoms and molecules become increasingly disordered until a final state of uniform disorder of atoms is reached. The final state is one of complete equilibrium.

The second law of thermodynamics is widely viewed as the law of entropy, to wit: disorder must increase over time until a uniform amount of maximum disorder is reached.

Let's return now to Humpty Dumpty.

Non-living Matter

If Humpty is a giant **non-living** plaster egg, the wisdom of the childhood fantasy is quite real. The laws of physics require that disorder will always increase. The plaster molecules of broken Humpty will not disobey the 2nd law of thermodynamics. Indeed the king's horses and the king's men will probably immediately add to the disorder by tromping on parts of Humpty in their efforts to reassemble him. Alas, all efforts cannot create Humpty's perfect order on top of the wall from the great disorder now lying below on the ground.

If Humpty is a jolly **living** soul sitting on top of the wall, his molecular fate will be determined by the outcome of the fall. Did the fall kill Humpty?

If Humpty died as a result of the fall, his molecular fate will be the same as if he were made of plaster. Immediately upon his death, now as non-living matter, his physical body would begin to decompose. It would proceed to achieve an external and internal temperature that is the same as the surrounding environment. On a molecular level the unavailability of energy would continue to increase over time and decomposition of tissues would proceed. Over eons of time, both plaster-egg Humpty and once-living Humpty would become indistinguishable from the dust of the ground upon which they fell. Disorder would increase until a maximum amount of disorder is reached. Such is required in accordance with the inviolate 2nd law of thermodynamics.

Living Matter

If, however, Humpty did not die as a result of the fall, and the king's men tendered him to the able care of a competent physician, his molecular fate would be quite different. His internal body temperature would remain constant and uniform. His broken bones would heal. The scrapes and cuts on his skin would mend as new. In time he could climb again to sit atop the wall. The increasing disorder required by the 2nd law of thermodynamics is avoided. How can that happen? Energy flows downhill, but, but…life runs uphill.

Life Runs Uphill

The **2nd law of thermodynamics** clearly provides that the unavailability of energy in a closed system - **entropy - always increases** until a state of complete uniformity or equilibrium is reached. As entropy increases, order decreases.

But, life holds entropy at bay. In fact **life reduces entropy**, contrary to the 2nd law of thermodynamics.

As a counter to the most-obvious observation that the processes of life reduce entropy, reductionist science says that is no problem, since living organisms are not 'closed systems'. Because living organisms take in matter and energy from the environment in order to maintain bodily processes, overall entropy in the entire universe (a 'closed' system) continues to increase even while it is held at-bay within the body itself. But, how can life do that?

Let's review some of the processes that occur within living organisms that provide a higher and higher degree of order and organization rather than a continuing state of disorder. They are quite remarkable and **exist only within living organisms**.

Life Requires Energy

The nucleotides that chain together to form the nucleic acids of RNA and DNA require energy to do so. The amino acids that chain together to form proteins require energy to do so. Both of these anabolic processes require energy in the form of molecules of ATP that are produced through the marvels of cellular respiration that we examined in Chapter 12.

Single-celled creatures first developed the wondrous process of photosynthesis and then green plants perfected it. The leaves of green plants: take in the Sun's energy; combine it with carbon dioxide and water; use those ingredients to make sugar; and then release oxygen into the atmosphere. Animals and plants then take in oxygen and combine it with the sugars produced initially by green plants when we eat them or other living things that had eaten them. We thereby extract the food energy (ATP molecules) that we need to power life processes. Through this wondrous process of cellular respiration, we nourish our bodies with usable energy and then rid ourselves of the carbon dioxide waste produced by the process. The carbon dioxide travels out our veins to our heart, is pumped through the pulmonary artery to the alveoli (capillary-type structures) in our lungs where we release it into the atmosphere when we exhale. The exhaled carbon dioxide is then extracted from the air by green plants, as the process of photosynthesis starts anew.

Without photosynthesis and cellular respiration all plant and animal life on Earth simply would not exist. Useful energy must be exquisitely extracted from the raw solar power of the Sun and then refined and further converted to power the 'uphill' processes of cellular life.

Life Requires Energy 'Finesse' and Recycling

After bacteria and green plants figured-out how to create usable energy by combining sunlight, carbon dioxide and water, animal life became possible on Earth. In order to live we, like all other animals, have to eat other living things or things that used to be alive.

We extract the energy from the things that we eat by kind-of reversing the process of photosynthesis. Through the processes of glycolosis and cellular respiration we add oxygen to glucose sugar and extract the energy held within. Carbon dioxide and water are expelled as our waste products and are then used as ingredients by green plants to continue the process of photosynthesis once again, and again, and again.

But things are not quite that simple. In order to extract the energy from the things that we eat (the catabolic process), living cells had to figure-out that just a little energy had to be used (the anabolic process) in order to begin the catabolic process in the first place. Otherwise, in accord with the physical laws of chemistry, we would simply spontaneously combust. A little finesse was needed to get things underway. Then, once a little energy is added to get things going, glucose (or other high-energy organic molecules) can be safely converted into usable energy molecules, called ATP. This is done through a controlled step-by-step process, with each step catalyzed by a different enzyme. The end result is that from one molecule of glucose we extract 38 useable molecules of ATP. That represents an efficiency rate of 40%. By comparison, the brilliant scientists employed by automobile manufacturers have been able to develop a car engine that is only about 25% efficient.

But, we are not through. Living requires an immense amount of energy. And, all that energy comes in the form of ATP. As we have seen, ATP is produced through glycolosis and cellular respiration. And, although these processes produce ATP most efficiently, they could never supply the enormous amount of energy required for living. Indeed, on an average run-of-the-mill non-athletic day, we consume our body weight in ATP. So, the cells figured-out that the vast majority of our energy needs had to be met through recycling.

When a cell constructs the information molecules of RNA and DNA it strings a series of nucleotides together to accomplish that construction. That requires energy. ATP supplies that energy by donating a phosphate group to an enzyme that catalyzes the chemical construction. When a cell constructs the 'workhorses' of life, that we call proteins, it strings a series of amino acids together in the manner prescribed by DNA. That requires energy. ATP supplies that energy by donating a phosphate group to an enzyme that catalyzes the chemical construction.

Whenever the ubiquitous energy molecule called ATP releases its energy to do work in a bodily cell, it does so by phosphorylating (donating a phosphate group). After ATP donates a phosphate group, ATP is transformed into a molecule of ADP, containing no energy. Then, through the wondrous process of cellular respiration a phosphate group is added to those molecules of ADP and they become an abundant source of new ATP energy molecules.

The elegant process of using a phosphate group to release energy and adding a phosphate group to recycle energy is amazing. Through this cellular energy recycling plant ATP continues to power all the cellular work required for life and living. It just so happens that quite a lot of that energy is used to keep us 'electrified'. That is a very good thing. For life itself requires electric charge.

Life Requires Electric Charge

Every living organism contains complex electrical circuitry that is used to communicate not only between cells, but also between the interior of the cell and the environment outside of the cell's plasma membrane. All living cells run on electricity and the electricity is generated by organic chemistry.

Maintaining cellular charge is a basic requirement for life. The failure to maintain proper electric charge in a living organism will result in organ failure and ultimately death. And, **cellular charge must be maintained proactively**.

Cells are constructed so that charged particles cannot pass directly through the plasma cell membrane. This feature lays the foundation for the electrical interactions. The inside of a living cell is constructed in such a fashion as to be polarized (have a negative charge). Basically, this is done by maintaining a high concentration of potassium ions within the cell compared to the outside of the cell, and a high concentration of sodium ions outside the cell compared to the inside. These ions move through specialized proteins that transverse through the cell membrane allowing for flow back and forth. They are known as **ion channels**. All of the ions naturally flow through passive ion channels in accord with the laws of physics from the place of highest concentration to the place of lowest concentration, as they strive to equalize concentrations. However, **if concentrations become equalized there will be no voltage** across the cell's plasma membrane.

A cell's membrane potential is the electrical potential difference (known as voltage) across the cell's plasma membrane. Voltage is caused by the separation of charges by the membrane. It is stored energy that can be used to do work. Importantly, if ion concentrations become equalized there will be no voltage across the cell's plasma membrane. So, something else is required in order to maintain the cellular polarity that is required for a multitude of wondrous functions to occur within a living organism that are necessary for life. That something else is the cell's construction of **active ion channels** transversing the cell's plasma membrane, **called sodium/potassium pumps.**

Each cell constructs literally thousands of these sodium/potassium pumps to actively transport sodium and potassium ions back and forth between the cell's interior and the outside of the cell's membrane. For each cycle of the pump, precisely three sodium ions are pumped out and two potassium ions are pumped in. All the pumps act the same way. Each time a pump operates, three sodium ions are pumped out and two potassium ions are pumped into the cell. The rate at which the pump operates is variable. The rate of pumping is regulated by hormones.

The sodium/potassium pump is an actual pump. It pumps the ions 'uphill', against their concentration gradients. And, that requires a lot of energy. It takes about one-third of our body's total resting energy to run these pumps. But, it is energy well spent. Without these pumps all living creatures would cease to exist. **Life itself depends on the maintenance of cellular polarity and our ongoing electrical circuitry.**

A Perspective Thought

The creation of multi-cellular life on Earth was wholly dependent on the ability of tiny single-celled organisms that are not visible to our naked eye to develop the process of photosynthesis over three billion years ago. In human history all of our scientific genius has never been able to accomplish that feat. At core, photosynthesis is the foundation that allows all plant and animal life to hold entropy at bay.

Living cells are tiny, tiny things. They are themselves, composed of organic molecules that are much, much tinier. Yet, somehow they managed to accomplish some most amazing things. Just for fun, let's compare some of their accomplishments to ours.

Dumb molecules figured-out how to combine sunlight, carbon dioxide, and water to produce a virtually inexhaustible energy source for all living organisms on Earth that are large enough for us to see. We intelligent human beings, including our most brilliant scientists, have never been able to figure out how to do that. If we could, today's energy crisis would quickly become a remnant of the past.

Dumb molecules figured-out an energy conversion methodology that first avoided the problem of spontaneous combustion and then:

- achieved a 40% energy conversion efficiency rate, and
- provided a virtually perpetual energy recycling factory.

We intelligent human beings, including our most brilliant scientists, have never figured-out how to do that. If we could, our energy usage in this country would be drastically reduced.

Dumb molecules figured-out how to create and maintain cellular electrical charge through chemistry. We intelligent human beings actually have figured-out how to do that. However, it took human scientific genius to do so. Doesn't it give one pause to wonder just how those tiny organic molecules figured-out how to do all this incredible stuff?

Elite and mainstream science steadfastly maintains that all this wondrous stuff simply evolved through the process of natural selection of random mutations of DNA molecules. I, for one, find that the discoveries of science make that hypothesis more and more difficult to even imagine.

A Most Fundamental Question

What then is the common feature of: photosynthesis; energy use and recycling; cellular electrical charge; and the other wondrous things within living organisms that allows entropy to be held at bay?

The answer is simple, yet profound. Life holds entropy at bay by virtue of the information residing within living things. In order to hold entropy at bay one ingredient is absolutely necessary. Information. **Life requires information.**

Just what is information, and where did it come from? Let's turn to the next chapter and take a look.

CHAPTER 20
The Nature of Information

Science has discovered a great deal about matter and energy and the forces of nature. As the fundamental forces are brought to bear on the fundamental particles, and fields of matter and energy, the wonders of the non-living world evolve. Those things evolve in strict accord with the laws of nature, which govern them. We can learn a great deal about those things but, in themselves, they contain no information. They represent only themselves.

Information can be transmitted or stored in matter and energy, but information itself is neither. Information is not matter. Information is not energy. What is it?

Information Requires Language

Information is an entity that always represents something other than just itself. The dictionary defines information as the communication or reception of knowledge or intelligence. And, the communication or reception of knowledge or intelligence necessarily requires a means of communication. That means is commonly called language. **All information is based on language.**

Language is the symbolic representation of something else (not itself). And **all language that we know of is designed** and contains several necessary features:
- alphabet,
- grammar (syntax), and
- meaning (semantics).

These features of language work together **to transmit messages**.

The alphabet is used to construct the words that represent things other than themselves.

Grammar is used to clarify things. Changing the word position and placing a ? at the end of a declarative sentence provides a major transformation. **The light is green. Is the light green?**

Semantics provides meaning. **You have the green light** could mean that:
- you are holding the green light;

- you should proceed to drive the car forward and not impede traffic; or
- your proposal for a new project at work has been approved.

The meaning depends on the context in which the information was transmitted and received. Meaning depends on context, and **all language aims to transmit meaning.**

Is DNA a Language?

The DNA molecule is an encoding and decoding system that represents something other than itself. The DNA molecule represents an entire living organism. DNA is most certainly a language.

The DNA genetic code language uses an alphabet of four letters. Each of the letters represents a specific nucleotide nitrogenous base: T (thymine), A (adenine), C (cytosine), and G (guanine). The nitrogenous base letters are grouped into words of three letters each. The three letter words string out into long sentences and paragraphs and chapters and books and encyclopedias in order to provide the instructions necessary for the construction of each and every protein and each and every part of a living organism.

Language is a systematic means of communicating ideas and includes rules for the formation and transformation of admissible expressions. A code is a system of signals or symbols (like letters, or numbers or words) used for communicating messages.

DNA contains a coded language. And, the DNA coded language is far more complex than our native English. The four nucleotide bases combine into three-letter sequences to form a word that describes one of 20 specific amino acids. For example, ACT*GTC*CAG* represents three coded genetic words (i.e., specific amino acids) strung together. Each three-letter word describes a specific amino acid. The exact sequencing of each of the three-letter nucleotide base groupings (called codons) is required in order for a specific protein to be built at the ribosome construction site in the cell. For most proteins the amino acids selected form a linear chain between 100 and 500 amino acids long, but some can number as great as 5,000.

The basic requirement of coded language is specificity. If the reading frame of reference is shifted to the right by one letter the sequence ACT*GTC*CAG* becomes A*CTG*TCC*AG; and, an entirely different sequence of amino acids is strung together to form an entirely different protein to perform an entirely different function in the living organism. Amazingly, a chemist, Dr. Robert E. Kofahl has recently discovered that a single-celled virus under study codes for more proteins than it has space to store all the necessary coded information by using just this mechanism. He provides this observation and challenge:

> "A string of 390 code letters in its DNA is read in two different reading frames to get two different proteins from the same portion of DNA. Could this have happened by chance? Try to compose an English sentence of 390 letters from which you can get another good sentence by shifting the framing of the words one letter to the right. It simply can't be done. The probability of getting sense is effectively zero."

Because of their entrenched belief that the mechanism for evolution of adaptive traits through natural selection **just has to be** random mutation, mainstream science has long-referred to the apparent non-coding portions of DNA as 'junk DNA'. With discoveries such as Dr. Kofahl's it becomes more and more apparent that science has only begun to understand the most-intricate informational structure of genetic language.

Language is Designed to Transmit Information

Information is the communication or reception of knowledge or intelligence. And, **language is designed to transmit knowledge or intelligence.** The transmission of knowledge or intelligence requires a thought process. It requires a sender and a receiver.

Language does not result from random mutation. No information is produced by random mutation.

Patterns Occur by Random Processes, Designs do Not

Nature randomly develops patterns in all sorts of things. Chaos forms patterns as tornados and hurricanes form, grow and abate. Snowflakes and crystals form beautiful, exquisite patterns, each unique and never to repeat. These things occur by the entirely random processes of nature, and they are never ever duplicated to be exactly the same. Patterns never repeat. Patterns are not based on language. Patterns are not designs. **Natural processes never produce coded language. Natural processes never produce information.**

Designs certainly look like patterns but they are very different. Designs can always be duplicated. **Designs are the product of coded language**.

Information is designed in a form that can be stored and then retrieved at a future time with no loss of information. Science has discovered that the information in DNA can be stored in other media with no loss of information. Yet, DNA itself is the most compact and reliable information storage and retrieval system ever developed. In its February 23, 2007 issue *Science Daily* reported that:

> "DNA, perhaps the oldest data storage medium, could become the newest as scientists report progress toward using DNA to store text, images, music and other digital data inside the genomes of living organisms....Data encoded in an organism's DNA and inherited by each new generation, could be safely archived for hundreds of thousands of years, the researchers state. In contrast, CD-ROMs, flash memory and hard disk drives can easily fall victim to accidents or natural disasters."

DNA seems to be the ultimate in designed information. Yet, elite and mainstream scientists are quite adamant that evolution by natural selection uses only the mechanism of random mutations for the development of adaptive traits. They completely eschew any design in living organisms.

The scientific fact remains that **all information that has ever been discovered is a product of language and design.**

Where Does Information Come From?

Information is a product of language and design. **Language and design are always, always preceded by ideas.** Information is the communication or message of knowledge or intelligence and the communication is accomplished through a designed language. **Meaning is imposed on a communication or message by the sender and the receiver**. The receiver may be pre-programmed to understand the message sent by the sender. But, always, **always the sender must intend or will the information sent. Information cannot be created unintentionally.** There is no evidence whatsoever that has ever been adduced by science that random mutation has produced information. **Yet, the very foundation of the mainstream version of evolution by** natural selection rests firmly on the assumption that not only can unintentional random mutation produce information, but that unintentional random mutation can actually increase information in an exponential manner.

Every living organism on Earth contains DNA information. DNA in the simplest living organism is a coded language message 500,000 characters in length. That is ½ million As, Cs, Ts, and Gs arranged in precise three-letter codons required to construct the complete organism. For human beings the number is three billion. That is, indeed, an immense amount of information.

Scientific evidence does not support the mainstream version of evolution by natural selection using only the mechanism of random mutations to account for adaptive traits. Some scientists maintain that there is, indeed, another mechanism, that has been documented again and again, which provides a much more rational explanation than natural selection through mistake and undirected mutation. They propose a mechanism for natural selection that points straight in the direction of **intentional organic development**. The mechanism is appropriately called **natural genetic engineering**. Let's take a look.

CHAPTER 21
Natural Genetic Engineering

Humans have been engaged in the engineering of domestic animals and crops for thousands of years. However, until quite recently such engineering was done in a strictly indirect manner through selective breeding. The advent of direct engineering through actual genetic alterations would not occur until the 1980's.

The first genetically engineered medicine to be approved by the federal government was synthetic human insulin. Approval came in 1982. The procedure used is commonly called recombinant DNA technology. Human insulin is a small, simple protein consisting of a string of 51 amino acids. To produce synthetic human insulin, the gene (segment of DNA) coding for human insulin is inserted and closed inside an E coli bacteria host cell. The host cell, now containing the gene for human insulin, then divides into new cells which are identical to the original. All of the E coli cells now contain the human insulin gene. The implanted DNA now induces the E coli cells to produce insulin, needed by humans who are afflicted by diabetes.

In 1986 the federal government approved the first genetically engineered vaccine to combat hepatitis B. Wondrous advancements in human medicine have quickly followed through further genetic engineering that has led to reductions in cancers, heart disease and other human illnesses.

In the last quarter of a century science has also made great advances in genetic engineering in agriculture. Gene splicing and implantation have been used to alter food crops and domestic meats. Disease-resistant crops have thereby been developed and genetic engineering has been used to genetically modify vegetables so that they stay fresh longer.

All of these marvels of genetic engineering are produced by:

- isolating a specific gene of interest (a specific section of DNA);
- inserting the gene into a vector (such as inserting the human insulin gene into the vector called E coli); and
- removing the gene product from the vector and transforming the cells of the organism to be modified.

Splicing genes has served to greatly improve the human condition. Of course, **the driving force behind such gene splicing in genetic engineering is *human intelligence.*** Moving cellular components from one location to another on a chromosome requires great skill and thought. Can the same result be obtained in nature by random mutation?

Jumping Genes

When Barbara McClintock first published her discovery of transposons (i.e., 'jumping genes') in Indian corn plants in the 1940's her work was scorned and ignored by the mainstream scientific community. The **central dogma of biology** demanded that only **undirected mutations** could cause adaptive change in organisms. Genes could not include transposable elements that could move from one location to another on a chromosome to turn-on and turn-off cellular change.

In the 1970's molecular biologists observed such transposition in bacteria and viruses as they gained resistance to antibiotics. Yet it took 35 years after McClintock's discovery of transposons for mainstream science to pay its highest tribute to a pioneer. In 1983 she was awarded the Nobel Prize for Physiology and Medicine for her discovery of 'mobile genetic elements'.

Today, the discoveries of science reveal that transposons are ubiquitous in living organisms. These sequences of DNA ('jumping genes') move to different positions within the genome of a single cell and thereby turn-on and turn-off or modify different cellular processes. Indeed, the magnitude of transposons in genome operation is enormous.

Retrotransposons are a subclass of transposons. They copy themselves to RNA and then reverse-transcribe themselves back to DNA. They seem to be ubiquitous components of plant and animal organisms. Almost **one-half** of the genome of mammals consists of transposons. Roughly 40% of the human genome is made up of retrotransposons. Every gene in the human genome contains about three retrotransposons.

While it is now common and accepted scientific knowledge that living cells contain controlling elements that govern cellular and organic development, **the central dogma of biology still requires that these controlling elements themselves are the simple products of random mutations.**

In a letter to a colleague in 1973 Dr. McClintock made this prescient observation.

"Over the years I have found that it is difficult if not impossible to bring to the consciousness of another person the nature of his tacit assumptions, when, by some special experiences, I have been made aware of them. This became painfully evident to me in my attempts during the 1950s to convince geneticists that the action of genes had to be and was controlled. It is now equally painful to recognize the fixity of assumptions that many persons hold on the nature of controlling elements.... One must await the right time for conceptual change."

Natural Genetic Engineering

In 2001 James A. Shapiro, Professor of Microbiology at the University of Chicago, published *A 21st Century View of Evolution*. In that paper Dr. Shapiro provides the following paradigm-breaking hypothesis for the adaptive-change-mechanism of evolution.

"Evolution is the history of organisms that have succeeded in adapting to changing circumstances. Over evolutionary time, this means altering the genome, the long-term storage organelle of all living cells, to provide the functional information needed to survive and reproduce in new conditions. Those organisms that have the most flexible computational capabilities, in particular those that have the best means of altering information stored in the genome, will have an advantage. Thus, it makes sense for organisms to possess crisis-responsive natural genetic engineering functions, and we should not be surprised to find them ubiquitous in contemporary organisms, all of whom are evolutionary winners. Indeed, it is now difficult to imagine how organisms that depend upon gradual accumulation of stochastic mutations could persist in the evolutionary rat race."

"The last half century has taught us an astonishing amount about how living organisms function at the molecular level, in particular about how they execute cellular computations through molecular interactions and about the systemic, modular, computation-ready organization of the genome.

We have come to realize some of the basic design features that govern genome structure. Combining this knowledge with our understanding of how natural genetic engineering operates, it is possible to formulate the outlines of a new 21st Century vision of evolutionary engineering that postulates a more regular principle-based process of change than the gradual random walk of 19th and 20th Century theories."

In 2003, the *Boston Review* published an article written by Dr. Shapiro entitled *'A Third Way'*. In that article Shapiro proposed that instead of strict adherence to either creationism or neo-Darwinism, science should more fully explore the fact that:

"...cells have molecular computing networks which process information about internal operations and about the external environment to make decisions controlling growth, movement and differentiation.... One can characterize this surveillance/inducible repair/ checkpoint system as a molecular computational network demonstrating biologically useful properties of self-awareness and decision-making....

We are learning that virtually every aspect of cellular function is influenced by chemical messages detected, transmitted and interpreted by molecular relays. To a remarkable extent, therefore, contemporary biology has become a science of sensitivity, inter-and intra-cellular communication and control. Given the enormous complexity of living cells and the need to coordinate literally millions of biochemical events, it would be surprising if powerful cellular capacities for information processing did not manifest themselves."

A Third Way prods an objective scientific evaluation of information command and control within living cells. Indeed, the command and control functions within living cells evidence an informational intelligence that seems to be far beyond that of human intelligence. Let's take a look.

CHAPTER 22
Smart Information in Living Things

All living things contain an immense amount of information within. That information is contained within the macromolecule known as DNA. **The only function of the DNA molecule is to process and preserve information.** And, DNA performs this data processing skill through a highly-complex language function.

DNA is at core simply the arrangement of atoms in accord with the laws of physics. Atoms of five of the 92 naturally-occurring elements of the universe electrically bond together in precise sequences to create a long strand of nucleotides that join together to make-up DNA. Amazingly, **while the specific sequence combinations obey the laws of physics, the laws of physics do not necessitate those combinations. In non-living matter the laws of physics necessitate elemental interactions. In living matter they do not.**

From all that science has discovered to date, each and every living organism that now exists or has ever existed on Earth has information-bearing DNA molecules. The blueprint for all information about each and every part of each living organism on this planet is contained in the DNA of that organism. **DNA is the repository of all life information and contains both the blueprint for the construction of each part of every living organism, and the recipe for how to bring the blueprint to life.**

The nucleotides making-up DNA are housed in **chromosomes**. The chromosomes are made-up of both DNA and various proteins. A chromosome contains 5 to 10 times more protein than DNA. However, only the DNA portion of the chromosome contains the blueprint and recipe information.

The nucleotides of DNA are **sequenced in a specific order** along the strands of DNA to make-up distinct segments called **genes**. Historically, **the central dogma of genetics** informed us that **each gene codes for** the exact-right information necessary for the construction of **one specific protein.** The number of nucleotides chaining together to provide the information necessary for the construction of each individual protein varies greatly. For example, the gene for the protein insulin is just over 200 nucleotides long, the gene for the protein hemoglobin is about 400 nucleotides long, while myosin, the gene for the protein that controls bodily muscles is about 4,000 nucleotides long.

The current scientific estimate is that there are about 30,000 different genes in the human genome, consisting of three billion nucleotide bases strung out along 23 pairs of chromosomes. Amazingly, 30,000 is not too far removed from the number of genes found in the genome of an earthworm. And, humans have even fewer genes than a rice plant. How can that be?

Until very recently the central dogma of molecular genetics has held that each gene codes for one protein. The portion of the gene that codes for a protein (i.e., polypeptide amino acid sequence) is called 'coding' DNA. But, more than 90% of DNA does not code for peptide sequences. This non-coding DNA has long been referred to by mainstream science as 'junk' DNA. But, with the completion of the human genome project it has been verified that all is really not so simple. During the transcription process from DNA to mRNA, different portions of the nucleotide sequence of a gene are removed. In short, it has been discovered that through the transcription gene splicing process each gene codes for an average of three different proteins. It turns out the elegant process of protein construction is even more elegant than anyone had ever imagined. The 'junk' in DNA turns out to be very purposeful.

The specific sequencing of nucleotides in a gene which remains after mRNA editing provides the information needed for the specific sequencing of amino acids necessary for constructing a protein. **The genes do no work at all. They simply supply the information** necessary to construct the proteins that do the work of living organisms. **DNA does no work, however, DNA activates the work. DNA supplies the information that directs the work to be done by the proteins. And, DNA serves not only to store and transmit the necessary information, it also serves to activate the information. The information in DNA itself serves as its own activating agent. DNA is smart information.**

Smart Information

We humans think a lot about a lot of things. Our intelligence allows us to gain information from the world around us. We process that information in the language we think in via our on-board computer - our human brain. While human thinking by itself cannot transform information into action it has resulted in our ability to create machines that can process information and convert it into useful, purposeful action. Human intelligence has succeeded in creating smart information.

Brilliant scientists have learned how to successfully program information into computers that then serves to automatically trigger information into purposeful action. Let's look at a few examples.

A machine has been designed and created through human intelligence that is capable of automatically implementing a missile defense system if it senses an enemy attack. The machine can be programmed to automatically monitor for incoming missiles and then automatically initiate and conduct a counter-strike with no human input required. Information has been programmed into the machine that endows it with the ability to monitor its environment and to implement appropriate action in response to environmental inputs and challenges.

A machine has been designed and created through human intelligence that is capable of automatically monitoring home security. The machine can be programmed to automatically monitor for home intrusion, smoke and fire and poison gas and then automatically initiate the specific action required. With no human input required the machine will initiate a call for police or fire services and can activate fire sprinklers and gas control devices. Information has been programmed into the machine that endows it with the ability to monitor its environment and to implement appropriate action in response to environmental inputs and challenges.

Many, many other machines have been designed and created through human intelligence that possess smart information. Smart houses, automated assembly lines, and satellite intelligence monitoring systems are but a few examples. But, human intelligence cannot begin to rival the intelligence that has designed and created the smart information that resides within the DNA of every living organism.

The smart information machines that humans create require us to figure out the physics needed to accomplish the intended goal, and then to physically assemble the machines from the elements of the Earth. DNA, on the other hand, first assembles itself into a coded language in order to supply the information necessary for the creation and maintenance of all living cells and cellular components. Then it commences an immensely-complicated construction process and proceeds to build a fully-functioning living organism. A single cell not only contains all of the information required to create the living organism, it proceeds to do the task with the natural materials that are readily at hand.

Human Intelligence versus DNA Intelligence

There is a vast chasm between the very nature of human intelligence and cellular intelligence. The intelligence residing within living cells is an animate intelligence while human 'thinking' intelligence is inanimate. The known zenith of inanimate human intelligence reposes in the minds of brilliant scientists. The known zenith of animate cellular intelligence reposes in DNA, the ubiquitous molecule that provides the blueprint and recipe of information required for all life.

In our human world we use our intelligence to extract information about the physical world around us and figure out how to use that information to assist us. Then, we need to activate that information. We must transform that information into useful, purposeful action. We do that through our intelligence and creativity. To transform information into action requires human action. Our thinking alone does not activate information. We can't do anything by just thinking about it.

In the cellular world the information within **DNA** figures out how to use the information that resides within itself and **then proceeds to activate that information**. DNA itself transforms the information within into useful, purposeful action. Is that not evidence of intelligence and creativity residing within the DNA molecule? DNA's intelligence combines thinking and doing.

The information that resides within our human thought (our intelligence) cannot do anything all by itself. To actually do something we have to physically transform that thought into action. We can think all day long about making a cup of coffee or cleaning out the garage but, at the end of the day, if thinking is all that we do we end up with no coffee and a messy garage. Human 'thought' alone does not result in action.

To the contrary, the information that resides within our DNA (our DNA library) not only knows how to do stuff - it does stuff. If DNA wanted to make coffee or clean out a garage it would not just 'think' about doing those things. It would physically cause them to be accomplished. DNA 'thought' alone results in action. **The information in DNA is 'smart information'.** Let's look at a few examples.

Example #1

Our modern society runs on electricity. Without the elaborate power generation, distribution and switching system that brilliant scientists and engineers have designed we would have no electric light, or computers, or TV or radio and on and on. The information for constructing an elaborate power grid resides in a library of hundreds of books. Let's do a simple thought experiment. Let's take all those books containing all that information and place them on the ground where we want the power grid to be built, and see what happens. Nothing. The books just sit there. Nothing happens. All the information in those books does absolutely nothing to transform itself into useful energy, into purposeful action.

Now let's look at what happens with DNA. The information necessary for constructing a far more elaborate power grid resides within the DNA molecule. Within the DNA library is the blueprint for constructing and the procedure manual for operating each and every part of a central and peripheral nervous system that provides virtually instantaneous transmission of useful electrical energy to every part of the body. But, DNA's information doesn't just sit there. Not by a long shot. DNA proceeds to transform that information into the useful transmission of energy. DNA transforms information into purposeful action.

Example #2

Modern cities need a complex food distribution system. A critical part of that system consists of the roadways. Food is transported to the local grocery by delivery trucks along a series of roadways, descending from interstate highways to state routes to city avenues to neighborhood streets. The information necessary for building and maintaining that complex roadway system resides in hundreds of books. Let's try another thought experiment. Let's take all the books in the library that contain the information necessary for mining rock aggregate, producing concrete and asphalt, and designing and building roadways, and place them on the ground where we want the roadways to be built, and see what happens. Nothing. Again, the books just sit there.

Now let's look at what happens with DNA. The information required for constructing and maintaining an elaborate food distribution system to supply nourishment to each and every cell of the body resides in DNA. Within the DNA library is the blueprint for constructing and the procedure manual for operating a blood supply network, beginning with the large aorta of the heart and ending with the minuscule capillaries that supply nutrients to each cell through a blood vessel system of some 50,000 miles in length. But, that information in DNA doesn't just sit there. DNA proceeds to build and maintain the elaborate system of arteries, alveoli, and capillaries to the exact specifications provided in the information library. DNA transforms information into purposeful action.

Example #3 (Then we won't further belabor the point)

The brilliant scientists of our society have discovered an immense amount of information about human anatomy and physiology. Scientific and medical texts contain hundreds of volumes describing how we are put together and operate, from toenail to brain. A final thought experiment. Let's put all those books together in a large library room, close the door, wait a while and see what happens. Much later, open the door and observe. Nothing. The books just sit there on the shelves.

Now let's look at DNA. DNA starts off and is contained in one single, solitary human zygote cell. That initial DNA molecule is composed of 23 pairs of chromosomes – ½ from our mother and ½ from our father. That initial DNA molecule contains all the information required to construct and grow and protect and maintain a unique living, thinking human being for an average lifetime in America of close to eighty years. That single molecule of DNA contains all of the information that has ever been written in books about human anatomy and physiology, and much, much more that is still missing from the library of human knowledge.

But, that is but a small part of the wonder of DNA. DNA 'thought' alone then proceeds to transform the information within into physical reality. The DNA in the initial zygote cell assembles the raw materials of nucleic acids and amino acids within the cell and then proceeds to use them to construct a number of embryonic stem cells capable of becoming most anything. It then proceeds to differentiate further cells with further and further and further specificity into a blood circulation system, neural electrical circuit, heart, lungs, kidneys, stomach, hands, fingers, feet, toes, eyes, ears, lips …. Nine months later we hear a newborn baby's first cry and see the first smile (probably gas). Then, growth continues through infancy and childhood and the teenage years until mature adulthood is finally reached. At each stage we are protected from disease by an elaborate immune system and an exquisite blood coagulation system. And, each part and each system was constructed strictly in accord with the information contained in our DNA.

The resultant adult human being now contains literally trillions of living cells. Each of those cells contains the same exact DNA molecule. Each of those cells (which contains the same DNA molecule) knows precisely what genes need to be 'expressed' from the proper segments of the DNA molecule in order to provide for constructing the appropriate proteins necessary for the exact action needed for the proper function of that particular cell. In each and every cell DNA not only 'knows' what raw materials to use in the construction project, it causes the construction to occur and to be maintained as part of a fully-functioning living organism.

Not a single human thought is required to accomplish all of this. The only 'thought' is that of DNA itself. And, DNA itself puts 'thought' into action.

Smart Information in Our Computer Age

At the start of the 21st century it is most appropriate to observe that we live in a computer age. The exponential rise in electronic capabilities leaves most of us in absolute awe. Yet all of the stuff that derives from computer advances is ultimately based on a pretty straightforward computer language.

Man-made Computer Language

Most of us acknowledge that the computers that we use each day are the product of scientific genius being applied to the physical world. Our wondrous computer information systems are the product of an immense amount of effort by brilliant scientists being applied over many years to the physics of the ubiquitous atoms and molecules of the mechanical universe.

All of the voice, data, photo, streaming video, as well as all the written information stored and accessed through our computers is evidence of advanced intelligence. However, all of that incredible information is the simple product of sequences of but two language characters: 1 and 0. Two bases only for a binary code.

While based on a binary code sequencing of 1 and 0, the intelligence required to provide a reliable and complex information system using that simple two-character sequencing was immense.

Genetic Computer Language

Remember the central genetic dogma that **a gene is simply a specific segment of DNA that contains the information required to construct a specific protein**. That information directs the specific sequence of amino acids necessary to construct a particular protein. But, the nucleotides defining the genes of DNA cannot talk directly to the amino acids that make up proteins. A code is needed. **Each gene codes for a protein**.

Nucleotides use a language containing but **four** base pairs of nitrogenous bases. Living proteins contain amino acid sequences always comprised of some of **twenty** specific amino acids. **A code is needed to translate the language of DNA nucleotides (four units) into the language of amino acids (twenty units).**

It took a brilliant scientist, Francis Crick, and his colleagues many years to first discover the amazing properties of the DNA double-helix and then to discover the three-letter genetic code necessary to translate the language of DNA and RNA nucleotides into the language of amino acids in order to construct proteins.

The genetic code is a **three-letter code** using **specific combinations** of **three nucleotide bases** called **codons**. Crick found that the three-letter code needed to uniquely identify sequences of the four possible bases (4x4x4=64) provided 64 combinations.

Each unique combination of three nucleotide bases is called a codon. Most of the **codons code for a single amino acid.** Of the 64 codons, 60 code for a specific amino acid. **Three** of the codons code only for a **stop signal. One** of the codons codes as a **start signal**, and thus provides an unambiguous marker for the first nucleotide base in a sequence of codons.

In short, DNA is a simple, yet highly complex coded genetic language.

No reputable scientist seriously contends that the binary language system that is the foundation for our modern computers developed by-chance without intelligent input. Yet, most scientists today contend that the far-more-complex coded genetic language residing within the cells of every living organism happened by chance, with no intelligent input required.

The Laws of Physics Don't Account for Information

The laws of physics explain how the electrical bonding of five natural elements combine to create an organic molecule of DNA. **The bonding occurs in accord with the laws of physics. But, the laws of physics do not require the long coded sequences to occur in a specific order. Yet it is the very specificity of such ordering that provides information.**

There is no law of physics that assembles atoms and molecules into information, let alone 'smart information'.

The laws of physics in no way explain where the information residing in DNA comes from. Yet the amount of information contained in DNA is nearly beyond human comprehension.

The information stored in the DNA of a single-celled bacterium would fill 100 printed pages of a book. The information stored in the genes of a single-celled amoeba would fill eighty 500-page books. The information contained in the DNA within a single human cell would fill a library of some **1,000 books, each 500 pages in length**. That is the amount of information contained in **each and every one** of our 100 trillion cells.

The human genome is the list of coded instructions necessary to make-up a human being. There are over three billion letters (base pairs) in the human DNA code. Those coded letters reside in the nucleus of each and every one of the trillions of cells that make-up the human organism. That information is tightly wound in the pattern of a double-helix on twenty-three pairs of chromosomes inside each cell nucleus. If the DNA information reposing within a single cell were unwound it would straighten-out to a length of about five or six feet. **If all the DNA information in a single human being were straightened-out and placed end-to-end it would reach from the Earth to far beyond the Sun.** That amount of information is, indeed, beyond human comprehension.

If no information is directed into man-made computers the computers will contain no information. The atoms and molecules comprising the computer hardware do not obtain the ability to create and process information without intelligent input.

If no information is directed into the cellular computers of living organisms how do they obtain such a massive amount of information? How do the atoms and molecules comprising the living computer hardware obtain the ability to create and process information? Where did that information come from? No one scientifically knows. The most truthful answer cannot be derived from science.

It should be stressed that at least one brilliant scientist disagrees with this analysis. Theoretical physicist Paul Davies in his book, *The 5th Miracle,* provides an explanation for the origin of all information that makes it no big deal. Dr. Davies believes that information was simply squeezed out of matter by one of the fundamental forces of the universe. Dr. Davies proposes that the primal source of information was gravity acting on hydrogen gas in the early universe. He offers this 'scientific' explanation:

> "In some as yet ill-understood way, a huge amount of information evidently lies secreted in the smooth gravitational field of a featureless, uniform gas. As the system evolves, the gas comes out of equilibrium, and information flows from the gravitational field to the matter. Part of this information ends up in the genomes of organisms, as biological information."

Dr. Davies' explanation is certainly imaginative and certainly not scientific. **There is no law of physics that assembles atoms and molecules into information. And there is absolutely no scientific evidence that discloses where the massive amount of information comes from that both informs and activates the assembly and maintenance and reproduction of all living things.**

Dr. Davies' analysis is certainly in keeping with the mantra of mainstream science that demands a naturalistic explanation for all phenomena. Dr. Davies' analysis is based not on science but, rather, on the **belief** of elitist and mainstream science. That foundational belief is **that the 'smart information' residing within the DNA of all living things just has to be the result of a simple, naturalistic process called '*random mutation*'.**

How and why did such a scientific belief become so prevalent in mainstream science? Let's turn now to examine that question in Part VI.

PART VI

SCIENTIFIC BELIEFS

A belief is a firm conviction in the truth or reality of something. Many beliefs are supported by evidence. Many are not.

'We are what we believe' is a timeless adage that contains the wisdom of the ages.

Our beliefs are the foundation for our values, our virtues and our morals. Our beliefs define who we are, both as individual human beings, and as a human society.

Our beliefs are important.

The bases for our beliefs should be examined with the highest scrutiny and care.

CHAPTER 23
The Historical Scientific Worldview

On a cosmic and planetary scale the universe is quite orderly and it is in constant motion. Without any scientific training or instruments, by use of our naked eye and our common sense, we can observe the heavens and come to some real insights about our universe. We see the daily progress and reoccurrence of the Sun and the Moon. We witness the orderly changes of the stars and the planets nightly, and with the seasons. There is heavenly order everywhere we look. And heavenly motion.

The insights we glean from these observations may or may not be right. They are dependent on the same method that our forebears used when they formulated the first 'laws' of science. And, the 'laws' of science proclaimed by mankind have indeed changed greatly through the ages.

The goal of science was for our forebears, as it is now, to first determine how things work, and then to figure out how to make things work for us, to assist in our lives. People observed things and were filled with wonder. Because movement was pervasive everywhere, the first intrigue was with things that moved.

In our western society the first 'laws' of science were postulated by the Greeks shortly before, during, and after the period known as the 'Golden Age of Greece' (circa 500 BC). The ancient Greeks were very philosophical and wondered a great deal about the nature of things. They developed an orderly way of thinking that assisted in figuring out solutions to life's problems and mysteries. They believed that human reason, based on observations about physical phenomena, would reveal solutions.

The Beginning of Western Science

So, the study of 'science' seriously began about 2,500 years ago in Greece. The elite of Greek society had the leisure, the time and the inclination to ponder the real nature of things. And, so they did. They introduced **philosophy** to the world. Greek philosophy included both **metaphysics** and **natural philosophy**.

Metaphysics is a division of philosophy that is concerned with **the fundamental nature of reality and being.** As you ponder the question of the ultimate nature of reality and being you embark into the realm of the metaphysical, a realm that includes both philosophy and religion. That embarkation, of necessity, occurs when we humans, and that includes the scientists and the laymen alike, run out of 'proof' and begin to conjecture. As we look for the best answers to ultimate questions concerning the nature of reality and being, and purpose and meaning, we should seek the answers that are supported by the best evidence available.

Natural philosophy is a division of philosophy that is concerned with the study of things physical, their characteristics, their relationship to each other, and their purpose. Today, natural philosophy is called **science.**

Aristotle was undoubtedly one of the greatest scientists and philosophers in history. Aristotle began his seminal work, *Metaphysics*, with the pronouncement:

"All men by nature desire to know."

In other words, science and philosophy should not be reserved for the great thinkers. They should be available to be used by all of us to assist all of us in gaining knowledge and wisdom. Another of the world's greatest scientists, Albert Einstein, provided this additional advice for exploring difficult scientific terrain:

"Things should be made as simple as possible, but not any simpler."

The Birth of Natural Philosophy

When the Greeks embarked upon metaphysics as a philosophy it was for a purpose. The purpose was to seek knowledge in order to find the best 'truth'. Truth will always be sought. There can be no end of searching.

Greek metaphysics led, in turn, to the 'scientific' study of physical things in nature. They called such study **natural philosophy.**

Contrasted to metaphysics, natural philosophy (known today as science) is the study of the physical properties and composition of things. The core assumption of science is that the properties and composition of things can become known by the exploration of the natural law and order that governs them. That core assumption is known as **naturalism.**

The collective genius of ancient Greece believed that the pursuit of all knowledge was good, but the ultimate good was the pursuit of the 'truth'. They concluded that only man could discover the truth for only man, of all the Earth's creatures, has the capacity for reflective and abstract thought. Only by such reflection and abstraction is truth discovered by rational reason.

This rationality allows man to discover the truth of underlying relationships that are eternal in the universe - the unchanging truths. This truth is that which abides with us forever, unchanging. And, the core of that truth we know to be relational, like the right triangle where the square of the hypotenuse is always equal to the sum of the squares of the other two sides. And the triangle always adds up to 180 degrees. Like the area of a circle can always be found by multiplying the square of the radius by the wondrous number called Pi (3.1416 and on and on with the sequence never repeating). No matter the size of the right triangle or the size of the circle these are absolute truths that will always exist. Even after the Earth has suffered a heat-death, when the Sun consumes all of its hydrogen, these relationships will hold true. Mathematical proportion, harmony, and balance was believed to be the 'truth' found in the nature of things.

Plato thought that such relational truths were true 'forms' that had some kind of actual existence - not a material, but nonetheless a real existence. Aristotle thought that such 'forms' were only abstract representations of that which can only exist in experience. In either event, the search for 'truth' becomes a search for the kind of truth that is akin to such mathematical relationship certainties.

Experience cannot teach us these truths that are covered-up by the multi-tasking of daily life. The senses cannot pick-up such universals. **These truths are 'a priori' - prior to experience. A priori knowledge is intuitive knowledge, a part of the very gift of human rationality.** The relational truths of the right triangle and the circle are not discovered by experience. They are discovered by the gift of intuitive rational thought uniquely possessed by the species Homo sapiens.

Plato thought that such relational truths could best be discovered by dialectical inquiry, a rational discourse of philosophical give and take. Aristotle disagreed.

Aristotle thought that such relational truths could best be discovered by studying the natural world. While the senses only pick-up particulars and not universals, they can get us well on our way to the discovery of such truths. You can't see a universal, but you can see an instance of a universal. With this stroke of genius insight, the serious study of natural science began.

Aristotle was the preeminent natural scientist of ancient Greece. He felt that our human nature provided us with an inexplicable impulse to know things. We should study nature and the way things work, for the information thus obtained will provide knowledge and insight.

Aristotle believed that nature produces nothing without a good reason. He proclaimed that **chance is not the operative principle** of the universe. Rather, the **universe is infused with purpose** and things are best understood in terms of the purposes that they serve. **To understand things we must begin with sense perception**, because sense perception is the first in a chain of causality that leads to the discovery of truth. **Things are best understood in terms of the purposes that they serve.**

In examining the natural sciences, Aristotle contended that all things could be understood in terms of what caused them. The famous example he used to make the point is a marble statue of a great personage, like Socrates. **Four causes** are at play. The **material cause** is the marble substance itself. The **formal cause** is the form of the statue itself, its recognizable shape. The **efficient cause** is the sculptor's craftsmanship. The **final cause** is the final thing realized in time; but it also includes the reason and purpose of the thing.

This rigorous study of causes of physical phenomena led Aristotle to scientific discoveries in fields as diverse as biology and astronomy. **He strongly felt that the things of the universe, both animate and inanimate, do instantiate purposeful design. 'Nature does not do things without a purpose'** was his foundational belief. Unless you have the intelligent design in the first place none of the other causal modalities will operate to any effect. This is regarded as a **teleological** explanation.

Does a thing exist? To what degree does it exist? What is its relationship to other things? What is it for? That was Aristotle's 'scientific method' for studies as diverse as biology and astronomy. His method was far different from the method of modern science. His method was a mix of physics and metaphysics. Some of his miscalculations led to a false understanding of both the cosmic and biological systems. But, Aristotle's work paved the way for scientific advancement in the ages to come.

Aristotle's Science

Aristotle was the foremost Greek scientist. He was a student of Plato and a tutor of Alexander the Great, and he lived, worked, and thought in the middle of the 4th century BC. All he had at his disposal to use in order to determine the nature of things was his reason. He could rely on very little previously recorded science and he had no scientific instruments. Aristotle was a genius.

Through his observations, Aristotle concluded that our planet consisted of four elements: earth (including all matter), water, air, and fire. **Each of the four elements had a natural state and that state was *at rest*.** And, each of the four elements moved to attain its natural state. Matter was heavy and moved to get as near to the center of the Earth as possible. Water was lighter than matter and attained its natural state by pooling on top of the Earth. Air was lighter than water and rose to its natural state of rest in the sky. Fire rose to its natural state above the air and presented itself from time to time as lightning.

Aristotle concluded that, since earth included all matter in the universe that had come together to form our planet, the natural state at rest of matter would require our planet to be a sphere. He verified his conclusion through scientific observation during lunar eclipses, which revealed to his naked eye segments of the Earth with a curved outline.

Aristotle postulated that everything beyond the fire in the heavens above (including the Sun, Moon, stars, and planets) was composed of a fifth element that he called **ether**. The **natural state of these heavenly bodies was one of perpetual circular motion around the Earth**. Heavenly bodies had to be perfectly circular and orbit the Earth in perfect circles, because the circle was the most perfect geometric shape. What else but a perfect shape would the gods employ?

Thus **the first scientific laws of motion were formulated on the natural states of matter and motion: *at rest* for earthly matter and motion; and *perpetual circular motion* for heavenly bodies.** These laws were determined by Aristotle, not by use of the scientific method based on induction and experimentation. Rather, they were derived by deductive reasoning stemming from intuitive assumptions that were themselves based on observations. They held sway for a lot of people for a long time. They formed the basis for mankind's belief in scientific laws of motion for almost two thousand years. **This worldview was founded on a firm belief that the planet Earth was a different kind of place, different from the rest of the universe, and subject to special forces and laws unique to this planet.**

Aristotle's Universe

Aristotle's treatment of astronomy and cosmology was a blend of both the physical and the metaphysical, with a strong emphasis on the later. He had no scientific instruments for observation. He had only his naked eyes and mind to guide him.

Aristotle's model of the universe was an amalgam of 55 concentric crystalline spheres to which celestial objects were attached. The objects rotated at different velocities but at a constant angular momentum for each sphere. The stationary Earth was a sphere and was located at the center. The Sun, Moon, planets, and fixed stars were attached, each to their own sphere, and rotated around the Earth. The sphere of the fixed stars lay beyond the planetary spheres, and beyond the stars was the outermost sphere which was the domain of the Prime Mover, or Unmoved First Mover. All the spheres were contained within nine perfect circles expanding ever-outward from the Earth. The chain of causality began as the Prime Mover set the outermost sphere to rotate at a constant angular momentum and this motion was then passed on from sphere to sphere causing the entire cosmos to rotate.

Aristotle's universe was unchanging, immutable, forever-existing. The Earth and the heavenly bodies had always existed. There was no creation event. And, **in Aristotle's universe everything was goal-directed and purposeful.** The Unmoved First Mover somehow put the whole organism of the universe, including the Sun, the Moon, and the planets, into motion. All the parts of the organic universe tend toward the Unmoved First Mover, which is also the Final Cause. Not very scientific, but pretty neat metaphysics.

Science Through the Middle Ages

Aristotle had a huge impact on the study of science for nearly two thousand years. He bequeathed a form of analytical reasoning known as Aristotelian logic. He bequeathed an approach of observing the natural world of plants and animals. But, he never experimented. He believed that simple observation, by reason alone, would allow the mind to perceive all the laws of the universe and understand universal systems. And, that is the very 'scientific' approach that he bequeathed to the worlds of the Roman Empire and medieval Europe.

With the rise of Christianity in Europe, Aristotle became the foremost 'Christian scientist'

Saint Thomas Aquinas (1225–1274)

A definitive melding of the Christian faith with the works of Aristotle was attained by the genius of a Dominican monk named Thomas Aquinas in the 13th century. Saint Thomas' **Aristotelian Scholasticism** dominated the schools of Christian Europe until the 17th century.

An intelligent God who leaves nothing to chance is the very foundation of Aristotelian Scholasticism. Nature is purposeful. God does everything for a reason. Natural knowledge is joined at the hip with Christian faith. The knowledge of ancient Greek society coupled with the knowledge of the Christian scriptures handed down from the Church fathers comprises the **presumptive authority of the past**. The people of the medieval world had become the beneficiaries of the Greek authorities and their interpreters. **The search for truth would be prescribed by the paradigms of the past.** A core belief was that the truth of the past had withstood the test of time and need not be reexamined.

Thomas Aquinas **systemized the universe** and all therein in accord with Christian doctrine. All was systematically explained in accord with Aristotle's four causes. And the ultimate explanation for everything was always in terms of the **final cause, the purpose or the end served**, called **teleology**.

As a **system builder**, Aquinas went on to align Christian virtues with the virtues of moderation espoused by Aristotle. We first need to have a deep knowledge of the system as a whole and its implications for the meaning of life. Aristotle believed that a well-lived life was a flourishing life, and that in order to flourish it must be in harmony with nature, and the fulcrum of the virtues of courage, prudence, and the like must be the 'golden mean'. To these Aquinas added the Christian virtues of **faith, hope and charity**. This ethical system proposed that **the purpose of life is to conduct ourselves in such a way that we will be worthy of God's grace and to gain salvation of our souls upon our death**. One should live in earthly moderation and model our life on the life that Jesus lived, abundant in faith, hope, and charity. Absent the Christian theological virtues of charity, hope and faith, salvation cannot be attained. Only through God's grace can we have faith. Only through God's grace can we hopefully and faithfully know that there is but one God in three persons, that we are made in God's image, that Christ died to redeem our sins, and that through our faith we may be saved. Scientific inquiry was rooted in this religious Thomistic system of scientific belief.

Within this Thomistic system, **the point of science is to better gain knowledge of the divine essence of God.** Since God is perfect he necessarily has perfect qualities. Love, wisdom, spirit (and not body) are those qualities that are best, for they are shared by God. Importance on the scale of being is determined by how close a being's qualities are to the qualities of God. The mind should dwell on higher, not lower, things. Unchanging Heaven is more like God than always-changing Earth. So, it is best to study the immutable things of the heaven than the mutable things of the material earthly world. Astronomy is revered while the earth sciences, biology and physics are not.

Aristotle's universe, formalized by Ptolemy in the 2nd century AD, was the understood model of the universe that reigned throughout the birth of Christianity and into its maturation during the Middle Ages. The fundamental rift between Aristotelian and Christian views was that Aristotle's cosmos was eternal while the Christian God created the world by speaking it into being:

"In the beginning was the Word, and the Word was with God, and the Word was God".

The earlier writings of Saint Augustine had explained to the Christian world that Jesus Christ is the Word of the universe. Now, the writing of Saint Thomas explained that Aristotle's Prime Mover is the Christian God. Science and religion were in lock-step.

The Historical Copernican Revolution

Thomas Aquinas synthesized Christian theology with Aristotelian thought at a time when the Roman Catholic Church was, by far, the strongest institution on the European continent. His synthesis became the backbone of Christian philosophy and science throughout the 14th, 15th, and 16th centuries. Thinking took a real upswing as university scholars became a mix of both clergy and laymen in a scholastic environment that sought knowledge in conformance with the **Christian duty to strive to better know God.** During this tumultuous period of reinvigorated thought, the Church tried to strike a balance between the preservation of the Christian faith and heresies of the faith that could arise from free-thinkers striving to better know God. The balance eroded as inquisitions were instituted by the Church to stop heresies and punish heretics by imprisonment or death.

The Roman Catholic Church had adopted as its own the Aristotelian/Ptolemaic model of the universe. Aristotle's Prime Mover had become the Christian God, and other Aristotelian universal concepts had been wedded to Christian doctrine. Some examples may be helpful. The **natural state of all matter on Earth was at rest** while the **natural state of all heavenly bodies was perpetual motion in perfect circles**. The 'stuff' of the heavens was the ethereal, incorporeal substance called **ether**. All of the corporeal 'stuff' of the universe had settled into our spherical planet, Earth.

The Catholic Church did not consider the Earth to be an 'honored' center of the universe as is popularly depicted today. Rather, all of the bad 'stuff' of the universe, the heavy, nasty 'stuff' had settled into the Earth. **Earth was a special place** with different laws than the rest of the universe, but **not in a highly-valued sense.** The heavens were light and beautiful and perfect. The Earth was heavy and imperfect and sinful and furthest removed away from God.

Nicolaus Copernicus (1473–1543)

A Polish free-thinker, Nicolaus Copernicus, directly contradicted the long-established official version of the Aristotelian universe formalized by Ptolemy.

Copernicus was an amateur astronomer, a viewer of the heavens. His naked-eye observations revealed to him, as they had to others before him, that the planets did not move in perfect circular orbits as required by the Aristotelian model. The planets would, at times, be seen to regress in their movements. Because the Ptolemaic universe presented such a strongly established paradigm, the regressive movements were explained in accordance with that theory as an epicycle (a smaller perfect circle moving within a larger perfect circle). As even more irregularities were observed, epicycles upon even more epicycles were required to produce the perfect circularity of orbits that the old universal model required. So, epicycles, inside other epicycles, inside yet other epicycles became the necessary explanation to maintain the established paradigm.

Copernican genius provided for **a mathematically-derived solution** contrary to common-sense observation. Copernicus proposed that the Earth rotated on its axis, and that it, along with the other planets, revolved around a stationary sun. The **Copernican model** retained Aristotle's perfectly-circular celestial spheres, and the Sun stood still. But, the Copernican universe was no longer Earth-centered. It was Sun-centered, or **heliocentric**.

Copernicus was a devoutly religious man who happened to be an amateur, naked-eye astronomer. He was a doctor of canon law and well-knew that his views could be seen by the Church as blatant heresy. So, his theory of the heliocentric universe was published as *On the Revolutions of the Spheres* as he literally lay on his deathbed in 1543.

Francis Bacon (1561–1626)

Until the dawn of the 17th century the presumptive authority of the past dominated European scholastic thought. The Charter of Trinity College of Cambridge University admonished attendees to pursue scholarship in reliance on the course of Aristotelian Scholasticism:

"All students and undergraduates should lay aside their various authors and only follow Aristotle and those who defend him."

When Francis Bacon entered Trinity College at the age of 13, he was further admonished that the Charter forbade:

"...all sterile and inane questions departing or disagreeing from ancient and true philosophy."

To say the least, a free-thinker like Bacon was not to be so constrained in his thinking. He felt that Aristotle's philosophy and the method of deductive reasoning by disputation was 'barren for the production of works' and did not serve for the benefit of the human condition here on Earth.

In the 40th year of Bacon's life, as the 17th century began, he produced profuse writings that extolled a new method of inquiry. He contended that much could be learned by the deliberate study of the nature of things. Instead of engaging in scholastic rhetoric that led nowhere, what was needed was a **philosophy of nature that would lead to the benefit of the human condition** - that would be aimed at reducing human suffering and increasing human well-being.

He did not for a moment suggest that science was at odds with religion. Quite the contrary. **He redefined the Christian purpose for scientific inquiry. Instead of concentrating on final causes, scientific inquiry should be aimed at acquiring better understanding and thereby better human control of natural things in order to improve the mortal condition of mankind**. The motivation for Christian science should not be to better know God. Rather, the motivation for Christian science should be Christian charity. The 'thought' system of the Aristotelian scholastics was devoid of Christian charity. **Bacon's proposed system was founded on Christian charity.**

Francis Bacon's paradigm-breaking system for scientific inquiry was based on four themes:
- knowledge is human power over the natural world;
- the only way to know the nature of God's creation is to study the nature of the created things themselves;
- **the inductive method is the proper method of scientific discovery** - observe the particular occurrences in the natural world, form a hypothesis, and test the hypothesis against the observed particular occurrences by experiment - using rationalism to form the hypotheses; and
- **discovered knowledge is forever cumulative and forever open to correction or falsification.**

Bacon appealed to the Christian ethic with his pronouncement that seeking knowledge must be in the service of charity. We should strive through science to discover things that are useful to mankind, that reduce suffering and enhance well-being. **Let us break with the authority of the past and gain the fruits of new knowledge.**

"There is no sign more certain and more noble than that of fruits. In religion we are warned that faith be shown by works. It is altogether right to apply the same test to philosophy. If it be barren, then let it be set at naught."

Through the influence of Francis Bacon and other free-thinkers, **the contemplative scientific view of scholasticism yielded to the dynamic view of modern science in the 17th century.**

Johannes Kepler (1571–1630)

Johannes Kepler was a German astronomer and mathematician who developed his first two laws of planetary motion before the telescope was invented. He developed the laws by pure mathematics. He literally mapped geometry into algebra, proving his findings. His first law states that planets moved in ellipses around the Sun and the second law provides that the closer a planet comes to the Sun as it orbits, the faster it moves. His laws strongly supported the Copernican universe.

After Martin Luther published his 95 points of challenge to the abuse of Papal dispensations, a Protestant religious movement spread rapidly in Germany. Kepler studied theology at a Protestant seminary near his German hometown and went on to study mathematics and astronomy at a German university where he was introduced to the works of Copernicus.

Kepler was deeply religious, firmly holding to a Christian duty to better know and understand God and his works. He was strongly influenced by the harmonies, first examined by Pythagoras in ancient Greece, and believed that God had created the universe in accord with a mathematical plan. He was convinced by his professors to abandon plans to be ordained in the church and to pursue a teaching career, which he did. At the start of the 17th century he became the court astronomer and mathematician for Holy Roman Emperor Rudolf II.

In addition to his famous laws of planetary motion, his mathematical theories were explanatory for observations that Mercury and Venus are never seen far from the Sun, explaining that they lie between the Earth and the Sun. This comported with Copernicus. He published the first textbook based on Copernican principles, titled '*Epitome of Copernican Astronomy*'. His work had a great influence on converting other scientists to Copernicanism (as modified by the Keplerian twist of replacing perfect circles with ellipses).

The Protestant revolution did not overthrow the traditional Ptolemaic model of the Aristotelian universe. Kepler's Copernican views were anathema to the Lutheran Church. He was excommunicated in the 40th year of his life.

Galileo Galilei (1564–1642)

Galileo was an Italian scientist and astronomer who was a contemporary of Kepler. They initiated the work that laid the foundation for Isaac Newton's discoveries of natural laws. Galileo was the first to use the newly-invented telescope to view the heavens. His observations revealed many wonders that supported the Copernican universe and dispelled many of the axioms of Aristotelian science. His discovery of sunspots and valleys of the moon were at odds with the Aristotelian notion of perfectly formed heavenly bodies.

He was educated at the University of Pisa and became a professor of mathematics. He was a scientific experimenter, using controlled conditions to confirm his mathematical hypotheses. He uncovered laws of physics that were again at odds with the Aristotelian idea that rest was the natural state of matter on Earth. He proclaimed that rest was not a more natural state than motion.

Galileo's astronomical observations revealed other physical phenomena that supported the Copernican theory of a heliocentric universe. In addition to sunspots and moon valleys he observed that the planet Jupiter had moons that circled it and that there were large variations in the brightness of the planet Venus, facts that could not comport with an Earth-centered universe.

Books that supported Copernican theory as 'true' were placed on the Index of Forbidden Books by the Catholic Church. Catholics could use Copernican theory to make calculations as long as they never claimed that it was in fact the truth. In this way, **Copernicanism was taught as a theoretical system** in the universities. Free-thinking begets free-thinking and leads to further discoveries.

Galileo took great efforts to conform his writings to the standards of the Catholic edict. However, he did predict that the Copernican theory was 'probable'. That was enough for the Church to try him for heresy. He was convicted and sentenced to confine himself to his estate for the remaining decade of his life. Despite the fact that Galileo renounced his Copernican beliefs before his Catholic tribunal, today his name evokes the image of a courageous scientist standing-up against the constraints of religious fundamentalism.

Isaac Newton (1642–1727)

Isaac Newton was an English mathematician and physicist who is widely-regarded as one of the greatest scientists who has ever lived. His genius discovered universal laws that explain a wide variety of phenomena here on Earth and throughout the universe.

Like Francis Bacon before him, Newton attended Trinity College at Cambridge University, where he studied mathematics, optics, and astronomy. He later become a professor of mathematics at Cambridge and labored to understand the causes of physical phenomena that did not comport with the wisdom of the ages.

Legend depicts Newton's discovery of universal gravitation occurring when an apple fell on his head. He did in fact discover that the force of gravity which caused the apple to fall was one and the same force as that which held the Moon and the planets in their celestial orbits.

At the age of 45, he published his seminal work on mechanical dynamics, including his three laws of motion and his law of universal gravitation. He explained the mysteries of the physical world in understandable mathematical terms. The picture of the universe became firmly founded on universal laws that governed motion and its effects on bodies, including heavenly bodies. Because of the author's prominence in his own time and throughout history his *Principia Mathematica Naturalis Causae* has become known to the ages simply as ***The Principia.***

Newton's laws of motion provided the backbone for belief in an orderly deterministic universe. The universal laws of motion could be used to determine precisely what physical effect would result from a physical cause. **All phenomena could be explained by cause and effect**. Just like the gears and flywheels and springs of a clock work together to keep precise time. **A Clockwork Universe**.

Newton's **first law of motion** is called **the law of inertia**. It posits: A material body remains at rest, or if already in motion, remains in uniform motion with constant speed in a straight line, unless it is acted on by an unbalanced external force.

Inertia only fully accounts for things in an un-Earthly world where no friction or air resistance maintains. Because in our Earthly world a body in motion is always affected by a force of some sort, Newton's laws set the groundwork for the study of those forces.

Newton's first law introduces the concept of **force**. A force is anything that changes the acceleration of a material body. His second law goes on to explain the effect that a force has on motion.

The **second law of motion** determines the strength of a force. It posits: The acceleration (a) produced by a particular force (f) acting on a body is directly proportional to the magnitude of the force and inversely proportional to the mass (m) of the body. In equation form the law is: $a = f/m$. Simple conversion provides $f = ma$. Example: Place two tennis balls and a bowling ball on the kitchen floor. Gently tap the tennis balls at the same time and they roll ahead for the same distance at the same speed. Apply the same gentle tap to the bowling ball and, if it moves at all, it will be very slow and but a small fraction of the distance the tennis balls traveled. That's the second law of motion in action.

The third law of motion is commonly called **the law of action and reaction**, but can most accurately be stated thus: Whenever one body exerts a force on a second body, the second body exerts a force on the first body. **These forces are equal in magnitude and opposite in directions.** Example: Place the palm of your two hands together and press hard. Now quickly move the left hand away and notice how the right hand surges left. That's the third law of motion in action.

Newton proclaimed that the three laws of motion applied to everything on Earth. And, more importantly, his mathematical calculations disclosed and he began to document, that **the three laws of motion were universal**. They applied not just on Earth, but indeed, throughout the universe. And that was a huge discovery.

Newton was a religious man who was not at odds with divine intervention. He believed that God started and sustained the motions of the stars and the planets. It was left to the successors of Newton to remove God from the picture and describe a purely mechanistic universe, an idea that is traditionally ascribed to Newton. In time Newton's idea of a lawgiver God yielded to the modern scientific concept of an impersonal mechanistic universe operating out of necessity in accord with the laws of nature.

Isaac Newton's discoveries both shocked the world and changed its viewpoint. The Copernican Revolution was finally successful, and firmly set the Sun as a star in the heavens above, around which we on Earth and the sister planets of our solar system revolve. And everything on Earth and everything above is controlled by the same universal laws. The Aristotelian Universe has been succeeded by the Copernican Universe with a Keplerian / Newtonian update - a Clockwork Universe.

Until Newtonian physics was discovered, all things scientific were grouped under the banner 'natural philosophy'. Modern science with its many intricate fields of specialization was founded upon the establishment of the Newtonian deterministic universe. But, **the worldview of Sir Isaac Newton and his followers was founded on a firm belief in a creative intelligent agent of the universe, which most scientists called God.** In the march of time that worldview has changed dramatically.

CHAPTER 24
The Modern Scientific Worldview –
Behold the Accidental God

Icons of scientific genius, like Aristotle, Isaac Newton, and Albert Einstein, have long-maintained a firm belief in a creative intelligent agent of the universe, in accord with an 'old' scientific worldview. Abdus Salam, a renowned scientist who was awarded the Nobel Prize for Physics in 1979 affirmed this 'old' scientific worldview concerning the source of the wonders of the universe, as he remarked:

> "Now this sense of wonder leads most scientists to a Superior Being - der Alte, the Old One, as Einstein affectionately called the Deity - a Superior Intelligence, the Lord of all Creation and Natural Law."

This 'old' worldview of the mainstream scientific community was shared by the mainstream religious community. Purpose was everywhere. But, within the last few decades that view has drastically changed. **That change in worldview grew not from scientific discoveries themselves but, rather, from the methodology that was used to make those discoveries.**

The Scientific Method
The scientific method is a disciplined regimen that is used to discover truths about how things work. When scientists use this method they:
- observe natural phenomena in the world;
- formulate hypotheses that predict how the things they observe will work the same way again and again; and then
- conduct experiments to test their hypotheses over and over again. If a hypothesis holds-true and continues to explain the hard scientific data again and again, it gains the status of **scientific theory**. This has proved to be an invaluable method for discovering scientific 'truths'.

Predictions from hypotheses are tested by experiment. If the predictions can be tested by experiment, and the predicted results can thereby be reproduced unerringly, then a hypothesis becomes a valid scientific theory. And the rigors of the scientific method require that **a truly valid scientific theory must be capable of falsification** (it must be possible that somehow, someway it can be proved to be in error).

For phenomena that are not capable of direct observation, the scientific method employs the rigors of advanced mathematics incorporating mathematical proofs to discover underlying truths. Again, such mathematics-based theories must ultimately be capable of proof and falsification in order to qualify as valid scientific theory.

In this way, the scientific method is used by the scientific community for building supportable, evidence-based explanations for the wonders of our natural world. If the scientific method is not thus-employed to explain natural phenomena, then the resulting theory cannot validly be labeled as a true scientific theory. It remains in the realm of the non-scientific, the stuff of science fiction.

Scientific Paradigms

Paradigms are outstandingly clear examples or archetypes. And, archetypes are original models of which all things of the same type are representations. Such **paradigms become the basis for worldviews**. And, such paradigms stoutly resist change.

The members of the scientific community, like all of us, are human. And, the human species has always had great difficulty in breaking out of a paradigm box that circumscribes our thinking. It is very comfortable inside the box. Inside the box is where our value-laden world resides.

From the height of the ancient civilization of Greece until Galileo's newly-invented telescope two thousand years later, the mainstream scientific community staunchly upheld the **paradigm of a Ptolemaic Universe**, with the Sun and the Moon and the stars revolving around the Earth. That scientific theory supported the Aristotelian worldview that the natural state of all things in the heavens was a composition of perfect spheres moving through the 'ether' in perfect circles around us.

If a square peg does not fit into a round hole, make it fit. Pound it in with authority. Historically, that has been the view of mainstream science. When the Ptolemaic scientists observed a retrograde motion of planets that did not follow the required perfect-circle pattern, they simply made the new facts fit within the paradigm. They explained that such retrograde motions evidenced, in fact, smaller perfect circles moving within the larger perfect circle orbits. They did not abandon the 'truth' of the Ptolemaic Universe until the invention of the telescope provided incontrovertible evidence to the contrary. Thereafter, with the discovery that the Earth was not the center of the universe, the scientific community became engaged in a quest for knowledge about all things physical. Thus began the historical Copernican Revolution of the physical sciences.

By the end of the 19th century, through the painstaking work of James Maxwell in the field of electromagnetism, and other scientists exploring ocean waves and sound phenomena, much had been discovered. All of the scientific theories in these fields required that all wave energy, sound energy and electromagnetic energy must travel through a physical medium. Ocean waves traveled through the medium of water. Sound waves traveled through the medium of air. Light from the Sun had to travel through some medium, for that was the scientific paradigm. The paradigm required all waves to travel through a physical medium. Light must therefore travel through the medium called '**ether**', for Aristotle's 'ether' in outer space remained a scientific assumption. That paradigm was broken only through the genius of Albert Einstein who pronounced, with the proof of discovered mathematical equations, that light did not have to travel through any medium at all. Thus was born Einstein's theory of special relativity.

The ancient paradigm of an unchanging and eternal universe had great staying power. Albert Einstein did not want to believe in an unstable universe even though the equations of his theory of general relativity required that the universe be either expanding or contracting. In order to maintain a stable universe he postulated an unsupported anti-gravity force that he called a '**cosmological constant**' to keep things in balance. Einstein was much discomforted when scientific evidence was discovered by Edwin Hubble that revealed an expanding universe. He later much regretted this 'unscientific' effort to maintain the existing paradigm of a stable universe.

The New Paradigm for Physics

Exploring a Newtonian deterministic universe of cause and effect by using Bacon's **inductive scientific method** led to great discoveries in the physical sciences during the 18th, 19th, and 20th centuries. New knowledge about the physical structure of the things of the universe grew and grew. Nuclear physics and quantum mechanics provided new insights into how physical things actually worked on a level that was forever hidden from the naked eye. Astronomy and cosmology provided new insights into an expanding universe of finite age.

In the early years of the 20th century:

- Albert Einstein published his earthshaking theories of relativity which go beyond the theories of Newton regarding the universal force of gravity and the physics of bodies at speeds near the speed of light, and

- Max Planck began the discovery of quantum laws of physics for sub-atomic bodies.

Further discoveries using the scientific method in physics led most physical scientists inexorably to a worldview based wholly on naturalistic explanations of physical phenomena. As astronomers and physicists made more and more wondrous discoveries of how the non-living things of the universe developed through naturalistic processes in accord with the laws of nature, **the physical sciences thereby became firmly wedded to expanded naturalism, to wit: there is continuity in nature; scientific laws alone are sufficient to account for all phenomena; the laws themselves need no explanation; and there is nothing special about the planet Earth or its inhabitants.** This **expanded naturalism** of physics, is known as the **Copernican Principle. For the mainstream scientific community of the 21st century the Copernican Principle has became the accepted scientific viewpoint for scientific discovery concerning non-living things.**

The New Paradigm for Biology

In the latter half of the nineteen century, Charles Darwin published his earthshaking theory of evolution by natural selection. **Until Darwin's publication** of *On the Origin of Species by Means of Natural Selection*, the **scientists of the western world who studied living organisms believed in** *purposeful* **evolution of living things.** They believed that living organisms evolved slowly over time, from individual to offspring, generation after generation. However, they believed that organic evolution was guided by a principle akin to Aristotelian teleology. In essence, the principle held that living things were purpose-driven. In other words, evolution of living things happened not-by-chance, that evolution evidenced purpose.

Darwin's theory of evolution by natural selection changed that viewpoint. As Darwin's brilliant theory gained widespread support, scientists of the western world who studied living organisms shelved 'purpose' and replaced it with 'natural selection'. Darwinian Evolution is not purpose-driven. Rather, Darwinian natural selection is the result of random chance physiological changes through inheritance that just happen to prove valuable for species survival. **The biological sciences became firmly wedded to expanded naturalism, to wit: there is continuity in nature; scientific laws alone are sufficient to account for all phenomena; and the laws themselves need no explanation. Expanded naturalism, embodied in evolution by natural selection of random mutations, and known as Darwinian Evolution, became the accepted scientific viewpoint for the discovery of scientific facts concerning living things**.

Darwinian Evolution has become the new paradigm of scientific thinking in the life sciences. The mainstream scientific community is very comfortable with the paradigm. It well supports the doctrine of continuity in all of nature. And, continuity in all of nature is a necessary prerequisite for making anything supernatural unnecessary. Continuity in all of nature is a fundamental axiom of the theory of emergent evolution to explain that the mystery of life happened by-chance-alone.

Scientism - the Overarching Paradigm for Mainstream Science is Atheistic

With the Copernican Principle and Darwinian Evolution, expanded naturalism became the accepted scientific viewpoint for scientific discovery. When you couple the Copernican Principle together with Darwinian Evolution the combined naturalistic mantra is called **scientism.** And, scientism **is nothing more than the worldview belief called scientific atheism.**

Brilliant scientists devoted to the use of the scientific method made gigantic advances in science throughout the twentieth century. Expanded naturalism, embodied in the Copernican Principle and Darwinian Evolution, enabled dedicated scientists to make incredible discoveries about natural phenomena and to use that knowledge for the betterment of the human condition. Most scientists came to believe that nothing in the universe was planned. All evolved in accord with unplanned natural law and natural selection of random mutations. No outcome was intended. All was, at core, accidental.

Behold the Accidental God

Charles Darwin's hypothesis of emergent evolution, coupled to natural selection of random mutations for the explanation of the origin of life and everything else in the living world, has been seized-upon by the elite and mainstream scientists of the early 21st century to form the basis of what can best be described as a **scientific religion**. The religion is **scientific atheism**.

Darwin's theory of evolution replaced the 'old science' characterized by **immutability of species**. That 'old science' required a creator in order to make each species of plant and animal immutable. That was the paradigm of the 'old' scientific religion of creationism. Darwin's theory of evolution through natural selection provided for the **possibility** that a creator was not necessary. And, that possibility thus formed the basis for the formation of the 'new' scientific religion that required no purpose. Things that lie beyond the purview of the scientific method could only be explained in the same manner as the things that lie within its purview. **The metaphysics of mainstream science became firmly rooted in an Accidental God.**

Mainstream scientists accepted Darwin's theory of evolution by natural selection of random mutations on faith. If evolutionary change through adaptation occasioned through natural selection of random mutations could explain how the beak of one species of bird (like a finch) became larger over time than the beak of another species of finch, the theory could be extended without proof to explain everything organic. If evolutionary change through adaptation occasioned through natural selection of random mutations could explain how the color of a moth's wings changed over time as a result of acclimating to English air pollution, then it could be extended to explain everything organic. And, if everything organic can thus be explained in this manner, why not further extend this 'scientific' theory to explain the transition from inorganic to organic as well?

That is precisely what the mainstream scientific community did. And, the rest of us non-scientists have simply relied on their **expert authority** to believe that there really was a 'scientific' reality to the evolution of life from non-life. There is not now, and never has been, any such thing.

Prior to Darwin's theory of evolution, the basic model for the science of biology was **fixity of species.** And, fixity of species meant that each species was immutable. Darwin's theory of evolution changed all that. The theory proposed that such **immutability of species was in error**. Species were in essence mutable. **All organic creatures evolved from common ancestors through natural selection of favorable random mutations**. Therefore, the ends of the evolutionary process are simply the result of a series of events that simply began by random chance. **Design by a creator was replaced by natural happenstance.** All can now be explained by chance. No need now for a creator. No need for a Purposeful God. **Behold the Accidental God.**

The theory of evolution by natural selection provided for a scientific belief in the continuity of nature, and **nature must be continuous in order to rule out any possibility of supernatural design**. Thus the seed was sown that gave rise to a new **scientific religion**. And, the fundamentalist scientific community has dogmatically adhered to the tenets of that religion into the 21st century. That religion is **scientific atheism.**

Darwin's theory provided scientists with the crude ammunition to mount a full-scale scientific revolution that **would result in changing a worldview of created order into a worldview of happenstance.** The mainstream scientific community thus adopted an **a priori evolutionary bias** for everything. Evolution became a scientific requirement for everything. Scientists proceeded to establish a credo of impregnable axioms and tenets in support of the new 'truth' that has now become firmly established as scientific atheism. Doctrine of the religion of scientific atheism proclaims that there is no need for a creative intelligent agent of the universe. Homage is dutifully given to an Accidental God.

Some few scientists today, outside of the mainstream, have had the courage to question the scientific validity of the new religion. Michael Denton, a physician and molecular biologist, in his book *Evolution: A Theory in Crisis*, observes:

"Philosophers and historians of science will probably be debating the nature of the Darwinian revolution for years to come, but whatever their final verdict on this event, the facts themselves were not sufficient to compel belief in the continuity of living nature or to establish beyond reasonable doubt that the whole drama of life on earth was generated by the sorts of simple random processes responsible for microevolution on the Galapagos Islands."

"As the years passed after the Darwinian revolution, and as evolution became more and more consolidated into dogma, the gestalt of continuity imposed itself on every facet of biology. The discontinuities of nature could no longer be perceived. Consequently, debate slackened and there was less need to justify the idea of evolution by reference to the facts.

…Thus the all pervasive affirmation of the validity of Darwinian theory has had the inevitable effect of raising its status into an impregnable axiom which could not even conceivably be wrong. It is not surprising that, in the context of such an overwhelming social consensus, many biologists are confused as to the true status of the Darwinian paradigm and are unaware of its metaphysical basis."

One of the preeminent standard-bearers of scientific atheism is renowned Oxford University biologist Richard Dawkins. In *The Selfish Gene*, he makes the following pronouncement:

"We are survival machines - robot vehicles blindly programmed to preserve the selfish molecules known as genes. This is a truth which still fills me with astonishment."

Then in his subsequent book, *The Blind Watchmaker*, he goes to great lengths to supply the 'facts' in support of this 'truth' and concludes:

"Each individual organism should be seen as a temporary vehicle, in which DNA messages spend a tiny fraction of their geological lifetimes."

And, then he provides us with this metaphysical position:

"Where are these facts leading us? They are leading us in the direction of **a central truth about life on Earth**.…This is that **living**

organisms exist for the benefit of DNA rather than the other way around." (emphasis added)

He further acknowledges the depth of the mainstream scientific belief about Darwinian Evolution by natural selection of random mutations when he affirms:

"Today the theory of evolution is about as much open to doubt as the theory that the earth goes round the sun, but the full implications of Darwin's revolution have yet to be widely realized. Zoology is still a minority subject in universities, and even those who choose to study it often make their decision without appreciating its profound philosophical significance."

Dr. Dawkins is a brilliant and highly-renowned scientist. And, his beliefs are now widely shared by many other brilliant and highly-renowned scientists. They proclaim as scientific fact that which is not supported by scientific evidence. Yet, they proclaim that the 'truth' is based on good science. That has become the quite smug atheistic belief of the elite and mainstream scientific community.

Tim Berra, professor of zoology at Ohio State University, provides a pointed example of the extent to which the doctrine of scientific atheism now permeates all scientific fields in the mainstream scientific community. In his book, *Evolution and the Myth of Creationism,* he nonchalantly includes this bold statement:

"**Since life evolved from non-living matter**, at some point we must arbitrarily draw a line and say that everything past that point is alive." (emphasis added)

Dr. Berra simply states as a given that life evolved from non-life. That is not a statement of scientific fact. That is a statement of atheistic belief. And, of course, no scientific proof is necessary to support such a belief. The problem for us non-scientists is that this belief is clothed in the garment of science. We tend to believe scientific authority. And, **the mainstream scientific community has firmly embraced and accepted the a priori truth of evolution** through natural selection of random mutations as fact in order to explain every aspect of life and even the creation of life from non-life.

Through such pronouncements **the theory of evolution through natural selection of random mutations** has obtained the exalted status of an axiom of truth. It has **ascended to the status of a priori knowledge**. A credo of the religion of scientific atheism is bluntly stated by Dr. Dawkins:

"Darwin made it possible to be an intellectually fulfilled atheist."

The New Goal of Science

The Copernican Principle and Darwinian Evolution that provided the foundation for scientific atheism redirected the goal of science. Historically, the goal of science has been to search for the most truthful explanation based on evidence from the physical world. But now, **the Copernican Principle and Darwinian Evolution has redefined the search to be restricted to the best naturalistic explanation.** If design appears everywhere in the natural world it is of no consequence or significance. No amount of apparent design can support a conclusion of actual design. Actual design is forbidden by the axiom of expanded naturalism, or scientism, which is at the core of the Copernican Principle and Darwinian Evolution.

The Modern Scientific Worldview

The modern scientific worldview developed and grew as an unplanned and unintended consequence of scientific devotion to the scientific method. The 'old' worldview of not-by-chance and abundant purpose was abandoned by the mainstream scientific community. It was replaced by the 'modern' worldview of by-chance-alone and no purpose. The 'modern' worldview now held that the natural world and everything therein evolved purely by chance without any design or purpose.

This 'modern' worldview maintains that a creative intelligent agent of the universe is no longer necessary. This worldview postulates that the universe and all therein was created by and has developed through the process of pure chance. Everything has evolved, at core, by accident. There is no special significance to our place in the universe at all. We are simply the product of chance and circumstance. As such, there is, at core, no real purpose and meaning to the universe or anything therein, including us. There is no real purpose and meaning in life at all. Our existence, in essence, is simply the result of some cosmic hiccup. That's what this 'modern' worldview of the mainstream scientific community says. And, that worldview, supported by so many brilliant scientists, has a profound effect on the rest of us.

A Lay Perspective of the Modern Scientific Worldview

In Darwin's day, about a century and a half ago, science knew very little about the inner-workings of life at the cellular level. Cells were thought to contain a kind-of mushy protoplasm. Nobody knew about DNA or RNA or cytoplasm or ribosomes or amino acids or how proteins are synthesized. Nobody knew about the cellular intricacies and the immense complexities of bodily organs and metabolic pathway systems. Science knew nothing about the 'Big Bang' theory of universe formation. Nothing was known of Einstein's relativity. Nothing was known of the quantum mechanics acting within atomic structure.

Yet, the feature of assumed continuity of nature, the feature of naturalism that underlies Darwinian evolutionary theory, has expanded to provide a paradigm for scientific research. There is nothing inherently wrong with that paradigm. It may in fact be true. But it does not provide valid scientific evidence to prove either how life began or, indeed, how the universe began. Those are one-time-only events that are incapable of falsification.

The explanatory bridge between the first cellular life on Earth and our civilization of the 21st century encompasses one enormous span after another enormous span, and another, on and on.

The brilliance of our scientists has, through genius and diligent hard work, discovered a great deal about some of those bridge spans. Scientific advancements have been made by the rigid adherence to the scientific method underpinned by an insatiable quest for truth. That is the nature of good science.

But, there is much more to be discovered. And, much more that may remain forever undiscovered. It may well be that some of the fundamental intricate laws of the universe are so intricate that mere human intellect may forever be incapable of discovering them. There is certainly no logical reason to assume that we humans, ourselves the products of stardust, should be able to factually discover the ultimate realities of the universe or of life.

Scientists know a lot of stuff. They have discovered a great deal about how things work on our planet and throughout the universe. They have thereby developed many wondrous applications and inventions that provide for our great comforts of modern life. Their work allows us to live longer and to prosper far better than our forebears.

In America, everyone is entitled to believe in any religion that they choose. But, religion is not based on science. It is based on belief. Belief in things that are not capable of scientific proofs. That is just as true for atheism as it is for theism.

Brilliant scientists serve as our knowledge authority figures for things physical. We trust that they tell us truthfully only the results of good science. When they delve into the world of metaphysics they owe a duty to the rest of us to tell us just that.

Science has not and cannot provide the answers to the most fundamental questions regarding the mystery of life, the mystery of the universe, or our role therein.

Science can only provide us with incomplete information. The journey of science is invaluable for ascertaining the best answers possible for ultimate questions. But, science can only take us so far. We will always have to base our conclusions regarding the most fundamental questions on incomplete information.

At the end of the day, there are two and only two choices for the foundation of all metaphysical 'truth' that lies forever beyond the realm of science. **Either the things of the universe and life itself happened by chance or they did not. If they happened by chance there is no purpose and meaning in life other than that which we invent. If they happened not-by-chance there is a higher purpose and meaning. We may never be able to factually discover what it is. But, each of us can then purposely live our lives seeking to discover ultimate purpose and meaning through our human reason.** That is the bottom line.

The majority scientific view is by-chance-only. The minority scientific view is not-by-chance. Both views will, of necessity, be extrapolations from observations of natural phenomena. Newton and Einstein observed natural phenomena and concluded from their observations that purpose was abundantly evident in the universe. The majority of scientists today seem to have concluded to the contrary. So, which metaphysical view is better supported by the discoveries of science as being truthful?

The answer to that question is the subject of the next chapter.

CHAPTER 25
Do the Discoveries of Science
Support Scientific Atheism?

Metaphysics begins where science stops. Science cannot explain one-time-only events like the beginning of the universe or the beginning of life. But the discoveries of science can go a long way toward a better understanding of ultimate reality.

Elite and mainstream scientists today support the metaphysical position that the ultimate reality is chance. No planning is involved. No intelligent agent is required. The ultimate reality - God - is accidental. That metaphysical position is called scientific atheism. Do the discoveries of science support that position? Instead of relying on the authority of elite and mainstream science to tell us, let's briefly review the fundamental scientific discoveries and decide for ourselves.

Creation and Development of the Physical Universe - By Chance or By Design?

The creation and development of the physical universe was either planned by an intelligent agent or it was accidental. One or the other. There is no in-between.

The discoveries of science provide the following facts concerning the creation and development of the physical universe.

- Einstein's Theory of General Relativity has been confirmed again and again by scientific observational evidence. The hot 'Big Bang' theory for the origin of the universe is a foundational postulate widely accepted by elite and mainstream science and rests solidly on Einstein's theory of relativity.

- In accord with the Theory of General Relativity, the universe began about 14 billion years ago with a 'Big Bang'. From a point of unimaginable density, called a singularity, the universe burst forth with unimaginable heat and expanded at an unimaginatively-precise rate in accord with the laws of thermodynamics. If the expansion rate were any faster there would have been too little gravitational attraction for matter to ever form into stars. If the expansion rate were any slower the universe would have soon collapsed under the force of gravity. The expansion rate was 'just right' to allow our universe to develop as it has.

- Scientific observations disclose that today the universe is spatially 'flat' and expanding at precisely the rate required to avoid collapse. Today's 'flat' universe could not have evolved from a 'Big Bang' unless the initial curvature of the singularity was confined to a very incredibly precise parameter on the order of ± 1 in 10^{50}. While there are several mathematical quantum theories, there is absolutely no scientific evidence to explain how that extremely-precise expansion rate could have occurred naturally. How could such precision have derived from natural causes? is one way to frame the question; but, renowned astrophysicist Steven Hawking provides much better phrasing:

"Why should the universe have started off at the big bang in just such a way as to lead to the state we observe today? Why is the universe so uniform, and expanding at just the critical rate to avoid recollapse?... Also, the initial rate of expansion would have had to be chosen very precisely for the universe not to have recollapsed before now. This means that the initial state of the universe must have been very carefully chosen indeed if the hot big bang model was correct right back to the beginning of time. It would be very difficult to explain why the universe should have begun in just this way, except as the act of a God who intended to create beings like us."

- The quantum uncertainty of quantum mechanics does not allow for a singularity to have occurred, but all of the observational evidence discovered by science supports the beginning of the universe in a singularity. This presents quite a conundrum.
- The first and second laws of thermodynamics are inviolate laws of the universe. They always, always maintain. The first law is known as the universal law of the conservation of energy and says that the amount of energy in the universe has always been and will always be the same. Energy takes different forms and energy may be converted into ordinary matter, but the amount of energy is always perfectly conserved. No energy can ever

be lost and no energy can ever be created. Energy in the universe always flows from a more-highly-ordered state to a less-highly-ordered state and will continue to do so until a complete equilibrium state is reached. This is the 2nd law of thermodynamics. The energy in the singularity that preceded the 'Big Bang' was necessarily in the highest possible state of order. The energy within the singularity could not possibly have been the result of natural physical processes, for such would violate the fundamental natural laws of thermodynamics. The energy that emerged from the 'Big Bang' was, therefore, not of a natural origin.

- Before the universe was created the physical laws of the universe did not exist. Out of that 'Big Bang' creation event the physical laws of the universe emerged intact. Creation of something from nothing is not a natural act.

- The physical laws of the universe control the interrelationships between the fundamental forces and fundamental particles of matter and energy that sprang into existence at that time. Those physical laws provide that at the sub-atomic level of matter there is a fundamental underlying 'uncertainty' whereby it is impossible to causally predict with accuracy the effect that will result from the interaction of energy and matter. However, this underlying uncertainty only maintains in the quantum world within atomic structure. The same physical laws require matter and energy to precisely perform in a 'cause and effect' manner when size increases beyond the tiny scale of atoms and molecules to form the ordinary matter we observe in our everyday world. No one knows why such a dichotomy maintains.

- To accurately describe the phenomena of the material universe, the physical laws of the universe require a foundation of some 20+ constants. Those universal constants cannot be predicted by scientific theory, they have to be discovered by experiment. Without the precise non-arbitrary values of those universal constants our Sun, our Moon, the sister planets of our solar system and our home planet would not have formed as they have.

- No natural physical law requires the universal constants to be the precise values that they are, but without the exact precision of each of those 20+ constants the universe would not have developed in a manner that allows life, including intelligent life, to exist on Earth.

Creation and Development of Life on Earth - By Chance or By Design?

The creation and development of life on Earth was either planned by an intelligent agent or it was accidental. One or the other. There is no in-between.

The discoveries of science provide the following facts concerning the creation and development of life on Earth.

From the 'Big Bang' start of the universe to the birth of our Sun and solar system a long period of time elapsed. Over the course of roughly nine billion years, ubiquitous electrons and quarks developed into the atoms of the 92 discrete elements of the natural universe in conformance with the universal laws and constants.

Then, on Earth, sometime between 3 and 4 billion years ago, atoms from only 6 of those 92 elements teamed-up to make the foundational components of all life - carbon-based organic molecules. Carbon, hydrogen, oxygen, nitrogen, phosphorus, and sulfur make-up the four (and only four) organic molecule compounds in living organisms: carbohydrates, lipids, nucleic acids and amino acids.

In short, all living organisms contain organic molecule compound building blocks that are themselves made-up of atoms from but 6 of the 92 naturally-occurring elements in the universe. Those organic molecules then teamed-up to make a living cell.

Every living organism is made-up of one or more living cells. The simplest living organism is a type of bacteria composed of a single living cell. The most complex living organism is a human being made-up of trillions of cells.

The atoms from 6 elements somehow combined together into discrete arrangements to form 4 organic molecular compounds. Then the 4 organic molecular compounds somehow worked together to perform the discrete and essential functions that must all come together at the same time to create LIFE.

In order for life to begin and evolve as it has on Earth the organic molecules figured-out how to do some amazing things. Science has discovered that:

- The molecules of nucleotides figured-out how to arrange themselves into alternating base-pair segments of DNA. No natural physical law requires such arrangements.
- The molecules of proteins figured-out how to work with DNA to coil into spiraling chromosomes in order to store a massive amount of information

in a tiny space in a cell nucleus. No natural physical law requires such arrangements.

- The molecules of nucleotides in gene segments of DNA figured-out how to carry all the information necessary to construct and maintain each and every part of each and every living organism. No natural physical law produces such information. Information is not a physical state of matter.
- The molecules containing the genetic information in DNA figured-out how to use a code. No natural physical law produces an information code. Coded information is not a physical state of matter.
- The molecules of nucleotides in mRNA figured-out how to disassemble, copy, edit-out 90%, and then reassemble the information in DNA. No natural physical law requires such editing and assemblage arrangements.
- The molecules of nucleotides in tRNA figured-out how to translate the language of nucleic acids into the language of amino acids. No natural physical law requires such translation.

The advent of complex life presented further problems to be solved by molecules. Science has discovered that:

- The molecules in single-celled bacteria and green plants figured-out how to hold the 2nd law of thermodynamics at bay and capture the energy of the Sun to supply the energy needs of living things through the process of photosynthesis. No natural physical law required the development of photosynthesis. No scientist has ever been able to create photosynthesis.
- The molecules in multi-celled animals figured-out how to solve the very real problem of spontaneous combustion that would occur if glucose sugar produced in plants through photosynthesis were ingested without further transformation. They figured-out how to do this, and then further developed the processes of glycolysis and cellular respiration. No natural physical law requires either the 10 discrete reactions catalyzed by different enzymes during glycolysis or the 8-step cyclical process of the Krebs cycle.
- The molecules of living cells figured-out how to supply the enormous amount of energy needed for life by recycling non-energetic molecules of ADP into an abundant source of new ATP energy molecules through the elegant process of oxidative phosphorylation. No natural law requires such arrangements.

- The molecules of living cells figured-out how to reproduce themselves. No natural physical law requires such reproduction.
- The molecules of living cells figured-out how to reproduce an enormously-complex, multiple-trillion-celled organism through sexual reproduction. No natural physical law requires sexual reproduction.
- The molecules of living cells figured-out how to use 'transcription factors' to 'express' the genes required for each individual cell in order for embryonic development to occur and for the living organism to perform the functions required for LIFE. No natural physical law requires such arrangements.
- The molecules of living cells figured-out that in order to do all this amazing stuff, it would be necessary to establish and maintain complex electrical circuitry in all living things. So, they figured-out how to generate electricity through organic chemistry and then developed sodium/potassium pumps within every living cell in order to maintain a constant electrical charge. No natural physical law requires such arrangements.

Scientific experiments and observations have determined the fact that life is a most complicated process. In order to be alive each living organism, from the single-celled bacteria to the multiple-trillion-celled human being, must be capable of doing three specific things all at once. It must be able to: (1) store and process information; (2) acquire and use energy; and (3) reproduce its cells and itself. Each of these discrete functions is essential for any living organism, from the most complex to the simplest. Science tells us that:

- The simplest living organism that has ever been discovered is a single-celled bacterium. In order to be alive the simplest living organism had to have DNA that codes for at least 200 different proteins. Noted agnostic astronomer Carl Sagan stated that the odds are 1 in 10^{130} for the DNA coded sequence necessary for the construction of a single protein of an average length of 100 amino acids to develop by chance. The total number of atoms in the entire universe is 10^{79}. So, the odds against constructing only one protein by chance are astronomically greater than the odds of selecting just one atom from among all the atoms in the whole universe. And the odds simply get much, much worse when you consider the same is true for each of the other 199 proteins required to bring just a single-cell to life. When you deal with such fantastic probabilities most statisticians agree that there reaches a point where the highly-improbable becomes impossible. The eminent statistician,

Emil Borel defined that point on a cosmic scale as anything exceeding a chance of 1 in 10^{50}.

- To survive, a living organism must be able to self-replicate, and self-replication requires both functioning nucleic acids and proteins. The process of natural selection by random mutations cannot proceed without something that exists to mutate and select from. That cannot happen unless the organism has the ability to self-replicate in the first place.

Theories of Elite and Mainstream Science in Support of 'By Chance'

Elite and mainstream atheistic scientists review these scientific discoveries and facts and conclude that all was accidental. No planning was involved at all. Some of their reasoning follows.

Life-friendly Universe

Elite and mainstream science believes that all of these facts, and many more 'just right' or 'Goldilocks' facts about the origin and development of the universe may give the appearance of design but no actual design is involved. Chance alone, acting in accord with the necessity imposed by the laws of the universe (laws that themselves were established by chance alone), are sufficient to account for all of these highly improbable occurrences.

In his text, *Elements of the Theory of Probability,* eminent scientist Emile Borel explains that, contrary to popular belief, given enough time, anything **cannot** happen:

> "We may be led to set at 10^{-50} the value of negligible probabilities on the cosmic scale. When the probability of an event is below this limit, the opposite event may be expected to occur with certainty, whatever the number of occasions presenting themselves in the entire universe.... *Events whose probability is extremely small never occur.*" (Note: 10^{-50} is the same as 1 chance in 10^{50})

Most scientists concede that the likelihood of all of these improbable events conducive to the development of intelligent life on planet Earth occurring by chance **in our observable universe** reaches the point of impossibility.

Our observable universe began in a 'Big Bang' some 14 billion years ago. The probability of the exactness of the universal constants and laws leading to the development of a planet in our universe with just the right conditions for the creation and evolution of intelligent life is, indeed, unimaginable. The scientific laws of probability do not support such a chance occurrence. It is in fact impossible. Nevertheless, the universe happened. Not to be deterred by such a fact, some scientists have 'created' the following solution to the impossibility problem.

In support of their position of 'by-chance-alone' the elite and mainstream atheistic scientists maintain that **the impossibility problem is solved by postulating that the observable universe is not the only universe**. They contend that there may be infinite areas beyond the observable universe that contain other universal laws and universal constants. Therefore, **given an infinite number of universes and an infinite amount of time, they contend that there is a reasonable probability that sooner or later a universe such as ours would form with the exact conditions that it has which are conducive to intelligent life on planet Earth.** Given an infinite number of universes and an infinite amount of time the answer 'by-chance-alone' becomes possible. This is known as **the multi-verse** or multiple universes postulate. It is based on purely mathematical calculations and **is not supported by any scientific evidence whatsoever**.

Science has discovered and documented again and again that the universal laws of the universe are indeed universal. Scientific evidence discloses that every particle of ordinary matter in the cosmos conforms to the same laws. No one has ever experienced or observed anything in the universe that is not subject to the same universal constants and laws. Yet, based on no evidence, elite and mainstream science believes that there 'just has to be' a vast number of other real universes that are subject to entirely different constants and laws. Such is simply required in order for them to maintain their steadfast belief in scientific atheism.

Creation and Evolution of Life

Most elite and mainstream scientists believe that the origin of life on Earth began by the evolution of non-living matter through the process of natural selection of random mutations. These scientists further believe that the evolution of the adaptive traits of all living organisms evolved by the simple process of natural selection of random gene mutations.

Most scientists concede the point that the laws of probability weigh heavily against the position of 'by-chance-alone' for the creation and evolution of life on Earth. Some argue that the laws of probability are simply not important when it comes to supporting their belief that life simply emerged from non-life by chance. They support that position by using the analog that it is *possible* for a marble statue to wave its hand. Just because the odds against such a feat are unimaginably great, they are not incalculably great. George Wald, a member of the National Academy of Sciences who was awarded the Nobel Prize for Medicine and Physiology for his work on how the eye passes messages to the brain, provided this viewpoint based on an infinite timescale:

> "One only has to concede the magnitude of the task to concede the possibility of the spontaneous generation of a living organism is impossible. Yet here we are - as a result, I believe, of spontaneous generation....Given so much time the 'impossible' becomes possible, the possible probable, and the probable virtually certain. One has only to wait: time itself performs the miracles."

The science of statistics tells us that in our universe (14 billion years old) anything is **not** possible. Some brilliant scientists simply choose to ignore that scientific fact.

Others who believe in scientific atheism maintain that necessary DNA and protein molecules may have arrived ready-made on planet Earth as bacterial spores carried by a meteor or comet and seeded by either intelligent aliens or chance. And, some simply concede that the information in the DNA and RNA nucleic acids necessary to begin sustainable life is too complex to have developed by chance and suggest that life be viewed as simply a given, like energy and matter - just givens. Nonetheless, they sincerely believe that such a given happened strictly by chance.

Most mainstream scientists contend that the first living cell was constructed as a product of emergent evolution from non-living matter 'by-chance-alone'. This theory of abiogenesis theorizes that smaller building-block structures of amino acids, called 'proteinoids' must have preceded the development of proteins and that the proteinoids themselves evolved through natural selection in a step-by-step process. Abiogenesis further contends that RNA and DNA as well simply must have evolved by natural selection from simpler structures. While there is absolutely no scientific evidence in support of the theory, such simply has to be the case in order to maintain the certainty of expanded naturalism.

Proponents of scientific atheism believe that all species of life on Earth have evolved through the process of natural selection, using the single mechanism of random mutations. All organic creatures thus evolved from common ancestors through natural selection of favorable random mutations. The ends of the evolutionary process are simply the result of a series of events that began by random chance and proceeded through the process of random chance. Over evolutionary time, favorable traits that are presented by random mutations are retained in the gene pool of the organic species while unfavorable traits are eliminated from the gene pool. **There is absolutely no scientific evidence that the evolution of any new species has ever occurred by the process of natural selection of random mutations**. For nearly two decades laboratory scientists bombarded fruit flies with high levels of radiation and chemicals in an attempt to evolve adaptive traits by the mechanism of random mutation. Nothing emerged from the experiments but hideous mutations of fruit flies. A fruit fly never mutated into anything else. No new traits or species ever evolved.

Yet, these Darwinian evolutionists believe that every aspect of every complex organ of every living organism, from plant roots, to the lateral line system of a fish, to the eyes of an owl, to the heart of a lion and the human brain, have all evolved through successive, slight modifications by means of natural selection, as pure chance has presented candidate mutations over evolutionary time.

Most scientific atheists concede that it is hard to imagine how all of the immensely complicated processes within all living organisms came about through this process that is founded on such incremental changes being made by random mutations within living cells. However, they boldly contend that such is the only 'scientific' explanation that can be used to explain organic change. Any other explanation would allow for the possibility of a creative intelligent agent. And the elite and mainstream scientists of the 21st century have declared that such an explanation is not 'scientific' and, therefore, cannot be true, in spite of the evidence.

Applying Occam's Razor

The answers to metaphysical questions can never be fully ascertained by the facts adduced by science. But, scientific discoveries and scientific principles can certainly help point the way.

Since the 14th century, when science was still called 'natural philosophy', a principle has been used by philosophers and scientists alike to help arrive at the best possible 'truth' regarding complicated things. The principle is colloquially called Occam's razor.

"Plurality should not be posited without necessity." With those words a Franciscan monk named William of Occam succinctly captured a compelling principle for deriving the best explanation for complicated things. In essence the most straight-forward, least-complicated explanation is the best. Occam's razor should be used to shave-off all but the most essential components.

For quite a long while atheists have used Occam's razor to argue against the existence of God. However, it seems today that the discoveries of science better support the metaphysics of the existence of a creative intelligent agent over the metaphysics of by-chance-alone.

The hypothesis of a creative intelligent agent is a straight-forward explanation guided by the discoveries of science. On the other hand, the hypothesis of by-chance-alone requires a complicated set of beliefs, including:
- the existence of multiple universes;
- the suspension of the laws of probability; and
- random mutations that produce not only biological improvements, but coded information as well.

Most importantly, none of the evidence discovered by science supports any of those beliefs.

Simply stated, **Occam's razor trims a purposeful shape for ultimate reality.**

Do the Discoveries of Science Support Scientific Atheism?

My conclusion is no. But, you should decide the answer for yourself based upon your own best judgment. If your conclusion is yes, you are in the company of the scientific elite of this country. **The truth remains that neither conclusion is a scientific fact. Both theism and atheism are 'metaphysical' beliefs.**

Proclaiming that science has discovered evidence which 'proves' that the universe and life itself began and evolved with no need for a Purposeful God is simply not true. Proclaiming that an Accidental God of scientific atheism is a 'scientific truth' is the real fallacy of that religion. And, that fallacy renders great harm to our society.

PART VII

THE ATHEISTIC RELIGION OF SCIENCE HARMS SOCIETY

Everyone in America is entitled to practice any religion of their choosing. There is certainly nothing intrinsically wrong with the religion of atheism that attributes the ultimate reality of the universe and the things therein to the chance unplanned events of an Accidental God.

However, a harmful societal wrong occurs when atheism is given the imprimatur of authority as being a scientifically-derived 'truth'. Scientific atheism is thereby given an esteemed place in society different than all other religions. It is a religion that is not only allowed to be taught, but rather, required to be taught as fact in the science classes of public schools. The effect of such teaching is to instill in the youth of America the factual idea that a Purposeful God does not exist. The elimination of a Purposeful God results in the elimination of any basis for objective human morality based on natural law. A Purposeful God is the ultimate source of such natural law.

The nobility of science is based on the search for the best possible truth. The discoveries of science will always be limited to facts that can be derived through the use of the scientific method. But, to insist that ultimate truth can only be discovered by use of the scientific method has no basis in scientific fact. To demean the profession of science itself by belittling contrarian scientists who do not toe the party line is a hallmark of bigotry. And, bigotry in this country is deemed to be abhorrent and unacceptable conduct.

In short, to insist that scientific atheism is a fact undermines the very principles of American society that assemble diverse peoples into one moral nation.

CHAPTER 26
Scientific Atheism Begets Brilliant
Fundamentalist Bigots

Each and every day of this 21st century, dedicated scientists are discovering more and more about the wonders that underlie both the inorganic and organic worlds. Yet the elite life scientists of our country maintain the atheistic worldview that all of the wonders of the living world **just had to** develop by accident. They insist that all of the wonders of the living world, in which we live and of which our very minds and bodies are a part of, cannot have a supernatural explanation. To these scientists unknowns are not really deep mysteries but, rather, are simply things to become known. They tell us that as random genetic mutations enhance the survival of a species' genome through genetic warfare, natural selection explains all. Only 'natural' explanations are necessary to explain everything. And, they staunchly contend that natural explanations are ultimately based simply on mistakes and accidents.

They maintain that they have discovered that the only 'natural' explanation for the evolution of myriad species of living organisms is founded solidly on **random mutations** that happen to assist in a species survival. The underlying 'truth' of this biological atheistic dogma is accident. No matter how intricate or exquisite or improbable an aspect of a living organism's being is, the answer **just has to be** the result of random mutations - accidental.

Many of the intricate workings of the living cellular world have been discovered by brilliant scientists. They observe and explain the methodology whereby DNA is replicated, and spliced and coded to provide instructions for the proper alignment of amino acids to make an incomprehensible number of discrete proteins and enzymes. Then they declare that they **just know** that the coding and transfer of such vast amounts of information within each of our trillions of cells came about strictly by random mutations - accidental.

The scientific understanding of the informational workings of living DNA is very, very superficial. Yet the elite and mainstream scientists maintain that DNA random mutations cause each and every part of each and every exquisite system within each and every living organism to operate without any direction. They

just know that the ultimate cause is pure chance - accidental. They insist that they have solved the mystery and the solution is **undirected** natural selection - accidental. That is akin to my understanding of how my television set works.

To me the TV set works by my use of the remote control. Once I find the remote and press the proper buttons all of the wonders of television appear. That's all I need to know. That is what makes the TV set operate. I have now solved the mystery. To scientific atheists, a living cell works by DNA converting information, that is **somehow** contained in nucleotides, into the proper sequences to exactly align amino acids to perform a vast number of discrete, amazing functions throughout the body. DNA is the cellular remote control that makes the living organism operate. That is all they need to know. They have now solved the mystery.

The fact is that identifying the remote control as the mechanism whereby the television set works in no way explains the phenomena of what I see and hear emanating from the box. Likewise, identifying DNA as the mechanism whereby an organic body works in no way explains the phenomena of the myriad wonders occurring within the body. **A rudimentary understanding of how something works is not the same as a complete understanding of how and why it developed to work in the first place.** I may know how to use the TV remote control expertly, but I really don't believe that what the remote causes to happen within the TV box occurs by chance, accidentally. And, I really don't believe that the remote control itself just happened by chance, accidentally, even though at times it does seem to magically appear from beneath a couch cushion.

Brilliant scientists are becoming more and more expert each day at splicing and manipulating the genetic sequences of DNA. They are learning how to use the coded DNA remote control expertly, but, amazingly, most of them adamantly believe that what the remote causes to happen within an organic body occurs by chance, accidentally. And, they believe that the coded DNA remote control itself also developed by chance, accidentally.

These atheistic scientists are wholly committed to material, naturalistic explanations being the final answer. There can never be a scintilla of planning involved in the construct of each and every part of each and every exquisite system of each and every living organism. They have found the remote control that makes the wonders of life work. And, they simply know that even the remote came about by accident. That is all they need to know. They have now solved the mystery.

Today the group-think of atheistic scientists has developed a paradigm of scientific belief that is akin to the group-think of theistic scientists past. At one point or another in human history the elite and mainstream majority of scientists believed in such things as the 'truth' of:

- all outer space beyond our atmosphere being composed of a mysterious 'ether';
- an Earth-centered solar system;
- 'pangenes' of a mother and father 'preforming' a tiny, tiny person that simply grew larger within the mother's womb; and
- spontaneous generation of life from non-living matter, whereby frogs grew out of mud, rats grew out of garbage and flies were born from rotting meat.

In our past history the elite and mainstream majority of scientists believed in these things and maintained that they were true. They were absolutely certain that they were right. And, they were dead wrong. Yet they belittled and chastised anyone who challenged their certitude.

The group-think of the elite and mainstream majority of scientists past caused them to actually see a smaller epicycle within a larger epicycle in order to maintain belief in an Earth-centered solar system. The group-think of the elite and mainstream majority of scientists past caused them to deny the wonders of embryological development and actually see a tiny preformed person residing within an egg or sperm sex cell and to actually see rotting meat giving birth to flies.

The group-think of the elite and mainstream majority of scientists today causes them to actually see mistake compiled upon mistake, again and again and again, resulting in the development of exquisite living systems that are capable of intricate communications and command and control functions. They are absolutely certain that they are right.

Absolute certainty is the stuff of fundamentalism. And, fundamentalism fosters bigotry.

A fundamentalist is one who stresses strict and literal adherence to a set of basic principles. A bigot is one who is intolerantly devoted to his own opinions and prejudices. Sadly, fundamentalist bigotry is alive and well in the membership of America's most elite group of scientists today.

The Fundamentalist Principles of Elite Science

Expanded naturalism, often referred to as 'scientism', is the overarching belief of America's elite scientists today and rejects out of hand any implication of purpose or meaning in the universe or in life itself. It is based on two primary principles:

- the Copernican Principle, also known as the Principle of Mediocrity, which states that Earth is an ordinary planet in an ordinary solar system and that there is nothing special about our planet or its inhabitants; and
- the Darwinian Evolution Principle, which states that all living organisms are the product of undirected natural causes; that life self-assembles and evolves from non-living matter through the process of natural selection of random mutations; and that all species of living organisms evolve from other species through the process of natural selection of random mutations.

Strict and literal adherence to the Copernican Principle requires a **belief in the actual existence of a vast number of universes containing different universal laws and universal constants than those governing the universe that we live in.** There is absolutely no evidence in support of that belief. That belief is founded on the brilliant 'thoughts' of brilliant scientific minds. That belief is contrary to all evidential scientific discoveries, yet that is a foundational belief of scientific atheism.

Strict and literal adherence to the Darwinian Evolution Principle requires a **belief in the suspension of the laws of probability and a belief that random mutations produce not only biological improvements, but coded information as well.** That belief is founded on the brilliant 'thoughts' of brilliant scientific minds. That belief is contrary to all evidential scientific discoveries, yet that is a foundational belief of scientific atheism.

The Copernican Principle and the Darwinian Evolution Principle can never be questioned. If a square peg does not fit into a round hole, pound it in with authority. Fundamentalism never lets facts stand in the way of belief. Fundamentalism requires that your own opinions and prejudices **just have to be** true.

The Bigotry of Elite Science

Bigots are intolerantly devoted to their own opinions and prejudices. They simply cannot tolerate anyone who believes differently than they do. Bigotry is not rational. Bigotry is visceral.

The elite membership of the Institute of Medicine of the National Academy of Sciences blatantly derides any scientist who believes that random mutation is not the sole mechanism for evolving adaptive traits. In the following passage from the book *Science, Evolution, and Creationism* they make their bigotry very clear:

"Some members of a newer school of creationists have temporarily set aside the question of whether the solar system, the galaxy, and the universe are billions or just thousands of years old. But these creationists unite in contending that the physical universe and living things show evidence of 'intelligent design'. They argue that certain biological structures are so complex that they could not have evolved through processes of undirected mutation and natural selection, a condition they call 'irreducible complexity'. Echoing theological arguments that predate the theory of evolution, they contend that biological organisms must be designed in the same way that a mousetrap or a clock is designed - that in order for the device to work properly, all of its components must be available simultaneously. If one component is missing or changed, the device will fail to operate properly. Because even such 'simple' biological structures as the flagellum of a bacterium are so complex, proponents of intelligent design creationism argue that the probability of all of their components being produced and simultaneously available through random processes of mutation are infinitesimally small. The appearance of more complex biological structures (such as the vertebrate eye) or functions (such as the immune system) is impossible through natural processes, according to this view, and so must be attributed to a transcendent intelligent designer."

It is true that some brilliant scientists do support the hypothesis of 'intelligent design', a theory that the NAS derides. These contrarian scientists have produced a well-reasoned and thoughtful hypothesis that is at odds with the party line.

Since they are so bold as to hold a contrary view, NAS demeans them and paints them with a broad brush as pariahs. How dare these contrarian scientists suggest that the intricate inter-workings of the myriad exquisite structures and systems within every living organism did not develop by random mutations? Don't they realize that the NAS speaks as the greatest scientific authority in America and that the NAS has closed the book on evolution theory? The NAS simply knows that accidental DNA mutations are the sole cause of the evolution of adaptive traits in all living organisms. If a scientist is to be reputable, that scientist must conform his belief to the NAS dogma.

Throughout the course of history, mainstream science has insisted that the party line be endorsed and the elite scientists of the times have always established scientific dogma. It appears that not much has changed in that regard over the span of the last several hundred years.

During the 16th and 17th centuries the elite scientists of the times joined with mainstream Christian clergy to insist on the 'truth' that the Sun orbited around the Earth while our planet reigned as the center of the solar system. For fear of retribution Nicolaus Copernicus waited to publish his radical idea that the Earth actually orbited the Sun until he lay on his deathbed in 1543. In the next century the elite of mainstream science of the times provided the Catholic church with the cloak of scientific authority when Galileo was tried for heresy after he proposed that the Copernican theory was 'probable'.

It appears that the NAS now follows that long-established tradition of insisting that current prevailing scientific beliefs be maintained at all costs. The inference in their book is that any proponent of 'intelligent design' is a closet proponent of a young-Earth biblical interpretation. That inference is a personal affront to the reputation of dedicated scientists who add great value to society. By denigrating the work of any scientist who happens to disagree with mainstream science, the NAS undermines the very objectivity of science. The fact is that the NAS is dedicated to the elimination of design from scientific consideration and will revel in bigotry to accomplish that goal. And that is a travesty.

Contrary to the inference of the NAS, the 'intelligent design' hypothesis is not tied to established religion. You don't have to be a born-again Christian to believe in the hypothesis. Rather, some proponents of intelligent design hypothesize that perhaps when life first appeared on Earth 3.5 billion years ago it was the work of a supernatural agent who designed a mechanism that allowed the first living single-cell organism to emerge on Earth, complete with the necessary components for life: DNA, RNA, and proteins. **The first living single-cell organism would have contained all of the information necessary to, over the course of evolutionary time, construct all of the intricate organs and complex biological pathways that are present in living organisms everywhere on Earth.** Perhaps, the first living cell, containing all of the information necessary for all organic development, proceeded to then engage in a Darwinian struggle for survival in a changing environment. **The information necessary to allow creatures to successfully engage in that struggle was contained in the information-bearing molecules of DNA.** As the molecular information system directed the construction of complete organic creatures, they thereby acquired the ability as individuals to deal with the vagaries and vicissitudes of physical existence in a changing external environment. That interplay, **over the course of evolutionary time**, could have resulted in the great diversity of living organisms on our planet in the 21st century. In short, **design may be an undiscovered self-organizing principle of the universe and design may be imbedded in living things as information**. Perhaps, just perhaps, science should not turn a blind eye toward exploring that hypothesis. After all, more and more scientific evidential discoveries of natural genetic engineering and information science seem to support that hypothesis.

Contrary to the bigoted characterization of the NAS, this 'intelligent design' hypothesis does not seem to be the radical view of someone who believes that planet Earth is only 6,000 years old. Based on the factual discoveries of science to date, 'intelligent design' seems to be a quite credible theory.

Evidence of the Fundamentalism and Bigotry of Elite Science

In their 2008 publication the NAS makes the following bold assertion:

"However, the claims of intelligent design creationists are disproven by the findings of modern biology. Biologists have examined each of the molecular systems claimed to be the products of design and have shown how they could have arisen through natural processes."

First Example

The first molecular system that the NAS then proceeds to examine is a flagellum. A flagellum is an organ of a single-celled bacteria that functions as a rotary engine that enables these bacteria to move through liquid.

The flagellum has a motor that is used to rotate the propeller. The motor is powered by the energy produced by a flow of acid through the bacterial membrane.

In order to properly perform the function of propulsion, the bacterial flagellum must have at least three main parts working in concert with each other: (1) a motor; (2) a rotor; and (3) a propeller paddle.

Those who are familiar with the whims of outboard motors well know that all of the parts must be working together from the outset or the boat just sits there or aimlessly drifts away. If any of the major components are missing or don't work, the entire function of propulsion does not work. The same thing is true for the flagellum of these single-celled bacteria.

The complete flagellum machine consists of at least three major parts working in concert to perform the single function of propulsion. At least 200 different kinds of proteins must be constructed by the single-cell of the bacteria to accomplish that result.

All of the proteins involved in this function are perfectly constructed and aligned to fit together so that their cooperative efforts produce but a single function for the bacterium. It can swim through liquid.

At the level of protein molecules, the nuts and bolts of the molecular machine must provide for nearly perfect fit if usable function is to be achieved. At the ribosome site in the cell specific amino acids are joined by peptide bonds to construct a chain of amino acids to form but a single protein. The single protein of a specific size and shape is thereby constructed in accord with the instructions provided by the organism's DNA. DNA further directs other amino acids, that serve to make-up all living things, to chain-together to make-up another protein, and then another, and another. Amino acids group together as instructed by DNA to form the remainder of the **200 proteins with unique shapes and sizes.**

The modern theory of Darwinian evolution has been updated to recognize that natural selection on the molecular level can only be achieved through modifying the information in the genes. Modified genes that favor the organism's survival become predominant over time in the whole population of the organism's species by becoming predominant in the species' gene pool. That is the essence of modern Darwinian evolutionary theory. And, the Modern Synthesis of Evolutionary Theory maintains that such gene modifications occur only through random mutations.

The modifications necessary for changing a heritable trait so that it may benefit the organism's function must be made to the organism's DNA, the information-bearing molecule for all living organisms. And, the DNA does nothing itself. It simply contains the information necessary to build the discrete protein machines necessary to perform any function of the organism, including the propulsion function of swimming bacteria.

The genes are simply segments of DNA that code for specific proteins. **The true work of evolution by natural selection on the molecular level must, therefore, provide information for 200 different proteins to be constructed in a manner whereby they cooperate with each other to perform the single function of propulsion via flagellum.** Each protein either becomes a part of, or catalyzes a part of, the to-be-completed **bacterial outboard motor.** In this manner 200 different proteins work together to form a tiny, tiny organ for bacterial propulsion.

The NAS believes that the flagellum evolved by random mutation of the bacteria's genes. Evolution by random mutation requires that this bodily organ was produced in a step-by-step evolutionary process of successful mutation upon successful mutation, and on and on. All 200 proteins are necessary to act in concert to produce the function that benefits the organism - swimming through liquid. Proponents of 'intelligent design' assert that evolution by random mutation would produce nothing but a series of non-functional parts, none of which would have been favored by natural selection. In fact, the evolutionary theory of evolving adaptive traits through random mutation would require that the non-functional parts would have been ejected from the gene pool of the organism's species.

To gain understanding of a complicated issue, one is well advised to follow the real evidence offered, eschew broad generalizations, and **take it slow**.

The NAS in their publication states that reputable scientists have shown how this flagellum was produced through the natural random mutation process. They then offer this specific evidence:

"There are many types of flagella, some simpler than others, and many species of bacteria do not have flagella to aid in their movement. Thus, other components of bacterial cell membranes are likely the precursors of the proteins found in various flagella."

That is the first 'scientific proof' offered in support of the contention that this complicated organ had to evolve in a step-by-step process of random mutations. That 'proof' is constructed entirely of 'hole cloth'. The facts stated provide absolutely zero information about how anything happened by random mutation. Let's take it slow:

- There are many types of flagella, some simpler than others - TRUE
- Many species of bacteria do not have flagella - TRUE
- Based on these two truths, the 200 proteins in flagella likely came from ancestor proteins - DOES NOT FOLLOW FROM THE FACTS STATED.

The second (and last) 'scientific proof' offered by the NAS in their publication is this:

"In addition, some bacteria inject toxins into other cells through proteins that are secreted from the bacterium and that are very similar in their molecular structure to the proteins in parts of flagella. This similarity indicates a common evolutionary origin, where small changes in the structure and organization of secretory proteins could serve as the basis for flagellar proteins. Thus, flagellar proteins are not irreducibly complex."

Again, the facts stated provide absolutely zero information about how anything happened by random mutation. Let's take it slow:

- Some bacteria contain proteins that are very similar to some of the proteins in flagella and those proteins are used to inject toxins into other cells as a means of protection - TRUE
- This similarity indicates a common evolutionary origin - NOT TRUE (This is a statement of BELIEF NOT BASED ON EVIDENCE)
- Thus, flagellar proteins are not irreducibly complex - DOES NOT FOLLOW FROM THE FACTS STATED

The only 'evidence' provided in support of the proposition that this flagellum was constructed through step-by-step random mutations of DNA is far too attenuated to ever be considered 'scientific'. Biologists have discovered that a sub-set of proteins (10 of them) that is used in constructing the flagellum (10 of 200) has been found in another cellular organism, wherein the 10 component proteins work together as part of a different function that has nothing to do with propulsion. That sub-set of 10 proteins also serves as a component of a biological process to kill-off cellular invaders - a function of protection, not propulsion. **The extrapolation that they then make is that this homology of a sub-set of 10 proteins helping to perform a different function in a different organic system is 'scientific proof' of evolution of adaptive traits through random mutations.** No explanation is offered for how the other 190 proteins evolved and no one can explain how the 10 proteins in the sub-set ever teamed-up in the first place. Without real evidence we are asked to simply believe the proposition that evolution of adaptive traits by the step-by-step process of random DNA mutations explains evolving sub-sets of protein groups that then further group-up to perform novel functions. That is at best superficial science. Without real evidence, that is not science at all. Yet, that is the total extent of 'scientific evidence' adduced to date in support of the proposition that bodily organs have evolved simply by the process of natural selection using solely the mechanism of step-by-step random mutations of DNA. Explanations including **'are likely'** and **'could serve'** are not the stuff of science.

Now consider the exponential orders-of-magnitude complexity involved in the development of our exquisite body organs like the heart, the lungs, the eye, and the brain. The 100+ billion nerve cells in our brain and nervous system all work together to control our every movement, sensation, thought, and emotion. And there is not a scintilla of scientific evidence in support of the proposition that evolution by natural selection using solely the mechanism of step-by-step random mutations of DNA is an adequate explanation for any of these wonders.

Second Example

In their 2008 publication the NAS then makes this further bold assertion:

"Evolutionary biologists also have demonstrated how complex biochemical mechanisms, such as the clotting of blood or the mammalian immune system, could have evolved from simpler precursor systems."

The second molecular system that the NAS then proceeds to examine is the mammalian blood clotting system. So let's first review the system and then look at the evidence provided by NAS.

Our blood clotting system is intricate and interrelated and serves to keep us alive as it 'cascades' to our rescue each time that we cut ourselves.

Blood is an essential organ of the body. We usually don't think of blood as an organ, but that is what it is. Blood carries nutrients and oxygen that are the essentials for life to all of the cells throughout the body. Oxygen is used by the cells for combustion that releases our source of energy for life.

About 45% of our blood is made-up of red blood cells and white blood cells. About 55% is made-up of **plasma.** Blood plasma is **a clear liquid** and it contains lots and lots of proteins called **clotting factors** that are used to form blood clots. One particular **protein, called fibrinogen, provides for the fibers that actually form a blood clot.** But, **by itself, fibrinogen is useless. Without the timing and placement necessary for a proper blood clot that is provided by numerous other proteins, blood loss resulting from a scrape or a cut would never stop. We would bleed-out and die.** All of the proteins working in concert together through an intricate **blood clotting cascade** are necessary in order for an adaptive trait (i.e., **the function of stopping blood loss)** to be successful.

Stopping blood flow when the body is cut requires the solidification of the blood, a wondrous process called **coagulation.** A solid barrier needs to be formed at just the right time and at just the right place or we will bleed to death. This function of stopping blood loss is done through the **blood clotting cascade.** And, the blood clotting cascade not only has to be turned on to start the clotting, it also has to be turned off at some point or else all of the blood in our body would continue to solidify. We would thereby die.

The process of coagulation is basically (like most of the wonders of the body) the work of proteins. The major proteins involved in human blood clotting, called the **clotting factors,** are **twelve in number**. Nearly all of them are produced in the liver. **None of these proteins are used for anything else. That is their sole purpose. If any one of them is dysfunctional, the blood will either not clot or else it will never stop clotting. The end result of either of those occurrences is death.**

Most of the major proteins in the clotting cascade spend their time just circulating in the bloodstream in an inactive state until some other protein signals them into action, to then turn on or turn off other chemical reactions making up the cascade. And, the cascade is very dynamic, with the initial stages of the clotting system feeding back information to make more of the initial activating proteins, and so on.

Let's take a closer look at what actually happens when we suffer a cut through the skin.

Upon being cut the blood vessel itself constricts to limit the flow of blood to the area of injury. Then a proteolytic protein enzyme called **thrombin activates platelets** that are floating around in the bloodstream. A proteolytic enzyme is an amazing molecule that splits protein bonds and adds elements of water, thereby forming simpler and soluble substrate biological products.

Platelets are cell fragments that look like tiny broken plates. When platelets are activated by thrombin they spring into action and become the **first line of defense** to stop the bleeding. They cause constriction in the smooth muscles surrounding the cut by releasing restriction chemicals that serve to narrow the hole. At the same time they aggregate together to form a sticky platelet plug to fill the hole. They complete their paramedic function as they **signal back to thrombin to begin the blood clotting cascade.**

You will recall that the fibers that actually serve to form the blood clot initially come from the protein called fibrinogen. **Fibrinogen itself is composed of six protein chains.** Upon receiving the signal from the platelets to begin the cascade, **thrombin slices-off** several pieces from the complex **fibrinogen** protein that itself is just **floating around in the bloodstream. The newly-trimmed protein thereby becomes active** and is thereafter called **fibrin**. The shape of the fibrin molecules causes them to cris-cross over each other to construct a **fibrin mesh** that entraps the escaping blood cells. The fibrin mesh ensures the stability of the platelet plug.

Finally, a plasma protein, called anti-thrombin, binds to the active forms of the various proteins involved in the clotting cascade and inactivates them. The blood clotting process thereby stops. And, after the wound has fully healed the clotting is then dissolved through the activation of **plasmin**.

The blood clotting process, that commences with the signal from the platelet, involves intricate pathways that connect, disconnect, reconnect, send signals back to the start for more help, connect, disconnect, reconnect again, and so on. It **is called a cascade** because as each chemical tumbles into the area of the wound it sets off another specific chemical, for all **the different steps in the cascade must occur in proper sequence or the function of blood clotting will not be accomplished.**

Each new step in the blood clotting process requires a protein enzyme that is just minding its own business in an inactive state. And it also requires another enzyme whose whole purpose is to activate or deactivate that inactive protein enzyme.

Hemophilia is a human disease that evidences a problem in the blood clotting cascade. Specific health problems routinely arise for patients with deficiencies in specific clotting factors. For example, hemophilia A evidences a deficiency of factor 8, hemophilia B evidences a deficiency of factor 9, and hemophilia C evidences a deficiency of factor 11. Without medical intervention to artificially inject missing chemicals or otherwise stop blood loss, serious health problems may lead to death. If any one of the twelve clotting factors in the cascade is dysfunctional the blood clotting system will fail.

Proponents of 'intelligent design' contend that the blood clotting system could not have developed by natural selection of random mutations. They contend that the exact sequencing requirements and the orchestrated feed-back features, that are an integral part of twelve major parts working together in cooperation to accomplish a single function, are **strong evidence against step-by-step evolution** of each of the component parts. Evolution by natural selection of random mutations requires that the selected adaptive trait **must be immediately useful for the organism,** not simply useful sometime in the future. **Natural selection by random mutations would work to eliminate the inactive proteins that, while essential for the cascade to properly function, simply do nothing except wait to be activated or deactivated by other proteins. By themselves, just floating along in the bloodstream, they are of absolutely no immediate value to the survival of the organism. The stepwise progression required for evolution through natural selection of adaptive traits through**

random mutations would simply remove them. They further contend that if an evolutionary step-by-step process were used to construct this intricate biological system each step would necessarily produce a fatal flaw. At each step the organism would die and that would lead to extinction.

NAS strongly disagrees and offers the following evidence to prove why 'intelligent design' is an incorrect hypothesis because the hypothesis of random mutation explains all. To gain understanding of a complicated issue, one is well advised to follow the real evidence offered, eschew broad generalizations, and **take it slow.**

The first 'proof' offered by the NAS in their publication in support of their position that they have shown how the blood clotting cascade was produced through the natural random mutation process is this:

"With the clotting of blood, some of the components of the mammalian system were present in earlier organisms, as demonstrated by the organisms living today (such as fish, reptiles, and birds) that are descended from these mammalian precursors. Mammalian clotting systems have built on these earlier components."

The second (and last) 'scientific proof' offered by the NAS in their publication is this:

"Existing systems also can acquire new functions. For example, a particular system might have one task in a cell and then become adapted through evolutionary processes for different use….Molecular biologists have discovered that a particularly important mechanism through which biological systems acquire additional functions is gene duplication. Segments of DNA are frequently duplicated when cells divide, so that a cell has multiple copies of one or more genes. If these multiple copies are passed on to offspring, one copy of a gene can serve the original function in a cell while the other copy is able to accumulate changes that ultimately result in a new function. The biochemical mechanisms responsible for many cellular processes show clear evidence for historical duplications of DNA regions."

These 'scientific proofs' are constructed entirely of 'hole cloth'. The facts stated provide absolutely zero information about how anything happened by random mutation.

The entire 'scientific proof' that is offered to explain how the exquisite process of blood clotting developed through step-by-step random mutations boils down to this:

- Some components of the mammalian blood clotting systems are found in fish, reptiles, and birds who preceded mammals - FACT
- The mammalian blood clotting system built on them - NOT FACT (this is a statement of BELIEF NOT BASED ON EVIDENCE)
- Existing systems also can acquire new functions - FACT
- Molecular biologists have discovered that gene duplication is an important mechanism by which biological systems acquire additional functions - FACT
- DNA makes a copy of itself each time a cell divides - FACT
- Segments of DNA are frequently duplicated when cells divide, so that a cell has multiple copies of one or more genes. If these multiple copies are passed on to offspring, one copy of a gene can serve the original function in a cell while the other copy is able to accumulate changes that ultimately result in a new function - NOT FACT (this is a statement of BELIEF NOT BASED ON EVIDENCE)
- The biochemical mechanisms responsible for many cellular processes show clear evidence for historical duplications of DNA regions - A FACT extraneous to the point.

NOTE: ALL OF THE FACTS STATED PROVIDE ABSOLUTELY ZERO EVIDENCE THAT ANYTHING HAPPENED BY RANDOM MUTATION.

Read again just some of the intricacies of the blood clotting cascade. The 'scientific proof' offered by the NAS in support of the proposition that blood clotting developed as an adaptive trait (that keeps us from bleeding to death each time we are cut) by the step-by-step process of random mutations provides no real evidence at all.

Without real evidence we are asked to simply accept the proposition that evolution of adaptive traits occurred through the step-by-step process of random DNA mutations. That is at best superficial science. Without real evidence, that is not science at all. Yet, that is the total extent of 'scientific evidence' adduced to date in support of the proposition that complex biological mechanisms, like blood clotting and our immune system, have evolved simply by the process of natural selection using solely the mechanism of step-by-step random mutations of DNA.

That's All Folks!

It is an amazing fact that **we have just finished the review of the totality of scientific evidence provided in support of the hypothesis that all of the complex organs and metabolic pathways of all living organisms have developed through the process of natural selection of random mutations of DNA. That's it.**

The classic example for explaining how a complex organ developed through natural selection was first provided by Charles Darwin and has been updated by modern biologists. It regards the evolution of the human eye. The explanation goes something like this:

- Some organism, simply by chance, developed a light-sensitive patch of skin. That somehow provided the organism with the ability to somewhat discern shapes in his environment. That ability provided the organism who possessed that trait with a better chance to survive than others of its kind who did not have the light-sensitive patch. Thus equipped for better survival, the trait of the light-sensitive patch was passed-on to more offspring and became predominant in the organism's gene pool.

- Some offspring who had inherited the trait, simply by chance, developed an improved light-sensitive patch that now provided the ability to 'barely see' through a 'cloudy' eye. That provided the organism who possessed the 'improved' trait with a better chance to survive than others of its kind who did not have the 'improved' trait. Thus equipped for better survival, the 'improved' trait was passed-on to more offspring and became predominant in the organism's gene pool.

- Over eons of evolutionary time further improvements in sight, developing simply by chance, provided a survival advantage and thereby became predominant in the organism's gene pool, and in the gene pool of the myriad species that evolved thereafter. Improvement after improvement was retained until the mechanism that provides us with three-dimensional color vision evolved.

Charles Darwin called the process 'evolution of species by means of natural selection'. He was not aware of the cellular processes of genetic inheritance but he was quite sure that the mechanism that controlled genetic inheritance had to be an undirected, 'by chance', mechanism. Modern biologists filled-in the missing piece and announced that the undirected 'by chance' mechanism was random mutation of DNA molecules that are part of all living cells. So, that is the up-dated explanation of the modern synthesis of biology.

Michael Behe is a dedicated scientist who offers 'intelligent design' as an alternative to random mutation. In his book, *Darwin's Black Box*, he challenges elite and mainstream science to actually look for evidence based on the discoveries of cellular biology. To illustrate the magnitude of the challenge, he provides the following biochemical sketch of the eye's operation:

"When light first strikes the retina a photon interacts with a molecule called 11- *cis*-retinal, which rearranges within picoseconds to *trans*-retinal. (A picosecond is about the time it takes light to travel the breadth of a single human hair.) The change in the shape of the retinal molecule forces a change in the shape of the protein rhodopsin, to which the retinal is tightly bound. The protein's metamorphosis alters its behavior. Now called metarhodopsin II, the protein sticks to another protein, called transducin. Before bumping into metarhodopsin II, transducin had tightly bound a small molecule called GDP. But when transducin interacts with metarhodopsin II, the GDP falls off, and a molecule called GTP binds to transducin. (GTP is closely related to, but critically different from, GDP.)

GTP-transducin-metarhodopsin II now binds to a protein called phosphodiesterase, located in the inner membrane of the cell. When attached to metarhodopsin II and its entourage, the phosphodiesterase acquires the chemical ability to 'cut' a molecule called cGMP (a chemical relative of both GDP and GTP). Initially there are a lot of cGMP molecules in the cell, but the phosphodiesterase lowers its concentration, just as a pulled plug lowers the water level in a bathtub. Another membrane protein that binds cGMP is called an ion channel. It acts as a gateway that regulates the number of sodium ions in the cell. Normally the ion channel allows sodium ions to flow into the cell, while a separate protein actively pumps them out again. The dual action of the ion channel and pump keeps the level of sodium ions in the cell within a narrow range. When the amount of cGMP is reduced because of cleavage by the phosphodiesterase, the ion channel closes, causing the cellular concentration of positively charged sodium ions to be reduced. This causes an imbalance of charge across the cell membrane that, finally, causes a current to be transmitted down the optic nerve to the brain. The result, when interpreted by the brain, is vision.

If the reactions mentioned above were the only ones that operated in the cell, the supply of 11-*cis*-retinal, cGMP, and sodium ions would quickly be depleted. Something has to turn off the proteins that were turned on and restore the cell to its original state. Several mechanisms do this. First, in the dark the ion channel (in addition to sodium ions) also lets calcium ions into the cell. The calcium is pumped back out by a different protein so that a constant calcium concentration is maintained. When cGMP levels fall, shutting down the ion channel, calcium ion concentration decreases, too. The phosphodiesterase enzyme, which destroys cGMP, slows down at lower calcium concentration. Second, a protein called guanylate cyclase begins to resynthesize cGMP when calcium levels start to fall. Third, while all of this is going on, metarhodopsin II is chemically modified by an enzyme called rhodopsin kinase. The modified rhodopsin then binds to a protein known as arrestin, which prevents the rhodopsin from activating more transducin. So the cell contains mechanisms to limit the amplified signal started by a single photon.

Trans-retinal eventually falls off of rhodopsin and must be reconverted to 11-*cis*-retinal and again bound by rhodopsin to get back to the starting point for another visual cycle. To accomplish this, *trans*-retinal is first chemically modified by an enzyme to *trans*-retinol - a form containing two more hydrogen atoms. A second enzyme then converts the molecule to 11-*cis*-retinol. Finally, a third enzyme removes the previously added hydrogen atoms to form 11-*cis*-retinal, a cycle is complete."

These intricate biochemical processes of vision result in the gift of sight. And, these biochemical processes operate in strict accordance with the laws of nature (physics and chemistry) discovered by brilliant scientists. However, nothing in nature requires these biochemical processes to occur in the first place. Yet, elite and mainstream science steadfastly maintains that these exquisite biochemical processes **just had to** occur accidentally through random mutations of DNA.

The elite scientists of the National Academy of Sciences belittle proponents of 'intelligent design' for not adhering to the 'truth' that each of these exquisite chemical processes that allows us to see the world **just had to** happen through the single mechanism of random mutations of DNA molecules. Elite and mainstream scientists seem to find the Darwinian explanation for the evolution of a 'cloudy eye' from a 'light-sensitive patch of skin' much more satisfying than attempting to explain the step-by-step evolution of the intricate biochemical processes that actually result in vision.

Fundamentalism Begets Bigotry

When you **'just know'** that you are right you also must **'just know'** that the other guy is wrong. Fundamentalists will promote their beliefs at all costs. If they find it necessary to degrade anyone who holds an opposing view in order to maintain the 'truth' of their position, no problem. Maintaining the sanctity of the belief justifies bigotry. Not much of a basis for pursuing 'truth'.

CHAPTER 27
Scientific Atheism Is A Religion and Should Not be Taught in Public Schools

For elite and mainstream science the answer to the question – 'by chance or by design? - is chance. That answer is deeply rooted in a core belief in expanded naturalism, often referred to as scientism.

There is nothing wrong per se with the naturalism principle. It works great for experimental science. But, when it goes beyond the realm of science to become a worldview for scientists, it must be recognized that such a worldview is based on a **metaphysical belief**, **not scientific fact.**

Basing the 'modern' worldview on expanded naturalism is quite understandable. Mainstream scientists have adopted the conclusion of the scientific elite that **the only possible explanation** for each and every aspect of the physical universe and all therein **is the same as the method** that they use to practice science, namely **naturalism**. Naturalism has served them so well in their profession it is easy to see why they would be wedded to it as the basis for metaphysical belief. But, that does not make it a fact.

The **'modern' worldview** of elite and mainstream science is based on a closed mind. It is based on this **over-arching axiom for 'truth': Truth can only be revealed through the scientific method which is itself based on naturalism.**

The 'Truths' of Science Have Natural Limits
The scientific method is the instrument of discovery, based on naturalism, that has consistently been used to make great scientific discoveries about the natural world. Science is thus rooted in naturalism, and rightly so. However, the use of the scientific method necessarily limits the range of scientific knowledge to the physical world. And, **science has not discovered that all of the things and phenomena observed in the physical world just had to be the result of purely chance natural causes.** Such a conclusion has no scientific basis whatsoever. It simply reflects a scientific atheistic belief. It is a metaphysical position that has no scientific basis.

Yet the elite scientists of the National Academy of Sciences are insistent that their metaphysical position is an absolute 'truth'. And, they insist, based on their scientific authority, that this belief should be taught as scientific fact in the public schools of America.

The creation of life is an ultimate question for every religion on Earth. Every religion has a creation story. If you have any doubt about what is being taught in science class **as fact**, those doubts can be laid to rest after reading the creation story of the elite scientists of the NAS in their publication *Science, Evolution and Creationsim*. Their narrative for the creation of life is certainly fascinating and, most certainly, not based on scientific evidence. It reads much like the articles of faith of a religion. The creed begins:

> "Figuring out how life began is both an exciting and a challenging scientific problem.... Nevertheless, researchers have been developing hypotheses of how self-replicating organisms could form and begin to evolve, and they have tested the plausibility of these hypotheses in laboratories. While none of these hypotheses has yet achieved consensus, some progress has been made on these fundamental questions.
> Since the 1950's hundreds of laboratory experiments have shown that Earth's simplest chemical compounds, including water and volcanic gases, could have reacted to form many of the molecular building blocks of life, including the molecules that make up proteins, DNA, and cell membranes. Meteorites from outer space also contain some of these chemical building blocks, and astronomers using radio telescopes have found many of these molecules in interstellar space."

Anyone reading these two paragraphs cannot avoid the clear impression that the writer intends to convey the message that life simply had to emerge from non-life. We are told that while there is not yet a scientific consensus, much progress has been made in showing how life evolved from non-life by natural causes. We are told that for over half-a-century laboratory scientists have been amassing evidence as to how the building blocks of life have formed, and that scientists have even found those building blocks in meteorites and in interstellar space. What we are not told (because this story is a creed) is:

- After more than 50 years of attempts in the laboratory, no scientist has ever been able to naturally self-assemble a chain of amino acids to construct even one simple protein.
- Nor have they been able to naturally string together a series of nucleotides to construct even one RNA molecule or one DNA molecule.

Those omitted scientific facts seem to be somewhat significant. Let's continue with the creed:

> "For life to begin, three conditions had to be met. First, groups of molecules that could reproduce themselves had to come together. Second, copies of these molecular assemblages had to exhibit variation, so that some were better able to take advantage of resources and withstand challenges in the environment. Third, the variations had to be heritable, so that some variants would increase in number under favorable environmental conditions."

Anyone reading this paragraph cannot avoid the clear impression that the writer intends to convey the message that science has discovered that these stated conditions are those that are required in order for life to emerge from non-living matter. We are told, in essence, that in order for a living cell to emerge from non-living molecules, those non-living molecules had to reproduce themselves and evolve in accordance with the same process of natural selection of random mutations that elite and mainstream science uses to explain the evolution of all living organisms. What we are not told (because this story is a creed) is:

- After more than 50 years of attempts in the laboratory, no one has ever observed any non-living molecules combining and reproducing themselves.
- The idea of non-living molecules exhibiting heritable variations, and that the favored variants increase under favorable environmental conditions, is simply the extension of Darwinian evolution principles to non-living things. Not only is there no scientific evidence in support of this idea, the extension of genetic mutations to molecules that have no genes requires more than a little bit of wishful thinking.

Those omitted facts seem to be somewhat significant. Let's proceed with the creed:

"No one yet knows which combination of molecules first met these conditions, but researchers have shown how this process might have worked by studying a molecule known as **RNA**. Researchers recently discovered that some RNA molecules can greatly increase the rate of specific chemical reactions, including the replication of parts of other RNA molecules. If a molecule like RNA could reproduce itself (perhaps with the assistance of other molecules), it could form the basis for a very simple living organism. If such self-replicators were packaged within chemical vesicles or membranes, they might have formed 'protocells' - early versions of very simple cells. Changes in these molecules could lead to variants that, for example, replicated more efficiently in a particular environment. In this way, natural selection would begin to operate, creating opportunities for protocells that had advantageous molecular innovations to increase in complexity."

Anyone reading this paragraph cannot avoid the clear impression that the writer intends to convey the message that life arising from non-life is a foregone conclusion of science. We are told that 'no one yet knows' how it happened, leaving the clear impression that it is only a matter of time. We are then told a tale of 'ifs' and 'coulds' and 'mights' and 'protocells' and 'self-replicators' in such a creative manner that it leaves the clear impression that something real is happening. What we are not told (because this is a creed) is:

- All of this is total speculation. The tale is woven out of complete 'hole cloth'.
- There is no such thing as a non-living 'self-replicator'.
- There is no such thing as a 'protocell'.
- Science has calculated that the simplest independent single-celled living organism must have at least 200 genes coding for at least that many proteins in order to survive. So, **the first life on Earth must have had at least 200 different proteins in that first single-celled bacterium.**
- The information for constructing each of those 200 different proteins had to be contained within the cell's nucleic acids, copied, transcribed, edited

and decoded into the language of amino acids, and then transported to the proper location within the cell for each protein to be built, and then further transported to the site in the cell for proper function. Not once. At least 200 different times **before the cell could be considered 'alive'.** All that information processing and transfer and protein construction had to occur before 'life' occurred on Earth.

These omitted facts seem to be somewhat significant. Let's proceed to complete the creed:

"Constructing a plausible hypothesis of life's origins will require that many questions be answered. Scientists who study the origin of life do not yet know which sets of chemicals could have begun replicating themselves. Even if a living cell could be made in the laboratory from simpler chemicals, it would not prove that nature followed the same pathway billions of years ago on the early Earth. But the principles underlying life's chemical origins, as well as plausible chemical details of the process, are subject to scientific investigation in the same ways that all other natural phenomena are. The history of science shows that even very difficult questions such as how life originated may become amenable to solution as a result of advances in theory, the development of new instrumentation, and the discovery of new facts."

Anyone reading this paragraph cannot avoid the clear impression that the writer intends to convey the message that science 'just knows' that life evolved from non-life. We are told that scientists 'do not yet know' all the details of how life happened but that life most certainly was an outgrowth of 'natural phenomena' and the job of science is to find out how life originated from natural processes. What we are not told (because the story is a creed) is that in order to be alive a living cell must be able to do three discrete things:

- A living cell must be able to obtain, store, and process all of the information necessary to first bring the cell to life and to then provide for its every function.
- A living cell must be able to extract energy from its environment and then convert that energy into a useful form that is required to power the processes of cellular metabolism that are essential for all life.

- A living cell must contain the information and possess the ability to replicate. It must be able to clone itself, and the living organism must be able to produce off-spring.

No scientist, indeed no one on Earth, knows how a living cell became endowed with the ability to do any one of these discrete things, let alone all of them together at the same time. That is a scientific fact.

Scientific Atheism is Not Based on Scientific Facts

All of this exquisite orchestration of information and usable energy and function has to be explained somehow. How did life begin? What is the best scientific answer? **The best scientific answer is—NO ONE KNOWS**.

Yet, the most distinguished group of scientists in America boldly espouse a belief, which they portray as scientific fact, that life began accidentally from non-living matter as specified in the 2nd edition of *Science and Creationism: A View from the National Academy of Sciences*:

> "For those who are studying the origin of life, the question is no longer whether life could have originated by chemical processes involving nonbiological components. The question instead has become which of many pathways might have been followed to produce the first cells."

That proclamation is not a scientific fact. It is a fundamental axiom of scientific atheism.

When Charles Darwin mused about life beginning by chance in a hot, watery ooze and pronounced that living organisms evolved through the process of natural selection, with adaptive traits occurring only through chance mutations, science knew absolutely nothing about cellular biology. Darwin and his colleagues believed that living cells were filled with a mushy protoplasm, not RNA and DNA and amino acids and proteins and organelles and nuclei. His understanding of human embryology was based on the belief that a preformed tiny person just got bigger and bigger and bigger inside the mother's womb. Yet, his theory of natural selection, with adaptive traits occurring only through random mutations, is still used as the template for modern biology, even in light of the amazing discoveries about the workings and inter-workings of biological structure and systems.

Brilliant scientists have discovered wondrous things about the structure and operation of living organisms. Incredibly, each discovery seems to unveil a more incredible intricacy residing within. And, most incredibly, elite scientists review all this incredible stuff and conclude that each and every part of each and every living thing, including life itself, began and evolved through undirected events called random mutations.

The creation of life 'by design' is the metaphysical belief of theism. The creation of life 'by chance' is the metaphysical belief of scientific atheism. The discoveries of science to date quite clearly reveal the fact that no one knows whether life began and evolved 'by chance' or 'by design'.

But, rest assured that the creed of the religion of scientific atheism, that we have just reviewed, is taught as fact in the science classes of public schools. The authority of the elite members of the National Academy of Sciences is so great that public school science teachers must consider the words of the NAS creed to be scientific facts.

To smugly contend that science has discovered that life arose from non-life through natural processes is simply not the truth. To smugly contend that evolution by natural selection has been scientifically proven to occur through the mechanism of random mutations of DNA is simply not the truth. These contentions of elite and mainstream science are based on **no evidence** and defy logic and common reason.

For the National Academy of Sciences to insist that their belief that life emerged 'by-chance' from non-life and that the evolutionary mechanism of random mutations should be taught as scientific fact in public schools flies in the face of logic and common reason. And, it flies directly in the face of the separation of church and state established by the American Constitution. The emergence of life from non-life and the evolution of organic species by natural selection of random mutations are core beliefs of the religion of scientific atheism. As such, they should be banned from the science classes of public schools.

Teaching Evolution and Intelligent Design in School

As of this writing there is a huge controversy in America over whether the theory of 'intelligent design' should be taught as an alternative to the theory of evolution by natural selection of 'random mutations' in the science classes of public schools. This issue presents a real dilemma for local school boards and courthouses, for the scientific community is not in agreement on the issue.

Scientists on both sides of the controversy generally agree that the beginning of life was a one-time-only event. And, both sides generally agree with the scientific hypothesis that all living things derive from a common ancestor. These propositions are based on physical evidence in support of organic evolution and the biological evidence that all living organisms contain a DNA information library. Thereafter the two sides part company.

The **elite and mainstream scientific community** supports the position that each and every adaptive trait of all biological life that has subsequently evolved from a common ancestor has resulted from natural selection of random mutations. Every tissue, every muscle, every organ, every hormone, every metabolic pathway, and every brain has evolved by means of random mutations of DNA. A random mutation that aids in adaptive response to environmental challenges is inherited by more progeny and thereby becomes predominant in the gene pool of the species. Each and every aspect of each and every complex biological organ and complex biological pathway has evolved in a step-by-step progression of inherited adaptive responses to environmental challenges in the same way. They further contend that, through the same mechanism, life itself evolved from non-life in a manner called emergent evolution or abiogenesis. They are the 'random mutationists'.

A gifted British author once wrote a fictional account of two kingdoms of tiny people who engaged in perpetual warfare over this enigmatic question: Should a boiled egg be broken on the little end or the big end? In honor of Jonathan Swift's brilliant satire called *Gulliver's Travels*, we will hereinafter refer to the mainstream scientific community, the 'random mutationists', as **Big Enders**.

The **minority scientific community** supports the position that complex biological organs and complex metabolic pathways are so complex that a step-by-step evolutionary progression to account for their development is simply not feasible. They contend that, in order for these organs and pathways to have any usable function at all, they would have to begin in a state of irreducible complexity that could not have evolved in a stepwise fashion. They conclude that such complex organs and pathways must, therefore, have been designed. They further contend that the beginning of life must have been designed, because the interactions of nucleic acids and amino acids necessary to provide for replicating DNA and cellular organisms is so complex that it could not have evolved in a stepwise progression from non-living matter. They are the 'intelligent designers'. Let's hereinafter refer to the minority scientific community, the 'intelligent designers', as **Little Enders**.

The little people that Gulliver encountered went to war over whether an egg should be broken on the big end or the little end. That issue was and still is incapable of being resolved by science. The Big Enders and the Little Enders in the scientific community today battle over other issues that are incapable of being resolved by science. Let's remove the battle from the science classroom.

Let me offer this solution to the current dilemma facing school boards and courthouses over whether the view of the minority scientific community should be taught as an alternative to the view of the majority scientific community in the science classes of public schools. **Neither should be taught in science class**.

Teaching Evolution in Science Class

The theory of evolution by means of natural selection is a valid scientific theory. It simply has limits. So, in science class, teach up to the factual limits then stop.

Teach all the stuff that is backed by good science, supported by valid scientific evidence, in the science classroom. But, **when the point is reached where scientific evidence is not available to support a position held by the scientific community stop teaching the position as science. Where science runs out of answers simply say so.**

Experts in science are renowned authority figures. We rely on them to provide factual information based on evidence. They owe society a duty to provide scientific truths based on actual evidence.

Scientists are human and exhibit the same foibles as the rest of us. Some scientists pronounce as 'truth' much that is not supported by the scientific method. Upon gaining acclaim and renown, they seem to become sprinkled with **'authority dust'**. The authority dust apparently reveals to them certain a priori scientific 'truths'. Sadly, we often don't know the difference between the 'truth' these scientists proclaim based on scientific evidence and the 'truth' that is revealed to them by authority dust. We trust them to tell us facts based on scientific evidence. Instead, they feel compelled to expound their beliefs veiled in the cloak of science.

This phenomenon is not restricted to scientists by any means. We observe the phenomenon quite frequently when politicians gain election to public office. Swept up by public acclaim, elected officials seem to instantly gain uncanny insight into how to best increase the public weal and decrease the public woe. They seem to have gained such a priori knowledge when they were apparently sprinkled with 'election dust'. And, the same holds true for members of the clergy who proclaim from the pulpit that they know the 'truth' regarding ultimate questions based on their holy insight. They too seemingly become sprinkled with 'authority dust'. Peter Pan and Tinker Bell do not discriminate between scientist or politician or clergyman.

Fundamentalists seem compelled to defend their worldview at all costs. Worldviews take on a life of their own. The avid supporters of a specific worldview seldom let the facts get in the way of their entrenched beliefs. They know the 'truth' and they are sticking to it.

The avid fundamentalist materialist knows the 'truth' that there is no God. The avid fundamentalist creationist knows the 'truth' that the world is only 6,000 years old. To him such has been revealed by the literal 'truth' of the holy scriptures. To the closed mind of the fundamentalist materialist, God simply cannot exist. To the closed mind of the fundamentalist creationist, discoveries of science that reveal that the world is over four billion years old simply cannot be believed.

No human authority knows the factual 'truth' to ultimate questions. So, **when authority dust pollutes the air of the science classroom it needs to be expelled.**

There is no direct scientific evidence - none - that the universe began and evolved as it has by chance alone. There is no direct scientific evidence - none - that life evolved from non-life. There is no direct scientific evidence - none - that complex biological organs or complex metabolic pathways evolved by step-by-step adaptive response to environmental challenges through the mechanism of random mutations.

On the other hand, there is no direct scientific evidence - none - that any of these things happened not-by-chance either. There is no direct scientific evidence in support of the intelligent design position.

So, in science class, simply state these facts as the truth that they are. Simply admit that no one knows the answers to ultimate questions.

The Nobility of Science

The history of human civilization, including the history of science, clearly reveals that we Homo sapiens have a very difficult time breaking-out of a 'paradigm box'. The 'truth' established by scientific authority has great staying power, and those vested with that authority go to great lengths to retain their 'truth'.

At the end of the 16th century, the scientific authorities of world-renowned Cambridge University proclaimed in the Charter of their Trinity College:

> "All students and undergraduates should lay aside their various authors
> and only follow Aristotle and those who defend him."

Quite clearly, the 'truth' of Aristotle's science, including the 'truth' of an Earth-centered solar system, had been resolved and that 'truth' was not to be questioned further.

More than 400 years later, at the beginning of the 21st century, the scientific authorities of the world-renowned National Academy of Sciences now proclaim:

> "For those who are studying the origin of life, the question is no longer
> whether life could have originated by chemical processes involving
> nonbiological components. The question instead has become which of
> many pathways might have been followed to produce the first cells."

Quite clearly, the 'truth' that life itself evolved from non-living matter through the natural process of random mutations has been resolved, and that 'truth' is not to be questioned further.

When the Greeks embarked upon metaphysics as a philosophy over two millennia ago it was for a purpose. The purpose was to seek knowledge in order to determine the best 'truth'. There can be no end of searching. Most unfortunately, it seems that by the adoption of scientific atheism the scientific elite of this country have now concluded that the search is over.

Science should be ranked among the noblest of professions. The nobility of science properly rests on the pursuit of the best possible truth. That pursuit requires a steadfast, yet open mind.

CHAPTER 28
Scientific Atheism Undermines Human Morality

The closed minds of our most elite scientists demand that the materialist 'truth' of an unplanned and unintentional universe be taught 'as fact' in the science classes of our public schools. Sadly, that 'truth' is not based on the discoveries of science. It is based on the 'wishful thinking' of atheistic scientists.

The Greatest Fallacy Ever Told

Scientific atheism is the most insidious fallacy that has ever deceitfully crept into western civilization. In America today it is nonchalantly reported with regularity as 'scientific fact'. Our most brilliant scientists have perpetuated the idea that they have actually discovered evidence that shows how the universe and life itself were created and evolved with no need of a Purposeful God. That idea has no scientific basis whatsoever. It is a scientific fallacy. It simply represents an atheistic metaphysical belief.

That metaphysical belief in an unplanned and unintentional universe is based on the unproven hypothesis that the material world is all that exists and that its existence 'just had to be' the result of unplanned and unintended chance occurrences. There is no scientific evidence whatsoever in support of that belief.

Belief in an unplanned and unintentional universe and living world requires that there must be a material explanation for everything and that our particular species is somehow capable of discovering that material explanation. Anyone who actually has the audacity to honestly believe that, cannot be endowed with too much common sense and certainly hasn't been paying much attention to life and living. The state that we call 'being alive' is not just a simple 'material' state. Being alive transcends that which is simply material.

All of the intricate and exquisite processes that constantly take place within our bodies and which serve to keep us alive operate pretty much on 'cruise control', with very little input from our conscious self. Living life is not just DNA replication and cell division and digestion and blood circulation and temperature regulation and neural synapses. Living life is **emotional.**

We love, we hate, we get upset, we rejoice, we sing, we laugh, we cry, we play, we sorrow. All of the experiences we call 'being alive' are 'beyond' the material. Yet elite and mainstream science insists that such life experiences are mere expressions of cellular chemistry. At core we are just living organisms whose only purpose for existence is to pass on our DNA - nothing more. That is the gist of scientific atheism.

When our children are taught in science class that scientific atheism is the 'scientific truth', the take home lesson for living life is the adoption of a philosophy called **nihilism** wherein:

- There is no basis for objective morality.
- There is no intrinsic value in human life.
- There is no purpose and meaning in life beyond that which we simply make up.

That is how scientific atheism undermines human society. And, that is why the religion of scientific atheism should not be allowed to be taught in our public schools as 'scientific fact'.

Since the beginning of civilization we, as a society, have been guided by human traditions that are based on a firm belief that human life has value and meaning and that there are objective moral standards. If we, as a society, are going to abandon those long-held traditions, we should most certainly insist that scientific atheists provide at least some actual scientific evidence in support of their metaphysics.

Tradition

There is a thread of continuity that defines civilization. It helps us in our daily lives. It provides a sense of things. It is called human tradition. It provides the basis for human values.

Tradition is made up of our thoughts and experiences and emotions. Human tradition in time evolves into religion and philosophy. The two fields are usually studied separately, but are so closely connected that we, as individuals, actually think of them together. One cannot really think too long of God without wondering why and how. One cannot really think too long about why and how without wondering about God. And, when we wonder about God we wonder about how we are supposed to act in this world.

Ever since our species gained the blessing (or curse) of reflection we have, as individuals, really been about the same as we are today. Each of us has questions, fears, doubts, and hopes. We are born. We live life. We wander. We wonder. We die.

With each age our opportunities to learn are enhanced by the discoveries and inventions of our most gifted peers. The rest of us benefit by the insights and genius of the few. We make their knowledge our own and, as a society, we progress.

Ancestral Beliefs

It's a pretty good bet that religion has been a part of mankind's existence right from the start. The start of mankind's existence being when the human species, Homo sapiens, became uniquely gifted with the power of reflection, the ability to intelligently think and ponder not only the wonders of the world, but the wonder of ourselves considering our place in the world.

Early insight into the nature of things led to the first animistic adoration and fear. Fire, water, earth and air were joined by the Moon and the Sun as objects worthy of worship and reverence. Religion was born.

Primitive man observed the world around him and marveled at it. And he marveled as well at the ghostly world that he experienced in his nocturnal dreams. The Sun rose, crossed the sky each day and disappeared, only to repeat the sequence again and again. The sky roared with thunder. Lightning bolted through the air and set the tree aflame, a fire he began to use for personal warmth but whose destruction and pain cowered him with fear. A bear that he had killed during the day returns to attack him again in his dream that night.

All the world was intensely alive and frightening, and mysterious. There seemed to be a hidden essence within all things. All things had souls, or hidden gods that could be worshiped. Such animism was the first religion.

The Sun god was the great giver of light and warmth. The Moon god controlled the weather and, along with the Sun god, gave us the measures for time. The sky itself was the god of rain. The Earth was the mother god. Animals of all types were worshiped as gods, because they were powerful and fearsome, and needed to be appeased. And so Sun, and Moon, and animal worship passed down to the early faiths of recorded civilization.

The appearance of the dead in dreams was fearful, and led to the belief in the continued life of the dead, which led to the belief in ancestor worship and gods of the dead.

Primitive religion made no real distinction between gods and men. The gods were ancestors. Men had been begotten from the gods and were, therefore, of the gods. God became our father.

The concept of god evolved into a pantheon of Greek and Roman gods, personified as willful and subjective deities. In time the omnipotent and righteous God of Jewish and Christian modern faith became possessed of goodness and 'right reason'.

Natural Law

Generations before Aristotle was born, a Greek playwright, Sophocles, in a drama about the city of Thebes, presented the concept of an eternal law, higher than man-made law, that governs human morality. The title character, Antigone, wants to provide her brother, who has been killed in a state rebellion, with an honorable burial. King Creon has decreed that those rebels like her brother must be left to lie on the battlefield and be consumed by the elements and the animals. Antigone proceeds to give her brother an honorable burial, which Creon, as the legitimate head of state, has declared is a crime punishable by death by stoning. King Creon asks Antigone if she understood the law and willingly undertook to violate it, and she replied:

> "Yes, for it was not Zeus that had published me that edict; not such are the laws set among men by the justice who dwells with the gods below; nor deemed I that thy decrees were of such force, that a mortal could override **the unwritten and unfailing statutes of heaven. For their life is not of to-day or yesterday, but from all time, and no man knows when they were first put forth**.
> Not through dread of any human pride could I answer to the gods for breaking these. Die I must, - I knew that well (how should I not?) - even without thy edicts. But if I am to die before my time, I count that a gain: for when any one lives, as I do, compassed about with evils, can such a one find aught but gain in death?

So for me to meet this doom is trifling grief; but if I had suffered my mother's son to lie in death an unburied corpse, that would have grieved me; for this, I am not grieved." (emphasis added)

The study of 'natural law' thus began with Antigone's pronouncement of the 'unwritten and unfailing statutes of heaven.'

While the concept of natural law predates Aristotle, it is he that provides its outline. The postulate is that the essential nature of any living thing can be determined by examining its inherent structure and typical activities and thereby discerning its distinguishing characteristic or nature. The distinguishing nature of birds is flying through the air; of fish is swimming through the water. The distinguishing nature of Homo sapiens is rational thinking. Rationality, made manifest in our abilities of speech and reason and introspection, is the hallmark of 'human nature'.

Humans are the only creatures who are possessed with the power of introspection and reason. Since the human 'essence' is 'reason' we are the only creatures for which natural law provides principles of right and wrong for our behavior. The concept of natural law applied to human nature is that there are **universal moral truths that govern the moral nature of human beings. Natural law presents the belief that human beings have intrinsic rights and intrinsic moral responsibilities and that our lives are governed by an intrinsic sense of justice.** Natural law is the descriptor of divine justice and equality that is the natural right of every member of the species Homo sapiens.

We can use our reason to reflect on what our common human nature is concerning ethics and morality. Because of our unique reasoning abilities it is believed that we can discern the 'right reason' that is the dictate of natural law that is promulgated by a Purposeful God. **The self-evident truths of 'natural law' are embodied in our human nature and 'written on our hearts'.** We can use our human reasoning abilities to develop a detailed code of human moral conduct based on a few self-evident truths of natural law. Reason, reflecting on our nature, will serve as the basis for promulgating codes of conduct, following the over-arching dictate to 'pursue good and avoid evil'. Since moral standards are rooted in natural law, the purpose of legal codes is to more precisely articulate the application of natural law to specific instances of conduct.

The 'laws of nature' are the universal laws that govern the interactions of the fundamental forces and the fundamental particles and fields of energy and matter. '**Natural law**' for the species Homo sapiens refers not to those things physical in the universe but, rather, to **universal moral truths**. **A Purposeful God is the ultimate source of natural law.**

The great religions of western civilization are based on belief in the existence of natural law that contains universal moral truths that should be used to govern the conduct of people in human society. All of the great religions of the modern world have some root in our ancestral beliefs and natural law. And the great religions of the modern world have many common threads that are an assemblage of wisdom traditions that have developed over many millennia of human civilization.

Wisdom Traditions

Our wisdom traditions are rooted in our religions. The great religions of western civilization are Judaism and Christianity. The moral teachings of those great religions developed over many generations of trials and tribulations. Let's take an abbreviated review of the history of both of these religions before we return to examine the common threads of wisdom traditions.

Judaism

About the same time that Hinduism was beginning to bud as a religion in India, circa 2,000 BC, the Jewish faith was born in what is today called the Middle-East. The birth was the result of the covenant made between the Hebrew God, Yahweh (also known as Jehovah), and the patriarch of the Jewish faith, Abraham. Jewish tradition provides this general outline of events.

Abraham (later renamed Israel by God) was 75 years old when God directly instructed him to leave, with all of his extended household, from his home Ur (modern-day Iraq) located by the Euphrates River, and relocate to the land of Canaan (modern-day Israel).

"The whole land of Canaan, where you are now an alien, I will give as an everlasting possession to you and your descendants after you, and I will be their God."

In return, Abraham agreed that he and all male descendants would be circumcised, and that Abraham and all of his extended family and descendants would render obedience to God and be righteous in all their actions:

"Abraham will surely become a great and powerful nation, and all nations on earth will be blessed through him. For **I have chosen him, so that he will direct his children and his household after him to keep the way of the Lord by doing what is right and just,** so that the Lord will bring about for Abraham what he has promised him." (emphasis added)

Abraham moved to Canaan. A son, Isaac, was born to Abraham and his wife, Sarah, in Abraham's 99th year. Isaac's eldest son was Jacob who had twelve sons. Their offspring became the twelve tribes of Israel. The youngest son, Joseph, was sold into slavery in Egypt by his brothers, but overcame to become a powerful figure in the Egyptian royal court. Joseph's forgiveness brought his father and brothers and their households to Egypt for relief from the famine devastating Canaan at that time. All lived prosperously in the land of Goshen in northern Egypt until the death of Jacob and Joseph. Thereafter, the Israelites were enslaved by the Egyptian Pharaoh for 400 years. And then baby Moses arrived in the bulrushes.

The prophet Moses delivered his people from slavery in Egypt, wandered with them for forty years in the Sinai desert while God provided them with manna from heaven to eat. They encamped at the foot of Mount Sinai for about a year. On one certain day on Mount Sinai God provided Moses with his Ten Commandments carved in stone. The Ten Commandments formed the basis of Jewish law and society henceforth from that day. At the end of their years of wandering, Moses was allowed to overlook, but not enter, the promised land of Canaan (Israel) before he died. The great warrior, Joshua, succeeded Moses, and took the favored people across the River Jordan into the promised land.

When Joshua led the Jews into Canaan, they recreated Yahu, one of the resident gods of their conquered enemy, as their own. Yahu was greedy, bloodthirsty and most fearful. The authors of the first five books of the Old Testament, known as the Torah or Pentateuch, transformed Yahu into the Jewish god Yahweh, the god of hosts, the god of war. At his command the Jews slaughtered whole nations of men, women, and children. Children were punished by God for the sins of their fathers and of their fathers' fathers.

Other gods persisted and were accepted in Jewish culture for another thousand years, during the reigns of the great kings, Saul, David and Solomon and the Judges thereafter. Indeed, wise king Solomon included in his palace, built next to the great Jewish Temple he had constructed in Jerusalem, alters to exotic foreign deities that his wives still worshiped. Yahweh remained a God of war and was capricious and arbitrary. The Ark of the Covenant, which contained the tablets of the Ten Commandments, went before them in battle, as a symbol of Yahweh, aloof, terrifying, and untouchable.

To believe in the polytheism of the nature religions is to conform to a view of the world that things really cannot be changed. The way things are in nature and human society simply represents the way things are in the world. Those things are beyond human change and are immanent in all. To the monotheistic Jews, however, their One and Only God was not immanent in nature, he was transcendent of nature and all. His will is above nature and the state of things. And, if we conform to his will we can change the way things are in the world for the better. We can do that by following God's law and exercising our free will in conformance with his law.

The beauty and hope of the great Jewish monotheistic religion of today sprang from the inspiration of the great Jewish prophets who emphasized this discovery. The great Jewish prophets began to 'speak out' about the eighth century BC. To be a prophet was not to foretell the future, but rather to speak out against idolatry, and corruption, and the inequities that existed at that time in Jewish society.

The prophets 'spoke out', over the course of several hundred years. They protested against the abuses of the religious rulers. They protested against social injustice. The grandeur of this religion was rooted in these individuals and the dramatic change in life's view that they brought about through this new approach that they espoused of individual equality and justice.

The Hebrew prophets recognized a fundamental difference between their God and the other personal gods of that era, particularly the gods of the Greeks and Romans. While the Greek and Roman gods were amoral and indifferent to mankind, their God was different. Yahweh was now righteous and just. And, since he had created everything in the universe as right and just, the natural and proper order of all things in such a moral universe had to be right and just. What's more, not only was God and all his world righteous and just, he loved and cared for each individual deeply. His loving kindness and tender mercy was the very foundation of a moral universe and human life. If the very nature of God's universe is righteousness and justice, a civil society also had to be righteous and just or it would be out of sync with God's will. Lastly, civil justice requires that every human being, as a child of God, possesses rights that must be affirmed by even princes and potentates.

God's nature and God's will for mankind was 'revealed' to the prophets, as it had been 'revealed' to Moses and Abraham. The revelations had occurred as direct insight from God. Some by visions and some by the actions of the Lord himself. God had designated the Jews as his chosen people, but they did not believe that they had earned such distinction. They had gained such status and knowledge of the truth of the nature of things simply through God's 'grace'.

The great Jewish prophets transformed Yahweh, the arbitrary, fearful, and vengeful God of war, into the righteous and loving God of goodness and Heaven. One Jewish prophet foretold the coming of a Messiah.

Christianity

Historians report that about 2,000 years after the founding of the Jewish religion one certain Jew, Jesus of Nazareth, was proclaimed by many followers to be the Jewish Messiah. Christian tradition provides this general outline of events.

God told Jesus' mother, Mary, that she would conceive and give birth to Jesus while still a virgin. Her husband Joseph would be Jesus' titular father, but his actual father was God himself. Jesus was the son of God.

Jesus was born, grew to manhood and, at the age of about 30 years, began to preach a message of peace and love throughout Israel. After three years of teaching and performing miracles he was arrested and tried by both Jewish and

Roman authorities and sentenced to death by crucifixion. It was claimed that he was the Messiah, the Christ. Jesus was crucified, died, and after three days arose from the dead. For some forty days following his resurrection from the dead he was seen to be alive among the populace. This was reported as witnessed by some 500 people.

After the crucifixion and reported resurrection, circa 33AD, the new religion of Christianity formed over a period of several hundred years. A common creed of beliefs of the religion was agreed to at the Council of Nicea in the 4th century AD. The teachings of Jesus were viewed as the word of God and were seen to supplement, not supplant, the Ten Commandments of the Jewish faith. The Christian religion thereby combined fundamental tenets of the Jewish religion (the Old Testament) with the teachings of Jesus Christ and his followers (the New Testament).

Common Threads of Wisdom Traditions

The word of God now formed a larger ledger to guide the conduct of mankind. The Old Testament told individuals that they should revere God and their parents, love their neighbor, and that they should not: steal; murder; commit adultery; give false testimony; or be covetous. The New Testament told individuals to humbly: love both friends and enemies; do unto others as you would have them do unto you; be righteous, merciful, peaceful; administer care to those less fortunate; avoid judging others; and forgive those who do you harm.

These things form the common threads of the wisdom traditions of the ages. Whether one believes that the prophets of these great religions actually communed with God is not really the point. The point is that these traditions came from a profound belief that a Purposeful God is transcendent of all and gives each of us fundamental rules whereby we should live our lives. God the Father is concerned with each individual personally and his will is, for each individual, and all mankind, to live in conformance with his divine law in peace, and happiness and freedom.

That represents the wisdom of the ages. It provides purpose and meaning in life.

The imperfections in mankind are legion. For the past two millennia the history of western civilization evidences breach after breach after breach of the wisdom traditions. In large part the breaches occur when human 'authorities' become filled with the hubris of actually believing that they have gained uncanny personal insight into the 'truth' of things. Holy crusades were undertaken to further God's 'truth'. Church authorities tortured and killed 'heretics' in the name of God. But, overall, western civilization learned from such tragic mistakes and returned to follow the wisdom traditions more closely.

As the twentieth century after the birth of Christ unfolded, America, as a nation, made great strides to more closely follow the wisdom traditions. Racial and sexual equality began to be actually realized. Neighbor helped neighbor deal with crises. American foreign aid and charitable contributions per capita reached all-time highs. A prosperous middle class flourished. By the start of the 21st century things were changing. A lot of reasons are proposed to explain the change. The reason which I feel is most compelling is that the underpinning of society, our wisdom traditions, are being eroded from within. This in large part is the result of the beliefs that have evolved among the scientific and intellectual authorities of American society. The foundational belief is scientific atheism.

Scientific atheists staunchly contend that all of the mysteries of science can be unraveled based on two fundamental 'truths':
* the actual existence of multiple universes; and
* random mutations of DNA.

The metaphysics of scientific atheism rests firmly on these two pillars of 'scientific truth'. Yet there is no scientific evidence whatsoever in support of these atheistic beliefs. Atheistic scientists smugly contend that the absence of evidence is not evidence for absence. That's quite true. And that quite simply places scientific atheism into the same category of 'beliefs', not fact, that is shared by all other religions.

The 'Scientific' Religion of Atheism

The use of the scientific method limits the range of 'truth' that scientific authority can tell us. Beyond that limit lies the metaphysical. The fact that the metaphysics of fundamentalist evolutionary naturalists is based on the use of the scientific method does not make their metaphysics a scientific fact.

The **metaphysics** of the 'modern' scientific worldview **begins when evidence stops** and over-reaching begins by requiring that each and every aspect of the universe, and all therein, must be explained in accordance with naturalism. **There can be no other explanation than a natural one. That is how science works.** Therefore scientists have become wedded to the

doctrine of naturalism as an a priori axiom of truth. They claim that, since intelligent design is beyond the scope of the scientific method, and since expanded naturalism is the centered heart of the scientific method, intelligent design has been disproved and by-chance-alone has been proven as a scientific fact. That proclamation is nonsense.

Naturalism, in and of itself, evidences no direction. It has no purpose. The 'modern' worldview of expanded naturalism says, in essence, that there is nothing special about our planet, or us, or our place in the universe. Design and purpose are denied.

The elite and mainstream scientific community has thus adopted a 'modern' worldview of expanded naturalism that eliminates the possibility of design and purpose in the universe. Ultimate reality is chance alone. Behold the Accidental God of scientific atheism.

The most outspoken proponents of this worldview either outright proclaim or strongly imply that this 'modern' scientific worldview is based on scientific fact. They 'just know' the 'truth' that every natural living and non-living thing in the universe has evolved purely by chance, with no design involved at all. That has been **revealed by the 'truth' of scientific extrapolation. They present the 'truth' of scientific extrapolation as scientific fact. And that is unconscionable.** Claiming as scientific fact extrapolations of the Copernican Principle and Darwinian Evolution, that go beyond the limits of scientific knowledge into areas incapable of falsification, violates the scientific method itself and does a great disservice to the search for truth.

To the fundamentalist evolutionary naturalist, discoveries of science that repeatedly disclose the wondrous appearance of design in the universe cannot be actual design. Actual design would indicate underlying purpose and, therefore, it cannot be believed. A **crusade has been mounted** by the fundamentalist evolutionary naturalists **to ensure that the metaphysical 'truth' of expanded naturalism**, that they present as scientific fact, **will prevail over the metaphysical 'truth' of a creative intelligent agent of the universe. They proclaim that the true God is the 'Accidental God.'**

The simple truth is that the discoveries of dedicated scientists have revealed a complexity in living things that cannot begin to be explained by application of the traditional rules of scientific naturalism. Yet, the most brilliant scientists in

America are insistent that the explanation 'just has to be' a natural one, because a supernatural intelligent agent is just not scientific.

The NAS tells us that such 'evidence for evolution can be compatible with religious faith'.

> "Science and religion are based on different aspects of human experience. In science, explanations *must* be based on evidence drawn from examining the natural world. Scientifically based observations or experiments that conflict with an explanation eventually *must* lead to modification or even abandonment of that explanation. Religious faith, in contrast, does not depend only on empirical evidence, is not necessarily modified in the face of conflicting evidence, and typically involves supernatural forces or entities. Because they are not a part of nature, supernatural entities cannot be investigated by science. In this sense, science and religion are separate and address aspects of human understanding in different ways. Attempts to pit science and religion against each other create controversy where none needs to exist."

Indeed, the only religious faith that is compatible with a belief that unplanned natural processes 'just has to be' the answer for the beginning and development of the universe and life itself is the atheistic religion of science. No amount of observations of 'apparent design' in living things can ever modify that scientific atheistic belief.

The great religions of the western world are solidly founded on an unshakable belief that the universe and life itself was created and evolved in accord with a purpose and reason known to God. The unshakable metaphysical belief of the NAS is that the universe and life itself was created and evolved by accident, with no purpose and for no reason. It is more than just a little disingenuous for these elite scientists to say that no conflict exists between science and religion and then proceed to hypothesize that the very foundation of the great religions of the western world is made of hole cloth.

One religion, and one religion only, is compatible with the NAS position. The National Academy of Sciences has become the prophet of the religion of scientific atheism.

The Spread of Scientific Atheism

Our western world of the 21st century is a world of specialists. That condition is actually a required one. With the amount of information available, unless we specialize, our information-based civilization will not work. There is simply too much information for generalists to succeed in our culture. One consequence of specialization is that we, as individuals, become quite limited in our range of knowledge. As we become expert in our chosen field we are forced to rely on the expertise of others knowledgeable in their chosen field. Some specialists become 'deep thinkers', others 'scientists', and others 'news reporters'.

The 'deep thinkers' of American society are usually found within the ivy-covered walls of our most prestigious universities. Most of them are not scientists. They are generally the professors who have earned doctorate degrees in philosophy.

The 'news reporters' of American society are generally not scientists either. Their background usually includes a bachelor's or master's degree in journalism or one of the liberal arts, such as history or political science.

It is widely recognized that the most brilliant scientists in America are members of the National Academy of Sciences. Many of these brilliant scientists, at some time in their careers, teach at the most prestigious universities in the country. They do not teach the future 'deep thinkers' or 'news reporters'. They teach scientists. They teach science in accord with the scientific method embodied in mathematics and natural reductionism. Because they are so wedded to the scientific method of reductionism they come to sincerely believe that ultimate 'truth' and reality can only derive from that source. The worldview that everything derives from a natural cause and nothing supernatural can possibly exist becomes firmly established as scientific atheism. Thus, the existence of a Purposeful God is actually believed to be disproved by science. Behold the Accidental God.

The most brilliant scientists thereby pass on their atheistic belief as established fact to gifted science students. Those students, in turn, go on to make marvelous scientific discoveries using the scientific method and become the leaders of the mainstream scientific community. The worldview of 'by-chance-alone' is passed on to science teachers and science colleagues throughout the country. Thereby scientific atheism becomes an a priori belief of mainstream science. Scientific peer review follows the party line and eliminates any contrarian viewpoint from the journals of respected science publications. Contrarians become pariahs and their work is belittled.

Both the 'deep thinkers' and the 'news reporters' are also highly intelligent. And, they generally share a common trait. They do not have a broad background in science, and they trust in the authority of science experts to tell them the truth about scientific discoveries. Since their background in science and mathematics is usually quite limited, as intelligent as they are, they rely on the authority of the elite scientists to tell them the 'truth' discovered by science. They accept as 'true' the pronouncements of the elite and mainstream scientific community that science has actually discovered that nothing exists beyond the natural world. The 'news reporters' widely announce through the print and electronic media the 'truth' discovered by elite and mainstream science. The 'deep thinkers' then meld into liberal arts the philosophical extension of that discovery and set upon examining the plight of mankind living in a godless world. The stage is properly set for the establishment of the 'truth' of the post-modern world that the only purpose and meaning in life is that which we make-up.

The foregoing cursory analysis of the impact of scientific atheism is, at best, a gross over-simplification of a highly complicated matter of influence in American society in the 21st century. Yet, it is pretty accurate.

The Foundation of American Society

The ultimate foundation of our American society is Natural Law. And, a Purposeful God is the ultimate source of Natural Law. The moral principles underlying our society are well articulated in the Declaration of Independence as 'the Laws of Nature and of Nature's God'.

"We hold these truths to be self-evident, that all men are created equal, that they are endowed by their Creator with certain unalienable Rights, that among these, are Life, Liberty, and the pursuit of Happiness."

The Declaration further proclaimed the Right of the People to establish a government founded on these principles.

A decade later that proclamation came to fruition as the Constitution of the United States of America.

The very conduct of our nation and its people is thereby underlain by the norms of natural law - what is morally right and what is morally wrong in order for us to live a happy life in conformance with the will of a Purposeful God. How well we accomplish the goal in fact is certainly subject to debate, but that is not the point. The point is that we, as a society, have always believed that we were capable of discerning those norms of natural law through our unique human reasoning ability granted to us by God.

Since the founding of our country we, as a people, have resolutely believed that there are self-evident truths of natural law embodied in our human nature and 'written on our hearts' that rests on the reality of a higher moral order of goodness and 'right reason'. We, as a people, have since the beginning of our history believed that we have a 'moral nature' that must govern our society.

The Effect of Scientific Atheism On American Society

People trust brilliant, gifted scientists to tell them the truth. When our elite scientists proclaim that their atheistic religion is supported by scientific facts they undermine society and science at the same time. Science is necessarily limited to the search for the best naturalistic truth. Science should not extrapolate naturalistic findings to require that ultimate reality must be based on the use of the scientific method. For really smart guys, that approach is just plain stupid. And, for society that approach is just plain devastating.

With the publication of the book *Science, Evolution, and Creationism*, the elite scientists of the National Academy of Sciences have delved deeply into metaphysical and religious territory. They clearly espouse a metaphysical belief of expanded naturalism, called scientism, whereby the only possible truth must derive from natural causes. And, thereby, they clearly espouse a religious belief called atheism.

When the really brilliant guys of science proclaim that scientific discoveries support the idea that the universe and life began and evolved without purpose or design they send to us the clear message that life itself is without purpose and devoid of meaning.

In America everyone is entitled to their own religious belief. That includes atheistic belief. The true danger for society presented in 'scientific atheism' is the inclusion of the word 'science'. The overwhelming implication is that it presents an authoritative 'truth' that derives from scientific discoveries. The tenets of 'scientific atheism' are now included in science textbooks and taught in the science classes of public schools. No other religious belief is afforded that opportunity, and rightly so. And, 'scientific atheism' should not be afforded that opportunity either, but it is.

Doing so places the religion of 'scientific atheism' in a favored class all its own. It provides scientific atheism with the imprimatur of discovered 'truth'. And that is simply wrong. **The insidious effect of 'scientific atheism' is to devoid the human psyche of the hope and inspiration of higher purpose and meaning in the universe and in life itself**.

Our American society is founded on the wisdom traditions of two great religions that provide a solid basis for the ethics and morals of righteous living. If 'scientific atheism' is believed to be the discovered 'truth', it necessarily follows that there is, indeed, no intelligent designer of the universe and life itself, that there is no Purposeful God. If there is no Purposeful God then both Judaism and Christianity become nullities and the underpinnings of our wisdom traditions for the ethics and morals of righteous living become nullities as well. And, if there is no Purposeful God there is no basis for any universal moral truth. Scientific atheism simply negates the concept of natural law that human beings have a moral nature. It denies that there are any objective moral standards for human beings who are simply the products of an unplanned and undirected cosmos. Human values are simply those that individuals choose for themselves. Moral relativism and situational ethics should prevail.

If, on the other hand, one believes that the discoveries of science better support the idea that the universe and life began and evolved not by accident, but rather, in accord with the design of a creative intelligent agent, then the great wisdom traditions underpinning the great religions of western civilization remain intact. And the precepts of moral conduct and individual rights pronounced by our Declaration of Independence and Constitution also remain intact.

There is no reason whatsoever to believe that mankind will ever be able to factually determine what God's purpose is. But, as long as the scientific facts point directly to the existence of a Purposeful God, every great religion on Earth has a completely valid reason to pursue their metaphysical beliefs with vigor. And, the moral foundation for our society stands intact.

Pillars of Faith

Atheists believe in an Accidental God. This metaphysical belief is based on two pillars of faith:

- belief that there are multiple universes beyond our observable universe which possess universal laws and constants that are different than those that obtain for all matter and energy in our universe; and
- belief that random mutations can create, organize, manipulate and transmit information that is contained in code in DNA molecules.

Theists believe in a Purposeful God. This metaphysical belief is based on two pillars of faith:

- belief that there is but one universe containing laws of nature that are universal (an observable universe of hundreds of billions of galaxies each containing hundreds of billions of stars seems quite enough); and
- belief that random mutations cannot create, organize, manipulate and transmit information that is contained in code in DNA molecules (information is neither matter nor energy, and all coded information that has ever been observed is the product of an intelligent agent).

The purpose of this primer has been to determine whether there is real purpose and meaning in life based on the best scientific evidence available. Did all the stuff of nature and life occur 'by chance' or 'not-by-chance'? Accidental God or Purposeful God? That is the bottom line. That is the basis of whether there is real purpose and meaning in the universe and in life.

By Chance or By Design? - What is the Most Probable Answer?

I have made an honest attempt in this primer to objectively examine the fundamental scientific facts of both the inorganic and the organic world that are relevant to answering this question in a manner that presents a knowledgeable and meaningful perspective. The reason for doing that was to enable the reader to decide the most probable answer to ultimate questions for himself. At the outset it was my intent to finish this book without revealing my own personal worldview. That, of course, has not happened. Yet, I am well aware that my views do not occupy a hallowed position. They simply represent what I believe. Intelligent people can certainly conclude differently than I.

Most scientific atheists whom I have met are a lot more intelligent than I am. But, higher intelligence does not necessarily yield a better belief. Quite often, common sense may be a better predictor of the more likely 'truth'. The icon of 20th century conservative intellectualism, William Buckley of Harvard University, once remarked that he would rather select a jury from the first 200 individuals listed in a telephone directory than from the directory of the university faculty.

The religious belief of scientific atheists is not based on scientific fact; it is based on metaphysical belief. Everyone should evaluate the facts and arrive at their own belief based on the best facts available. If further scientific facts are discovered in the future that point directly to 'by chance', I will reevaluate my beliefs based on the new facts discovered. But, until the time that such discoveries are made (if they ever are, which I doubt) I will continue to intelligently believe in a Purposeful God. And, I will **never accept the *belief* of scientific authority as scientific fact.**

If intelligent people are armed with basic factual knowledge about the many wonders of the world we live in, the genius of common sense can prevail against the crusades mounted by biased human authority. If the fundamental basics can be extracted from the overwhelming mass of information detail and can be presented in a straight-forward manner, then intelligent people can discover and decide their own core beliefs based on the best facts available. Human intelligence and dignity are ennobled when we examine the bases for our core beliefs through our own human reason, independent of the bias of authority figures.

The best and most meaningful answer to the question rests solely with you.

CHAPTER 29
A Disclaimer and Overview Perspective

I acknowledged in Chapter 1 that I do not have any credentials in either religion or science. My formal education in both arenas is quite scant. My knowledge is primarily self-taught. Because of these facts, it is most obvious that there may be factual mistakes in this book. Indeed, it would be most surprising if there were not.

In the science sections I have likely misunderstood, and thereby mis-described, some of the myriad details and processes of many complicated things, like how a dying star explodes as a supernova, or how neurons fire at synaptic terminals in the brain, or how the complement system serves as a bridge between the innate and acquired immune systems of the human body. Brilliant scientists who have an extensive scientific background and credentials devote a lifetime of arduous effort to learn the details of just one of each of these most-complicated things.

In the religious section of the last chapter, my capsulizations of the development of two great religions and the wisdom traditions that they embody are, at best, gross over-simplifications. Brilliant theologians who have an extensive background and credentials devote a lifetime of arduous effort to learn and understand the teachings of these most-complicated traditions.

I fully accept responsibility for all factual mistakes. All errors are mine and mine alone. They may be properly attributed to my lack of detailed understanding of the accurate facts contained in the works of others that I read and studied and then misunderstood.

The wider the scope of information presented in a book, the greater the chance for mistakes. That is a fact. Yet it is an equally important fact that we must examine a vast array of information about a vast number of very complicated things and then make generalizations from our examinations in order to gain a proper perspective of the whole, whereby we may arrive at our own independent assessments and judgment. In order to grasp an understanding of the larger picture lesser attention must be given to the details of the component parts. The idea is much better expressed in the ageless adage: 'Sometimes you can't see the forest for the trees.'

This primer has been aimed at a better understanding of the forest. And, I am most confident that any mistakes or misunderstandings on my part do not diminish one iota from an accurate understanding of the main inquiry of this primer - Did all this stuff happen by-chance or not-by-chance? Accidental God or Purposeful God?

Whether there are 20 or 25 or 30 fundamental parameters and constants of the universe that operate with the exact precision necessary to allow the natural world to evolve in the precise manner that permits intelligent life to evolve on Earth is not important to our central inquiry. What is important is understanding that if any one of these parameters or constants of the universe were only slightly, slightly more or less than it actually is, then intelligent life would not have evolved on Earth.

Whether the process of glycolysis whereby the cell breaks down a molecule of a six-carbon sugar into a three-carbon sugar involves 10 discrete reactions or 9 (each catalyzed by a different enzyme) is not important to our central inquiry. What is important is understanding that the processes of metabolism whereby each living organism on Earth acquires and uses energy, originally derived from the thermonuclear reactions within our Sun, are a composite of extremely complicated and elegant things that must work together in concert in order to sustain life and hold the 2nd law of thermodynamics at bay.

Whether the 12 unique proteins that are called into action to perform the blood clotting cascade, in order to keep us from bleeding to death each time our skin is cut, are produced in the liver or the pancreas is not important to our central inquiry. What is important is understanding that this wondrous process requires extreme precision and the utmost cooperation by molecular components in order for the function to occur at all.

Whether I have presented the details of these things with flawless precision is not the central inquiry. The central inquiry is 'how did these intricate things evolve to perform the functions required with the flawless precision that they exhibit?' By-chance or not-by-chance?

Whether I understand how each of the wondrous things examined in this primer works is not the point. The point is 'how did they come to work as they do?' By-chance or not-by-chance?

The information in this primer is most condensed. Many brilliant scientists may most justly view this information as too simplistic to provide a proper understanding of the scientific issues. That is certainly true. But, that is not the point. The point is: What evidence do these brilliant scientists actually have to support their metaphysical belief that the universe and life itself began and evolved accidentally and contains no real purpose and meaning?

Our perspective on and in life is dependent on our worldview - on how we see things. Do we see the glass as being half-full or half-empty. The half-full versus half-empty perspective is important. It provides a 'default setting' for our understanding of the things of this world.

The proponents of an 'Accidental God' adhere strictly to a naturalistic explanation for the ultimate reality of everything. The full glass represents the totality of understanding that they will ultimately obtain when they discover the naturalistic explanation for every part and particle of the physical universe. They always will see the glass as partially empty and they will always be sure that their efforts will one day result in physical fullness.

The proponents of a 'Purposeful God' adhere to a belief in ultimate reality that contains both the naturalistic world and that which lies beyond. The full glass represents the totality of understanding that they will never be able to fully obtain. They always will see the glass as partially full and will always be quite certain that they will never be able to fully know the reality represented by ultimate fullness. They delight in their belief that they are a part of an ultimate fullness that has moral purpose and meaning, and that their lives should be lived in accord with the teachings of natural law and wisdom traditions that have guided human civilization through the ages.

The ultimate fullness may never be known by mere mortals. But the ultimate fullness is always there. It resides in our hearts and our minds as the belief in ultimate fullness - in higher purpose and meaning in life.

www.ingramcontent.com/pod-product-compliance
Lightning Source LLC
Chambersburg PA
CBHW031815170526
45157CB00001B/62